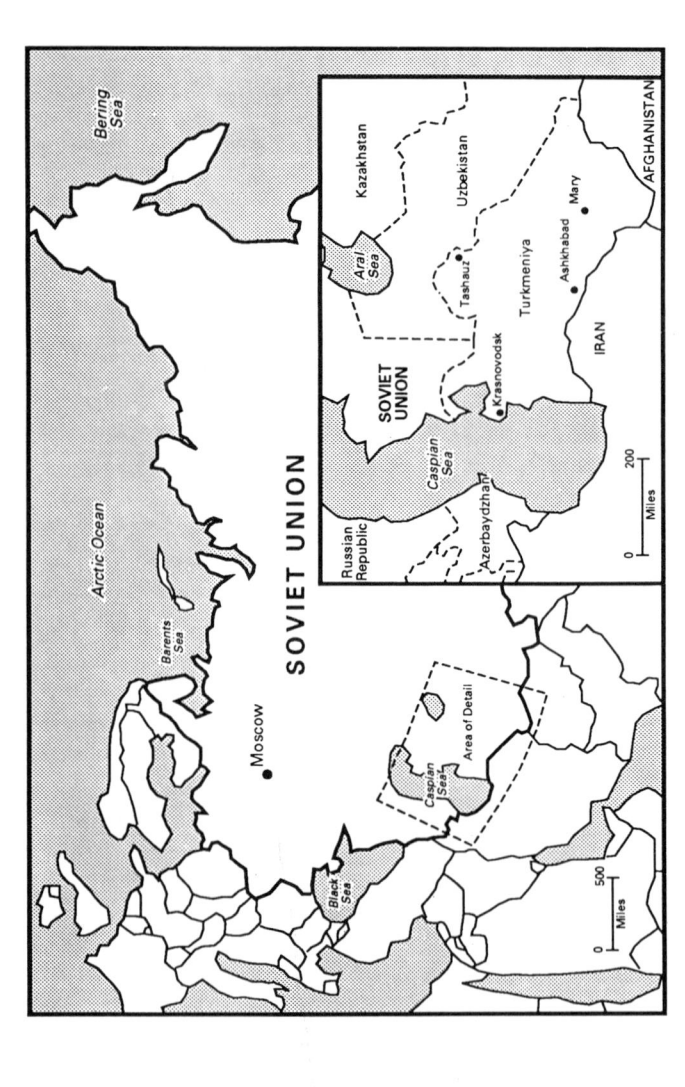

Geographic
Perspectives on
Soviet Central Asia

STUDIES OF THE HARRIMAN INSTITUTE

Columbia University

Founded as the Russian Institute in 1946, the W. Averell Institute for Advanced Study of the Soviet Union is the oldest research institution of its kind in the United States. The book series, *Studies of the Harriman Institute*, begun in 1953, helps bring to a wider audience some of the work conducted under its auspices by professors, degree candidates and visiting fellows. The faculty of the Institute, without necessarily agreeing with the conclusions reached in these books, believes their publication will contribute to both scholarship and a greater public understanding of the Soviet Union.

Some titles in the *Studies of the Harriman Institute Series*

In Stalin's Time: Middleclass Values in Soviet Fiction by Vera Dunham. Duke University Press (forthcoming)

The Soviet Agrarian Debate by Susan Solomon, Westview Press, 1978

Education and Social Mobility in the Soviet Union, 1921–1934 by Sheila Fitzpatrick, Cambridge University Press, 1979

The Soviet Union and the Third World: An Economic Bind by Elizabeth Kridl Valkenier, Praeger, 1983

Geographic Perspectives on Soviet Central Asia

Edited by Robert A. Lewis

Cartography by **Robert R. Churchill**
and Amanda Tate

London and New York

First published 1992
by Routledge
11 New Fetter Lane, London EC4P 4EE

Simultaneously published in the USA and Canada
by Routledge
a division of Routledge, Chapman and Hall, Inc.
29 West 35th Street, New York, NY 10001

Typeset in Linotron Palatino by J&L Composition Ltd, Filey, North Yorkshire
Printed and bound in Great Britain by
Biddles Ltd, Guildford and King's Lynn

British Library Cataloguing in Publication Data
A catalogue reference for this title is available from the British Library.

ISBN 0–415–07592–0

Library of Congress Cataloging in Publication Data
has been applied for.

ISBN 0–415–07592–0

To the memory of our colleague,
Theodore Shabad,
*for his contributions to the study of
the geography of the USSR.*

Contents

Part III

Economic Geography

Part IV

Social Geography

Tables

Maps and Figures

MAPS

FIGURES

Part I

Introduction

Chapter 1 —————————————————————————

Introduction

Robert A. Lewis

T hese essays on Soviet Central Asia are from a geographic perspective. Academic disciplines by necessity present a simplistic view of the world, and geography is no exception. Reality is extremely complex, and theory simplifies by generalizing, in an effort to explain and to make reality comprehensible. Thus, to the degree that we draw upon geographic, demographic, and other theories, categories, and concepts to explain processes in Soviet Central Asia, we present a simplified view of this region, but we greatly increase our understanding of its societal processes. Moreover, the phenomena that a discipline considers relevant comprise a further simplification of reality, because no field attempts to cover all aspects of reality. Yet, geography is not a narrow field; such traditional geographic subjects as the natural environment, the spatial organization of society, population, resources, ethnicity, and economic and political geography provide a broad framework and essential elements for regional study, history, and international affairs. Thus, the geographic approach provides considerable breadth and explanatory power. Whether one views geography as the study of the interrelationships between man and the natural environment, or spatial distribution and structures, or regions as coherent entities, a sense of geography is crucial to the understanding of human behavior, the society around us, and the world. By a sense of geography is meant a sense of spatial variation and distribution and relative location or spatial context. Thus, even though the geographic perspective can be considered as only a window on reality, much can be seen through this window. In fact, it is impossible to achieve an adequate understanding of the world or any of its regions without a knowledge of geography. Moreover, the profound socioeconomic transformation that has occurred in the USSR in this century has resulted in a drastic change in the geography of that country. If modernization continues and predicted climatic trends occur in the coming decades, this geographic change will accelerate, and become even more essential in the understanding of

Soviet society. The proper study of the USSR requires a knowledge of geography, particularly its regional geography, because of the great diversity of the country. Increasingly spatial and regional differences in such factors as nationalism, socioeconomic development, fertility, mortality, migration, growth of the total and working-age population, age structure, and regional investment are becoming crucial to the understanding of the course of events in the USSR and constitute strong forces shaping Soviet society.

Soviet Central Asia is a very important region in the USSR (see frontispiece map). It is the most populous of the economic regions of the USSR, and it has the highest rate of population growth in the USSR, especially that of the working-age population. This is resulting in a major spatial redistribution of the Soviet population and its working-age component. Although most of the population and infrastructure are in the western part of the USSR and most of the resources are in the eastern part, most of the growth of the population and its working-age component are in the southern non-Slavic tier, primarily in Soviet Central Asia. Furthermore, this region contains the major portion of the Turkic-Muslims in the USSR. In the postwar period, Soviet Central Asia has probably experienced more socioeconomic change then any other major Soviet region. The products of its irrigated agriculture, especially cotton, make a major contribution to the Soviet economy. The relative location of Soviet Central Asia is also significant, in that it is on the frontier with the Middle East, an area with which it shares a common culture. Thus, from many perspectives, Central Asia is an important Soviet region. Despite these important attributes, it is probably the rapid growth of the indigenous population of the region relative to economic development, and its potential for social and political disruption, whether real or not, that has attracted the most attention to Soviet Central Asia, and will impede perestroyka to the degree that resources are required to solve these problems.

An important consideration for the future of the USSR and Soviet Central Asia is the degree to which this region will be integrated economically, demographically, politically, ethnically, and socially into Soviet society. Probably the most important aspect of this integration would be the out-migration of surplus indigenous workers from Soviet Central Asia to labor-deficit areas elsewhere in the USSR. Such an out-migration would be economically beneficial to the USSR, but probably ethnically troublesome, in that the world over the geographic interaction of ethnic groups, particularly in the context of ethnic homelands, has led to considerable ethnic tension.

The purpose of this anthology is to survey the geography of Soviet Central Asia and to highlight geographic issues in the region. Thus, this book includes chapters on the major divisions of geography, in which basic trends are established and analyzed. Geographic issues include regional population growth, the prospects for Turkic-Muslim out-migration from Central Asia to other parts of the USSR, the degree to which Turkic-Muslim culture is immune to the forces of modernization, the extent to which investment in Central Asia has been equalized relative to other parts of the USSR, the role of large water diversion projects in the future development of the region, the declining level of the Aral Sea, irredentist tendencies in political integration, geographic segregation by ethnic groups and its effect on ethnic processes, the social and economic integration of Turkic-Muslims into the modernized society, the effect of the dominance of investment in primary activities, and the demographic effect of agricultural mechanization.

The book is organized around two unifying themes: integration and the prospects for out-migration of the indigenous population to other parts of the USSR. Economic, social and ethnic integration is investigated at various scales from state to Central Asian, to republic, to national integration. Given the surplus labor and rapid growth of the working-age population in Central Asia, and the labor shortage elsewhere in the Soviet Union, one would expect, in accord with the universal experience, that there would be considerable out-migration of the Turkic-Muslim population to the north. As yet, this has not occurred. The authors examine this issue from a variety of perspectives.

The participants in the project, with but a few exceptions, are affiliated with the Geography Department of Columbia University as faculty or graduates. The advantage of this common affiliation is unrestrained and close interaction among the participants. The major weaknesses of most anthologies are a lack of focus, duplication, and considerable variation in the quality of the various contributions. It is hoped that this close interaction afforded by a common affiliation will avoid these weaknesses. A workshop at Columbia University on 13 and 14 November 1987, funded by the Harriman Institute for Advanced Study of the Soviet Union, facilitated interaction among the authors and others interested in Soviet Central Asia. The authors are Soviet specialists whose work has to varying degrees included Soviet Central Asia and a topical speciality.

For the purposes of this book, Central Asia is defined as the four southern republics excluding Kazakhstan, but because of logical necessity we frequently include southern Kazakhstan in our spatial framework. This is because, as is common knowledge among geographers, there is no all-purpose set of regions for the geographic study of the Earth, and there are compelling similarities with Soviet Central Asia in the natural environment, economy, and culture of the southern oblasts of Kazakhstan. As to the temporal framework, our emphasis is the postwar period, although once again out of necessity some authors give consideration to the prewar period. Turkic-Muslim is defined as Turkic and Muslim, thus including the Tadzhiks in the Central Asian context.

The transliteration system is that of the US Board on Geographic Names.

One volume clearly does not suffice to cover the geography of Soviet Central Asia, because the topic is vast in terms of space, time, and subject matter. That there has been relatively little geographic study of Soviet Central Asia published in English or other Western languages further complicates the problem of geographic coverage. Therefore, it was decided that the period since World War II would be stressed, and the volume would provide more basic geographic information than one might ordinarily expect. This emphasis excluded more coverage of the relationships of Soviet Central Asia to other parts of the USSR and the world, which are of course geographically important subjects. Some subjects are not fully developed, because they lie outside the central interests of geographers, and others are omitted because there is a lack of information for adequate coverage. This does not mean that we underestimate the importance, of, say, religion or culture or aspects of history that have been covered in sufficient detail elsewhere. However, the subjects that we have emphasized are central to the understanding of the geography of the region.

This book is based on a number of assumptions that affect its scope and content. The first assumption is the importance of testing the universality of societal processes in the Soviet context. The most reasonable approach is first to test general societal concepts in the USSR to determine if, or to what extent, they explain the processes under investigation, and then, when necessary, to examine conditions specific to the USSR to explain further these processes. In fact, logic should compel one to demonstrate that the processes under investigation are not universal, before one accepts a unique approach. In short, as a matter of scientific policy, one should logically concede some peculiarity or uniqueness to the societal processes of a nation, country, or ideology only if data require it. Since there are gross signs that modernization and related societal change are occurring in more or less the expected manner independent of political boundaries, the most reasonable attitude would be to hypothesize that the causal determination is not "culture bound." Implicit in our regional approach is the contention that the forces of modernization are dominant in social change, and one cannot adequately understand societal processes in the USSR solely by studying the USSR, because a knowledge of the generality of these processes is necessary for their analysis in the USSR. Thus, our chief working hypothesis is that people throughout the world tend to react in the same manner to the socioeconomic forces that influence their behavior—whether they live under "capitalism" or "communism" or anything in between.

The importance of avoiding ideology, dogmatism, and wishful thinking in the analysis of societal processes in the USSR is a corollary of the above assumption. Stark differences in ideology and superpower rivalries promote ideological and dogmatic interpretations and wishful thinking about societal processes in the USSR. The best way to achieve a more objective, realistic view of the USSR is to analyze it in terms of existing socioeconomic theories and concepts and from a comparative perspective. In short, the best way is to test the universality of societal processes in the USSR. In doing this, however, one must avoid discredited theories and models. For example, the totalitarian model has been widely discredited as a gross oversimplification, but yet it continues to be influential in thinking on Soviet society.

A major constraint on the application of theory in the Soviet context would be effective government policy. Theory attempts to explain practice, and policy attempts to control practice. Therefore, a successful policy would control and explain practice, and theory would be redundant. Some aspects of societal change in the USSR are closely related to policy, such as the centrally planned economy. Economic change is, of course, the driving force behind modernization, so it has a major indirect influence on other forms of social change. However, there is no comprehensive, effective set of policies that directly govern social change in the USSR. For example, there is no government plan with respect to the divorce, urbanization, fertility, migration, or mortality rates in the USSR, and there is no comprehensive, effective population or ethnic policy, despite the authoritarian nature of Soviet society. The control of social change is generally very difficult, because in most instances it is complexly interrelated with a variety of socioeconomic trends and is poorly understood, particularly in the USSR, where Marxist-Leninist dogmatism has adversely affected social theory and policy. However, theory can be applied to many aspects of social change in the USSR, and if there is a considerable universality of these processes, one can make guarded forecasts based on the universal experience as to future societal trends. Clearly, a conceptual knowledge of the economic, demographic, political, geographic, and social forces shaping a society is necessary in the appraisal of future trends.

However, in appraising the impact of modernization on a society, one should avoid various brands of one-factor determinism, whether it be environmental, economic, demographic, or cultural, because the processes of societal change are interrelated and very complex, and should not be oversimplified. For example, the exaggeration of the dominance of culture or religion can mask other important cleavages in a society, and the growth of a population or relative decline of an ethnic group does not necessarily herald crises.

Another assumption of this book is the importance of empirical research, which is to be expected because geography is a heavily empirical subject. Data, theory, and method constitute the essential elements of research, and a rough balance in these elements is required in order to avoid the common academic inclination toward excessive and tedious description, method for its own sake, obfuscation, and theory unconfirmed and unrelated to reality. However, whereas theory and method are the subject of considerable academic attention, data consideration, especially data evaluation, is generally neglected and considered mundane, even though virtually all socioeconomic data contain errors. Thus, data skepticism is of the utmost importance and is the mark of scholarly maturity, because the mindless acceptance of published data as being accurate or representative of the phenomena under investigation often results in erroneous interpretation and bad scholarship. Some questions can be answered with relatively bad data, but others cannot. The most reasonable question to ask is not if the data are accurate, but how accurate they are and whether they are sufficiently accurate to study the subject under investigation. Of course, conclusions not based on data or those that go beyond available data must be avoided. In addition to its crucial importance in analysis, the disaggregation of data is very useful in evaluating accuracy, because there are frequently geographic differences in the quality of data, especially in large diverse countries such as the USSR.

Another data problem is that of data comparability. The basis of the problem of data comparability is that most data are artifacts, and thus there are few uniquely correct definitions, territorial units, or temporal periods for which data can be collected. This is so because most statistical data are continuous and their demarcation into categories is, in most instances, arbitrary. As to definitional comparability, if there were uniquely correct, universal, and constant definitions for socioeconomic phenomena, there would be no problems of definitional comparability. Likewise, the territorial units in which data are collected are also artifacts, and can bias data. Regionalization is merely the spatial aspect of the classification problem; just as there is no all-purpose statistical or taxonomic interval or class, there is no perfect set of regions for a country. Unfortunately, data are not collected at the required scale in the USSR to permit the construction of an ideal set of regions for each analytical purpose. Consequently, the researcher is usually forced by circumstances to use some set of existing units, but should be aware of their limitations. Time intervals between censuses should also be constant in order to facilitate temporal comparisons, but in many instances this is not the case.

The geographic scale of analysis is also an extremely important consideration in empirical analysis, because results often differ depending on the scale of the territorial units. For example, data by union republics in the USSR are not suitable for most purposes because of the great differences in their size, population, and socioeconomic diversity. The Russian Soviet Federative Socialist Republic (RSFSR) is about twice the size of the United States; Armenia is about the size of Belgium; and the city of Moscow has a population greater than that of ten of the republics. Thus, whenever possible it is necessary to disaggregate data geographically to below the republic level to minimize spatial variations as much as possible.

What follows is the selective application of these assumptions where applicable. We start with a regional overview as a background to the region and progress through chapters on the various branches of the discipline of geography. We attempt to stress the key geographic trends and problems in the region. Our hope is that this research will provide a foundation for the study of the geography of Central Asia and will further the understanding of this important region in general.

CONCLUSIONS AND IMPLICATIONS

Chapters dealing with historical, political, and physical geography comprise Part II, "Regional Background." Ralph Clem provides the necessary historical-geographical background by investigating the frontier and colonialism in Russian and Soviet Central Asia. He traces the expansion of the Russian Empire into Central Asia and compares its development with models derived from European colonial expansion. He concludes that there was a considerable universality in the expansion into Central Asia, and the colonial heritage continues to affect society and economy in Soviet Central Asia.

Lee Schwartz covers the political geography of Soviet Central Asia by focusing on the integration of the Central Asian Frontier of the USSR. After surveying the historical political geography of Soviet Central Asia, he investigates political integration from three perspectives or scales of analysis: macro (the region), meso (the republics), and micro (tertiary political-administrative units). He concludes that both integrative and disintegrative forces are influential, depending on the scale of analysis, but disintegrative predominate. At the regional level historical and cultural factors tend to be disintegrative, but at the republic level there is a high degree of institutional integration, which is negated by the growing nationalism of nations in their legalized homelands. At the oblast level, disintegrative forces predominate in the form of anti-Soviet sentiments, interrepublic conflict and localism.

Peter Sinnott provides an overview of the physical geography of Soviet Central Asia, and then focuses on a major environmental problem, the declining level of the Aral Sea. He concludes that natural conditions have limited settlement and agriculture in Soviet Central Asia to irrigated oases along the major rivers and to the piedmont, and that the expansion of irrigation and cotton monoculture has led to an overuse of water and a breakdown of the ecosystem. The rapidly declining level and area of the Aral Sea and associated problems are the best examples of this ecological disaster. Sinnott questions whether the planners who created the problems can restore the ecological equilibrium and a more rational utilization of the water resources.

Chapters in Part III deal with the economic geography of Soviet Central Asia. Ronald Liebowitz provides the necessary background to this section by surveying the geographic imbalances of resources, infrastructures, and work force increments in the Soviet Union, which have adversely affected the growth of its economy. He relates these imbalances to economic developments in Central Asia, with its rapidly growing work force, primarily by analyzing investment flows in the postwar period. The picture that emerges is that there has been, and continues to be, insufficient investment and economic development to accommodate the rapidly growing population and work force, and that most investment, which has been declining per capita, has been directed to resource development and agriculture and not to labor-intensive industries. Liebowitz concludes that the general lack of capital in the USSR and recent investment policies indicate that the problem of insufficient economic development relative to the rapidly growing population will not be addressed in the 1990s, and that economic conditions should continue to deteriorate, unless there is more invest-ment, especially rationally directed investment, or significant out-migration. He also concludes that Central Asia is not well integrated economically with the Soviet economy, because of the colonial nature of Central Asian development.

Peter Craumer investigates agricultural change and conditions in rural areas in Soviet Central Asia. Although there has been considerable agricultural development in the postwar period, he documents deteriorat-ing agricultural conditions in recent years: a rapidly growing rural population, little rural out-migration, increasing labor density per unit of agricultural level, limited water, little further expansion of the irrigated acreage, stagnating or declining cotton yields, declining labor productivity, declining mechanization, lagging agricultural wages, ris-ing labor costs, extensive salinization, considerable water pollution, and increasing inability of the private plots to absorb excess labor. These deteriorating conditions in rural areas, in conjunction with rapid popula-tion growth by world standards, should eventually constitute a power-ful economic push out of rural areas for migrants, although pull factors are generally dominant in migration.

Michael Sacks investigates work force and social change in Central Asia. He concludes that there has been a dramatic shift of workers out of agriculture, and thus the economy was not stagnant, although his anlaysis was limited by the lack of data after 1970. Russians and other nonindigenous groups, however, predominated in the modern sector of the economy, although indigenous participation in the modern sector has been increasing. Turkic-Muslim men outside agriculture tended to be in occupations that were dominated by women in the RSFSR, but there has been increasing female competition in these occupations. Female occupational segregation was particularly evident in rural areas, where female work force participation rates were low, and Sacks attributes this to the strength of patriarchy, which he relates to the low rates of rural out-migration, high fertility, and male resistance to entry into the modern sector.

Part IV is devoted to relevant aspects of the social geography of Soviet Central Asia. Ozod Ata-Mirzayev and Abdukhakim Kayumov, our Uzbek contributors, document the rapid growth of the Central Asian population, largely as a result of high fertility, which has declined somewhat recently, and moderate mortality, which has resulted in low crude death rates because of the young age structure (70 percent of the population is below age 30). They also discuss the high infant mortality rates, high marriage rates, low divorce rates, and low levels of urbanization, particularly for the indigenous population. They conclude by listing the primary factors which explain the high fertility rates.

Richard Rowland investigates in some detail such topics as relative population shares, fertility, mortality, age, migration, population distribution, population growth, and urbanization. He concludes that Central Asia is characterized by rapid population growth, an increasing share of the Soviet population, a highly concentrated population with high rural densities and vast uninhabited areas, a relatively low level of urbanization, and a drastically distorted age structure in rural areas. Rowland focuses on migration and concludes that the indigenous population is relatively immobile, and that there has been relatively little rural or northward out-migration. Thus, there is relatively little demographic integration, because indigenous population characteristics differ sharply from the northern areas. He further concludes that, given the rapid growth of the working-age population, the lack of investment, and deteriorating economic conditions, out-migration to the north is the most realistic option.

Robert Kaiser investigates social mobilization and integration in Central Asia in terms of education, Russian bilingualism, and occupational structure, with special reference to the status of women. He concludes that with respect to education there is a rapidly growing indigenous elite with equalization between the sexes, that there has been increasing Russian bilingualism among both sexes, and that there has been considerable upward occupational mobility. Thus, there has been considerable social integration, and the indigenous nations have achieved an intermediate position in the modernization process, even though female equalization has lagged, especially with respect to occupation and in rural areas. Kaiser believes that without more investment and a restructuring of the Central Asian economy, the continued pressure for indigenous upward mobility will not be satisfied, and that perception of relative deprivation will probably result in increasing national tensions.

In the final chapter, Robert Kaiser investigates the interrelationship of nations, homelands, and international integration in Soviet Central Asia, and concludes that the considerable social integration and mobility and socioeconomic development that have occurred have not resulted in appreciable international integration. Rather, because of the concentration of the nations in their homelands, their emotional attachment to them, and the associated benefits, social mobilization has resulted in increasing national awareness and consolidation, and the nation has become the primary allegiance in the region. Thus, the chief problem faced by the state is to accommodate the rapid population growth, increasing social mobility, and rising expectations in order to dampen national tensions. With respect to migration, Kaiser concludes that the advantage of residence in one's national homeland raises the economic threshold of out-migration, but does not preclude migration. He further argues from a theoretical perspective that as socioeconomic development proceeds, national problems will accelerate.

The picture that emerges is grim: a rapidly growing population, especially the indigenous population which is concentrated in rural areas, labor surpluses, little rural or northward out-migration, deteriorating economic conditions especially in rural areas, environmental degradation, significant social development, and thus rising expectations and national awareness. At current growth rates, the indigenous, the rural, and the total and rural working-age populations will double about every generation or less. There is relatively little rural out-migration; in the past decade it constituted only about 6 percent of the rural natural increase in Tadzhikistan and Turkmeniya, and only about 25 percent in Uzbekistan and Kirgiziya. The urban areas and socialized agriculture are not absorbing the rapidly growing rural population, and the urban population is growing more slowly than the rural population. Thus, the region is becoming more rural, and the urban share of the total population is declining, an unprecedented occurrence in a developed country. There is no mass exodus from the region of the nonindigenous population. Russian net out-migration in the past decade, for example, was about equal to Russian natural increase, so the Russian population remains roughly stable. However, since 1989, it appears that Russian out-migration has increased. In short, a demographic caldron is brewing in Central Asia.

Yet, to date there is no concerted effort on the part of the central government to accommodate this growing population. On a per capita basis and in terms of productive investment, investment flows to Central Asia have been declining, and much investment is not directed toward providing local employment, but toward resource development for export. It is unlikely that these investment patterns will change in the near future, because the current program for economic revitalization calls for investment to be directed toward areas that maximize the return to capital and labor, which is not Central Asia, but the European west. Moreover, the recent program of enterprise self-financing would not seem to produce the necessary capital, because of the generally low productivity in Central Asia.

Economic conditions in rural areas are deteriorating, as labor surpluses grow. Furthermore, there is little water for additional irrigation, and past extensive irrigation development has resulted in an ecological disaster: the declining level and area of the Aral Sea and associated problems, and extensive soil salinization and water pollution. It appears highly likely that the Aral Sea will disappear, even if fertility continues to decline, the young age structure constitutes a considerable fertility and growth potential, although further slow decline in natural increase can be expected, especially in urban areas. Thus, in rural areas where the population is growing the fastest, conditions are worsening and are likely to continue to do so. If these conditions persist, the standard of life could be halved in rural areas in the next couple of decades.

There are, however, other options. The least plausible would be that the central government would support a large welfare population on the land in Central Asia, because of the current economic squeeze in the USSR. The most plausible would be significant out-migration, which would at least alleviate the deteriorating conditions. In the past, economic equalization and rising nationalism have probably dampened out-migration, despite abundant job opportunities to the north. Clearly, the pressure for out-migration from Central Asia is building very rapidly. At the current growth rates, the rural working-age population will increase 40 percent by the year 2000, and the total working-age population will grow only slightly less. This growth without more investment should be sufficient impetus for considerable out-migration. If, however, there is a major increase in economic efficiency and a release in labor in the northern areas, as is planned, there may not be many jobs suitable for young educated people, and rising expectations will be frustrated. Moreover, if large numbers of Central Asians migrate, the universal experience would indicate that this would create ethnic tensions at both the origin and the destination of the migrants. At the destination the migrants are considered alien intruders, in multinational, multihomeland states, much like the Russians in the Baltic areas and the Meskhetian Turks in Uzbekistan, and at their origins they are considered a loss of national blood, especially if the migrants are driven out by economic conditions controlled by the central authorities (Lewis and Rowland 1979: 422–4).

Our conclusions in a previous study seem appropriate, even though it was less detailed than this present study.

> We maintain that there is great utility in applying universal formulations in the Soviet context. Our guarded forecasts on the prospect of Central Asian migration are based on this assumption. What we have done here and elsewhere, essentially, is to investigate those initial or determining conditions in Central Asia that universally have promoted migration, such as the growth of the native working-age population, labor surpluses, mechanization of agriculture, social and cultural change, investment, wage differentials, the ability of the nonagricultural sector to absorb the growing rural population, the expansion of irrigation, job availability elsewhere in the USSR, and so forth. We do not deny that there could be conditions specific to Central Asia that might impede migration, at least in the short run, but so far we can see no particularistic factors in Central Asia that are sufficiently strong and resistant to change related to modernization to counter the strong demographic, economic, and social forces that are intensifying in Central Asia and that elsewhere in the world generally have resulted in substantial out-migration.... We cannot determine precisely when significant out-migration will begin or the numbers involved, because we have insufficient information on a number of factors, and future conditions, as usual, are impossible to assess. However, we can predict with some certainty that unless trends in the USSR change drastically one can expect considerable migration from one ethnic territory to another in the USSR and the universally associated ethnic problems. (Lewis and Rowland 1979: 423–4)

Previously we pointed out the importance of geographic differences in demographic processes, socioeconomic development, and so forth, to the understanding of developments in the USSR. To the degree that it is necessary for the central authorities to direct funds to Central Asia, an area of relatively low return to capital and labor, it will impede the economic restructuring that is considered to be crucial to the revitalization of Soviet society. Further economic neglect of Central Asia with its rapidly growing population should result in considerable out-migration and associated ethnic problems. Regardless of the course of events, Central Asia will present serious problems to the Soviet government.

The concept of integration, the other theme of this book, is difficult to define and is somewhat ambiguous in that it differs with each aspect of society. In the most general sense, it is the process of becoming a part of a whole, or a region becoming an integral part of the Societ Union. It implies the sharing of values, traditions, beliefs, and way of life, and participation in a common endeavor such as an economy with complex ties, common markets, and labor pools. Convergence of measures or homogenization is a part of the process, and the easiest to describe and analyze. Although the elements of stability are not well known, it would seem that a high degree of integration would contribute to the stability of a country, and would be a force that would counter territorial disintegration. Our efforts to investigate integration must be considered preliminary, because the definition of the process has differed from author to author. Yet, we can state with some confidence that in general Central Asia is not well integrated into the Soviet Union. Although there has been appreciable social and some political integration, there has been relatively little international, demographic, or economic integration, if one considers convergence important to the process. Current policies associated with perestroyka are not designed for the further integration of Central Asia, so the process may slow in this non-Slavic colonial frontier. About all one can say is that the relatively low degree of integration does not enhance the territorial stability of the USSR.

In the wake of the August 1991 coup attempt and subsequent developments, the grim economic picture in Soviet Central Asia becomes even grimmer. Given the highly interdependent Soviet economy with much concentration of production, the dissolution of the central government and the command economy, the political disintegration of the USSR should result in a further rapid deterioration of economic conditions in Central Asia. Even if a loose political confederation with a common internal market were formed, economic relations with the other republics will be disrupted, and there very probably will be no more subsidies, income transfers, cheap energy and resources, or investment funds for water projects or environmental reclamation. Self-financing will be immediately imposed on Central Asia, as well as world-market cotton prices, and there will undoubtedly be no more effort to equalize production in backward non-Russian areas as former policy directed. Because of the unbalanced development, low productivity, and rapid population growth of Central Asia, the effect on the standard of life could be catastrophic, especially with regard to the necessary food imports. According to Goskomstat, even before the April 1991 price reforms, about a half or more of the populations of the Central Asian republics fell below 100 rubles per person per month, which is well below the current poverty level. Clearly, in the absence of extraordinary resources and development or considerable outside support, people living in deserts with limited water cannot procreate at will without becoming impoverished. This region could well rejoin the Third World in the not too distant future, and a point will be reached when local leaders realize that uncontrolled population growth contributes to their societal problems. At least for the near future, there are no countervailing forces or trends that could alleviate these current and impending conditions.

As to some of the consequences of these developments, economic adversity should lead to further indigenization of the Central Asian republics with the associated economic competition and ethnic conflict which, in conjunction with declining living standards, should intensify the current trend of nonindigenous out-migration from Central Asia. A declining standard of life in the region should be a major impetus for the out-migration of the indigenous population, although the lack of economic opportunity in the north may limit this out-migration, because historically pull factors have been more important than push factors in migration. Yet, there should be major differences in income between Central Asia and the north and thus considerable out-migration and ethnic conflict. There probably will be more economic cooperation among the republics, but one Muslim nation hostile to the Russians will not form, despite the considerable Western literature emphasizing the Muslim menace to the contrary. Their primary identification will continue to be to their respective nations and increasingly so, judging from the universal experience. At present, the Central Asian republics are relatively more politically stable than the other republics of the USSR; in fact, a worst case scenario is a union consisting of the RSFSR, Kazakhstan, and the Central Asian republics. If this highly unlikely event occurred, the 1989 population of this union would be 196.2 million, of which the southern republics would account for 25.1 percent and the Great Russians 66.0 percent. But deteriorating economic conditions very probably would lead to deteriorating political conditions, as elsewhere in the world. Eventually, the Center or the RSFSR or whatever is left of the USSR will come to realize the economic burden of these underdeveloped, overpopulated republics, and appreciate the advantages of political independence from them.

REFERENCES

Lewis, Robert A. and Rowland, Richard H. 1979. *Population Redistribution in the USSR*. New York: Praeger.

Part II

Regional Background

————————————————————————————

Chapter 2

The Frontier and Colonialism in Russian and Soviet Central Asia

Ralph S. Clem

> There is more land than you could cover if you walked a year, and it all belongs to the Bashirs. They are as simple as sheep, and land can be had almost for nothing.
>
> Tolstoy, "How much land does a man need?"

It is by now commonplace to describe the history of Russia, as did Klyuchevskiy, as a singular and continuous process of migration and colonization, something quite apart from events elsewhere and with a quality all its own. Like most sweeping generalizations, this one contains an element of truth, but the shortcomings of such a simplistic view are only too evident on closer examination. To begin with, although the historical expansion of the Russian state was inexorable, it was nevertheless temporally and geographically differentiated in terms of causes, circumstances, and implications. In fact, much of the complexity of the current ethnic situation in the USSR derives from the timing and conditions through which a given group and area were absorbed into its predecessor state, the tsarist empire.

Secondly, the notion that Russia was "a country which colonized itself" is clearly wrong; Russia was actually an aggressive state that *aggrandized* itself territorially at the expense of neighboring, non-Russian peoples, and then colonized *their* lands. Migration, taken usually to mean a large-scale spatial movement of population, in the Russian case followed military action and political incorporation and should not be seen as a benign trend, because it entailed conquest and domination of one group by another.

Moreover, an approach to the study of the formation of the tsarist empire which treats that experience as *unique*, with this distinctiveness either stated explicitly or implied, fails to relate the Russian pattern to that of other societies or to ideas in history and the social sciences.

Although there are some noteworthy exceptions in the literature to the Russocentric paradigm, as a rule the majority of scholarly work in this field is bereft of references to the universal model of colonialism and of the genesis of multiethnic societies. This issue is vital to our purpose here, because the modern history of Russian and Soviet Central Asia is made so much more meaningful if it can be compared to situations where diverse peoples have gone through the sequence of encounter, settlement, conflict, and accommodation. Unfortunately, so much of the academic literature in history and the social sciences is rooted in the parochial tradition that theoretical work is rare. It should be recognized, however, that the general or comparative method requires knowledge of other areas and hypotheses which are open to refutation. That we do not yet have definitive, overarching theories by no means rules out the attempt to reach higher levels of generalization. Indeed, by such means our knowledge of complex historical socioeconomic processes is furthered and our appreciation of a specific case is enhanced (Price 1950: 1–3).

Accordingly, I will endeavor to place the incorporation of Central Asia into the Russian Empire and its retention in the USSR within the larger conceptual framework of frontier settlement, colonialism, and modern ethnic group relations. In other words, I have tried to distill the far-ranging body of scholarly work in these subjects into a sequence which fits the course of Russian and Soviet history. I certainly do not pretend that this is anything other than a first approximation of an understanding of historical developments in Central Asia, but in the attempt we might gain some valuable insights into events of the recent past in this important region, the manner in which they correspond to similar situations elsewhere, and the prospect for change over the near term. This discussion should also provide a necessary historical-geographic background to the course of events in Soviet Central Asia in this century.

FRONTIERS, COLONIES, AND MULTIETHNIC STATES

Probably for as long as humans have existed on this planet, people have transgressed on the lands of others. Certainly within the last millennium there has been very little empty space suitable for habitation or economic exploitation. Rather, the prevailing tendency has been for one group to expand into lands occupied, however sparsely, by those of different tribal or ethnic background. Without doubt, the dominant feature of this trend for the last 500 years has been "the outreach of Europe and the impress of its political power for some considerable period upon very nearly every part of the non-European world. . . a myriad of lesser movements [merging] into a single grand process' (Meinig 1969: 213). The difficulty is to see the abstract "forest" of European hegemonism through the "trees" of the many different and varied individual events, locales, and circumstances which that complex process encompassed (Fieldhouse 1967: 3–10). To deal with this complexity, the use of typologies is often adopted, either as an heuristic device or as a concession to inadequate theory. I will briefly define and review two of the several forms which the European diaspora has taken and then discuss the manner in which these situations relate to the formation and maintenance (or dissolution) of the multiethnic states which grew out of the frontier and colonial era.

The frontier

The first type of expansionism that I will consider is the *frontier*. Thompson and Lamar define the frontier

> not as a boundary or line, but as a territory or zone of interpenetration between two previously distinct societies. Usually, one of the societies is indigenous to the region, or at least has occupied it for many generations; the other is intrusive. The frontier "opens" in a given zone when the first representatives of the intrusive society arrive; it "closes" when a single political authority has established hegemony over the zone. (1981: 7)

Again, although frontiers as broadly conceived have existed as long as dissimilar peoples have been in contact, our attention here will focus on the most prevalent form such contact has taken since the eighteenth century: the movement of sedentary cultivators of European origin into vast prairie areas which were previously the domain of nomads. The sequence of events in such instances has followed a consistent pattern, whether in Eurasia, North and South America, Southern Africa, or Oceania (Price 1950: 1–3). Generally, the key elements in these situations are: (a) the geographical attributes of the frontier, especially the potential for agriculture; and (b) the level of technology and sociopolitical organization of the interacting groups (Thompson and Lamar 1981: 8–9).

Typically, frontier history has progressed through stages. Initially, only a small number of adventurers, traders, miners, and/or missionaries penetrated the contact zone; these individuals were not necessarily seen as a threat by the indigenes—although their influence, material or otherwise, was often destabilizing. If the borderland was suitable for settlement, the usual result was for a much larger number of invaders to follow as agricultural pioneers, thereby overwhelming the indigenous population. It was this relative numerical preponderance of intruders over indigenes which set the agricultural frontier apart from other forms of European expansionism, such as the commercial empires of the Portuguese and Dutch, or plantation colonies.

The heavy influx of settlers brings to the fore another critical point: the differences by which the idea of land "ownership" is conceived in nomadic as opposed to sedentary agricultural societies. As a rule, nomadic or pastoral peoples considered *territory* to be important and defensible; they required large expanses for hunting, herding (often on a seasonal cycle), and sometimes shifting cultivation, but ordinarily they did not lay personal claim to specific parcels. On the other hand, the European intruders who began settling the vast grasslands of the world viewed *land* as a commodity, something which individuals could own outright through legal means. In many instances, therefore, the indigenes ignored the granting or sale of property rights to settlers by the intrusive authorities as the frontier advanced, simply because nomadic peoples had no appreciation of the right to *exclusive* control of a particular area (Pollock 1980: 85). Inevitably, conflict arose when traditional hunting and grazing areas or transhumant and migration routes were foreclosed by settlement.

Largely because the European intruders usually enjoyed military-technological superiority, the outcome of the conflict in most cases was a "final solution" to the problem of local resistance, as the indigenes would be expelled, confined to reservations, or exterminated. This military "pacification" of frontier areas by Europeans had many common features, and was not achieved solely by technological advantage. For one thing, the intruders often found that they could exploit internal factionalism within or between indigenous societies, forging alliances with and then turning against their local confederates in the classic divide-and-conquer strategy. Also, the Europeans viewed expansion as a long-range goal, and maintained pressure through a series of campaigns and the construction of fortified lines. To be sure, there were occasional setbacks in this process, as illustrated by the fate of the US Seventh Cavalry at the Little Big Horn in 1876 and the British Twenty-Fourth Regiment at Isandhlwana in 1879. In most such cases, the defeat of the Europeans owes as much to cultural arrogance as to tactics; how else to explain how an experienced soldier such as Custer could have attacked 2,000 heavily armed Dakotas with about 600 men other than by the fact that he held the indigenes in contempt? These reversals were only temporary, however, and after such a debacle an aroused public opinion on the European side would bring down the full weight of the intruders on the indigenous people with predictably catastrophic consequences for the latter (Morris 1965; Connell 1984).

As a rationale for these aggressive actions the intruders would "justify" their takeover of the frontier region on the basis of making more "efficient" or "productive" use of the land. The nomadic or pastoral way of life would be characterized by the intruders as wasteful, primitive, and inferior. Thus, Arnold J. Toynbee could write that "the descendants of West European colonists had occupied effectively vast habitable spaces which previously had been virtually empty or else had been occupied ineffectively and utilized inefficiently by their former inhabitants" (Webb 1951: viii).

Indeed, the propaganda of the times is replete with ethnocentric and inflammatory polemics. A common practice was the use of crude ethnic stereotypes and exaggerated accounts of misdeeds by the indigenes in fiction and journalism to legitimate the notion that the "only good Indian is a dead Indian" (Miller and Savage 1977: 109–37). Thus, establishing the indigenous peoples as wild, subhuman beings in the European mind served the purpose of validating the moral correctness of their subjugation, their extermination, or—at the very least—the expropriation of their lands.

Beyond this popular genre, there developed academic literature about the frontier and its meaning for the larger expansionist state, as epitomized by the work of Frederick Jackson Turner (1920). Turner, his disciples, and his critics have argued at length over the influence of the frontier on American society. This discussion is voluminous and is very difficult to summarize beyond the main points, which are: (a) the American frontier experience was (or was not) unique, and (b) it promoted (or did not promote) egalitarianism and rugged individualism (Elkins and McKitrick 1954: 321–53; Mikesell 1960: 62–7). It is probably safe to say that frontier environments everywhere were—certainly at first—hazardous and fraught with risk, and no doubt attracted the more and filtered out the less hardy types. The marchlands may also have served as a safety valve for a burgeoning population or as a haven or place of exile for malcontents who otherwise would have been a threat to the established order; the extreme case of the latter was the deportation of over 150,000 convicts from England to Australia from the late eighteenth to the mid-nineteenth centuries (Hughes 1987). One of Turner's devotees, Walter Prescott Webb (1951), expanded the scope of the frontier to include the totality of European imperialism, with Europe as the "metropolis" and the remainder of the world (except Asia) as the "frontier." The safety valve or "global frontier" aspect remained popular until quite recently, with the call for settling the last available lands still heard in the 1930s (Bowman 1931, 1937; Eidt 1971).

The denouement of the political status of European frontiers took two principal forms. The first type, as illustrated by the United States, Argentina, and the Russian Empire, involved expansion by sovereign polities into geographically contiguous areas which were eventually incorporated into an enlarged state. The other model, examples of which would be Australia, New Zealand, and Canada, were longtime overseas colonies which gained independence after the frontier was enclosed. The distinction is, of course, mostly an academic one, and the point would be considered moot by the indigenes who were deprived of their territory and way of life. Ultimately, the flood of intruders into frontier regions swamped the locals and their culture and left them as minorities in multiethnic states dominated by the outsiders.

Colonialism

The second form of European expansionism relevant to the Russian experience is *colonialism*. In the sense in which it will be used here, colonialism is virtually synonymous with the classic form of "imperialism," if by this latter term we mean "the reaching out by one people to impose some degree of political control upon another" (Meinig 1969: 214). Beyond political control, however, one must also consider social, cultural, and economic factors as elements of colonialism. The general thrust of colonialism was described succinctly by Meinig as the last period of European imperialism (a phase he termed "nationalistic imperialism")

> to identify it with the outreach of virile national states, each highly conscious of power and position, strongly driven by the prestige of empire under the pressure of intense international rivalries. It was a drive also intimately related to the vigorously expansive stages of modern industrialism, with its omnivorous demand for materials, its need for expanding markets, and its development of ever more efficient tools for overcoming distance and for conquering peoples. (1969: 231)

Colonialism had attributes which set it off from frontier settlement. First, unlike the frontier setting, colonialism normally involved a relatively small number of intruders who could employ superior "technology and military power ... [to] subjugate a large native population and hold the area as an economic complement to the mother country" (Shibutani and Kwan 1965: 126). To enhance this economic complementarity, agriculture in the colonies was reorganized along plantation lines to provide grist for the mercantilist mill. Mining was frequently an important economic activity, as minerals were required by industries in the imperial homeland. Transportation routes were established to link the extractive area of the colony with outlets to the imperial power; railroads were especially important in this regard (Meinig 1971: 38–52). The intruders frequently established their rule through military authorities, and, when prudent, concluded treaties with local leaders to secure their cooperation (for as long as such cooperation was necessary). Violent resistance to the spread of colonialism or rebellion in reaction to the imposition of objectionable policies by the intrusive authority was not, of course, unknown (Horne 1977).

The intruder presence in the colonies was concentrated in the major towns and cities (often in cantonments or highly segregated quarters of urban areas), instead of being dispersed as in frontier areas. These colonial cities were typically cosmopolitan, as other outsiders moved in to fill the economic interstices between intruders and indigenes; hence the spread of Lebanese, Armenians, Jews, Chinese, and Indians throughout the European empires (Meinig 1969: 234).

Another important difference between the frontier and colonial scenarios is that in the latter the culture and way of life of the indigenous population—although subordinated to those of the dominant intruders— were not eradicated. Rather, the intrusive and indigenous cultures existed side by side; the traditional language and customs endured, at least in a modified form. Thus, colonialism is a situation where the intrusive is a veneer over the indigenous, and it consequently proved to have much more of a tenuous hold than the frontier model. The once great European empires dissolved for the most part in the face of rising nationalism after World War II, whereas the frontier areas continue to be firmly in the hands of peoples of European origin.

Ethnic stratification

Once established, the European frontier and colonial settings evolved from the contact and conquest stage into a new, relatively stable form of multiethnic society. In examining such situations, we have found it fruitful intellectually to refer to the idea of *ethnic stratification* as elaborated by Shibutani and Kwan (1965; Lewis et al. 1976: 83–93). In brief, they suggested that within ethnically plural societies the main socioeconomic cleavages take on an ethnic dimension; that is, there tend to be differences in status, occupation, education, and standard of living among ethnic groups. These differences result from formal or informal sanctions imposed on minorities or vanquished indigenes by the dominant group. Once in place, an ethnic stratification system reinforces itself through overt discriminatory measures (at the extreme, *apartheid*); stereotyping of social roles (such as "natural" occupational tendencies); geographically exclusionary policies and residential segrega-tion; unequal treatment before the law; and preferential standing for the dominant group's language and culture. Individuals from lower-ranking groups or nations who might present a threat to the system are co-opted, and certain minorities can benefit from "middleman" activities (Shibutani and Kwan 1965). In many cases, the conquering powers also co-opted indigenous peoples into "native" military formations and used them to consolidate their hold over subjugated territories (Enloe 1980).

Although ethnic stratification tends to be inertial, Shibutani and Kwan believe such scenarios "should be viewed as an ongoing *process* rather than as a *structure*" (1965: 131, italics in original). The principal agent for change in such systems has been modernization, which opens new avenues for ethnic group socioeconomic development (particularly through education). When the arrangement begins to weaken, one often finds a determined resistance to change by the established dominant group, a reaction which often turns violent. Both the opportunities for modification to the system and the tendency for the privileged to want to maintain their status lead to a strengthening of ethnic identity as individuals seek to acquire power through their respective groups. The ultimate form of ethnic identity is the nation-state, the seeking after which is nationalism, a goal pursued by many but achieved by few groups. Short of that, nations and ethnic groups in mature multiethnic societies must content themselves with regional autonomy and interest group policies such as affirmative action. In the realignment of ethnic stratification engendered by modernization, it is to be expected that the dominant group will retain its relative advantage, but other groups may benefit disproportionately (the "achiever group" phenomenon). Thus, the outcome of modernization in multiethnic societies of frontier or colonial origin is an alteration but not an elimination of the *status quo ante*.

FRONTIER EXPANSION AND COLONIALISM IN RUSSIAN CENTRAL ASIA

Although there are some arguable points about—and, of course, interesting local peculiarities in—the territorial expansion of the Russian Empire into Central Asia, the process should be seen as part of the outward movement of European peoples discussed above. Meinig, in his general survey of European imperialism, thought that "The Russian conquest of the pastoral and oasis societies of Central Asia was an exhibition of nationalistic imperialism: the capture at great military cost and political strain of a vast realm full of alien recalcitrant peoples and of little discernible economic value" (1969: 232). Actually, according to his typologies, it would be more appropriate to classify the "conquest of the pastoral societies" as another of his forms, "settler expansion" (what I referred to earlier as a *frontier*) which entailed "encroachments upon relatively primitive, semi-nomadic peoples, thinly and loosely spread upon the land, starkly different in culture, intractable to European control" (1969: 228). Thus, the advance of the Russian Empire in this region had two distinct, but not unrelated, stages: the extension of an agricultural frontier across the Kazakh steppe and foothill zones (Meinig's "settler expansion"), followed by a classic colonial scenario in the southernmost part of the area, or Central Asia proper ("nationalistic imperialism"). This spatial sequence of frontier and colony has all of the essential characteristics of the two models presented earlier; it can be shown that exceptions to the universal experience are minor and claims of uniqueness for the Russian case are unsupportable. Regardless, the incorporation of Central Asia into the tsarist empire was the last of a succession of Russian territorial acquisitions stretching back at least 350 years.

Russian imperial expansion: the early stages

In the introduction to his excellent study on the Russian conquest of Bashkiria, Alton S. Donnelly stated that "the growth of the tiny principality of Muscovy into a gigantic world empire is one of the wonders of modern times [and that] a survey of the steps of Russian expansionism reveals a series of stories, not just a single tale" (1968: 1). Furthermore, Donnelly saw Russian expansionism as "an example of European imperialism that coincides in time with the imperial movement of the Western powers which began with the great age of exploration" (1968: 1).

One might argue over an historical benchmark for the beginning of Russian imperialism, but the year 1552 is better than most in this connection. In that year, Russian forces led by Tsar Ivan IV ("the Terrible") captured the Tatar fortress at Kazan', thereby breaking a three-centuries-long containment begun when the Mongols invaded the Slavic lands in the mid-1300s. After taking Kazan', the Russians swept down the Volga, capturing the strategically important city of Astrakhan' (1556) near the point where that river empties into the Caspian Sea. Relatively secure behind the Volga, the Russians earnestly probed far beyond the Urals and thence across Siberia to the Pacific; only 64 years elapsed between Yermak's conquest of the Siberian Khanate and the founding of Okhotsk on the far coast in 1647, places separated by over 3,000 miles of forbidding country. This rapid eastward extension of the Russian domain through the boreal taiga was fueled by an insatiable demand for furs ("soft gold") and expedited by the lack of large-scale resistance from the comparatively few indigenous inhabitants, who were quickly overwhelmed by superior Russian military force (Gibson 1969: 4–24; Stebelsky 1983: 144–53). In so many ways, this Siberian expansion resembled the American and Canadian West experience, both in tangible manifestations (such as the construction of forts and

entrepôts and the use of riverine trade routes) and in the mercantile
network, at the vanguard of which were adventurers, trappers, and a
collection of freebooters and brigands. Thus, Yermak, Vlassiyev, and
the Cossacks moved through time and space much as did MacKenzie,
Bridger, Carson, Fremont, and the "mountain men" and *voyageurs*
(Lobanov-Rostovsky 1957: 79–87; Meinig 1969: 224–6). In another way,
the use of the new lands as a penal colony, the history of Russian Siberia
foreshadowed that of Australia.

The push across Siberia had skirted the northern edge of the vast
Eurasian steppe, penetration into which would necessitate a much more
concerted effort to establish an agricultural frontier in the face of
opposition from the nomadic peoples of the grasslands. Through the
sixteenth and seventeenth centuries Russian settlement of the steppe to
the west of the Volga has progressed slowly southward and southeast-
ward out of the mixed forest zone by means of the construction of
successive fortified defense lines ("zaseki") by which the nomads of the
western steppe at the time (the Tatars, Nogays, and Kalmyks) could be
kept at bay (Shaw 1983: 117–42). Gradually, cultivation of the fertile
chernozem proceeded behind the defense lines and, as Robinson so
eloquently put it:

> Here was a world so different from the ancestral forest of the Russians, that
> one may also see them blinking in the sunshine of these open plains; and
> certainly they must have been astonished at the easy turning of the rich
> black soil before the plow, and the manifold reward of the harvest. (1932: 14)

By the mid-1700s the Russians had advanced their defensive lines
east from the Volga and had in place a string of forts linked by post-
roads and screened by outposts and patrols, stretching about 2,500 miles
from Gur'yev on the northern shore of the Caspian Sea north to Ural'sk,
then east to Orenburg, Omsk, and southeast along the Irtysh to
Pavlodar, Semipalatinsk, and Ust-Kamenogorsk at the foothills of the
Altay Mountains (Stebelsky 1983: 149). Thus, the "Ural gates" (the
corridor between the southern terminus of the Ural Mountains and the
northern edge of the Caspian Sea) were closed to the nomads, and the
Russians were poised for their *Drang nach Süden* into the Asian steppe
region (Map 2.1).

The Kazakh frontier

The political-military situation in the Asian steppe region in the
early eighteenth century was extremely unstable. Tsarist expansion in

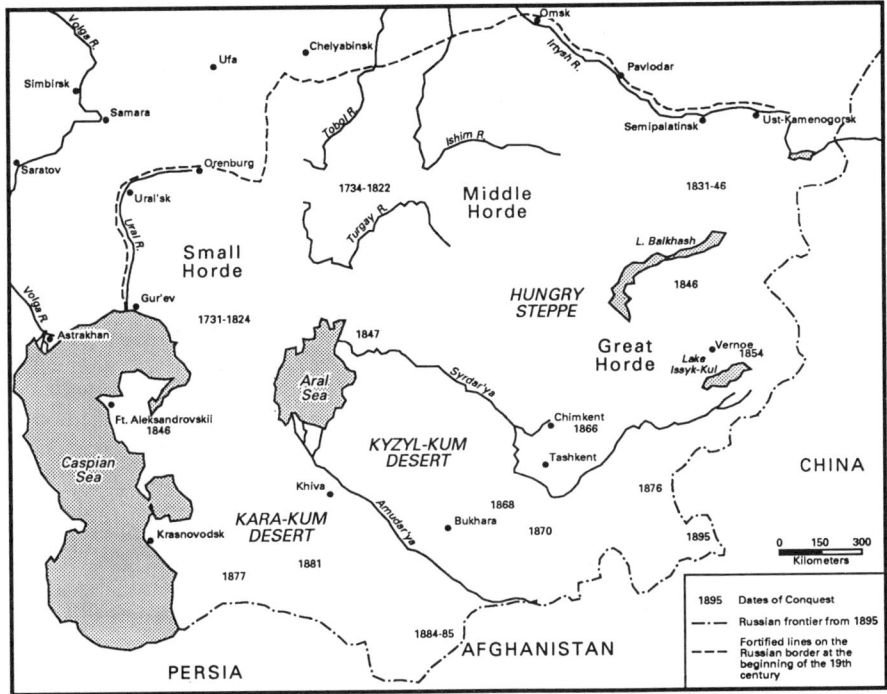

Map 2.1 Russian conquest of Central Asia

the Trans-Volga area had set in motion a series of moves by nomadic peoples retreating from Russian aggression, and forced the realignment of grazing and migration routes. Complicating this fluid scenario were shifting alliances and periodic conflict among and within the non-Russian groups, as Nogays, Kalmyks, Bashkirs, and Kazakhs marched and countermarched across the steppe and semidesert. Also, during this period the indigenous people of the steppe were under serious pressure from the Dzhungars to the east, and maintained a largely symbiotic, but not infrequently hostile, relationship with the settled population to the south—that is, in Central Asia (Allworth 1967: 10; Donnelly 1968: 41–5).

Once the long struggle to subdue the Bashkirs was concluded, the Russians could turn their attention to the Kazakh steppe and, beyond that, to Central Asia. The Kazakhs, numerically the largest of the nomadic peoples, were divided into three principal groups (hordes) which functioned independently and within which there existed considerable disunity. The Small Horde was found to the west, ranging from the northern littoral of the Caspian to the east and southeast toward the Aral Sea; the Middle Horde controlled the vicinity of the Tobol, Ishim, and Irtysh rivers; and the Great Horde was located along the mountain foreland to the south stretching from the Chu River in the west past Lake Balkhash east to the vicinity of Lake Zaysan. The Russians recognized opportunities to split off one group from another through trade and diplomatic maneuvering, as epitomized by the agreements with the Small Horde in 1731 and with the Middle Horde in 1732 and again in 1740 (Olcott 1987: 31–53). These agreements were probably seen as temporary expedients by both the Russian and Kazakh sides, and did not prevent periodic outbreaks of bilateral violence or attacks on allied groups; through such diplomacy, however, the Russians gained intelligence about and eventually influence over internal Kazakh affairs.

The early nineteenth century witnessed a major acceleration in tsarist actions toward the Kazakhs, as dissension within the Small and Middle hordes offered the Russians the chance to exploit internecine strife and annex steppe territory directly. Promulgating a new administrative structure in the Kazakh lands, the Russians backed up the political change by introducing strong military forces and constructing new fortified lines deep within the Kazakh frontier. By this time, the Russians had also proven adept at employing non-Russian irregulars to supplement their own military forces (Baumann 1987: 492–3). Traditional Kazakh pastures and herding routes, especially along the piedmonts, were cut off, and thousands of Russian and Ukrainian settlers moved south from the Siberian line to cultivate the virgin steppe. To add insult to injury, various taxes were imposed on the indigenous peoples and restrictions were placed on activities such as fishing and cutting timber (Demko 1969: 51–123; Stebelsky 1983: 154–5).

These and later Russian actions prompted repeated rebellions by the Kazakhs, most notably that by Kenisary Qasimov from 1837 to 1844. Such revolts, usually on a small but occasionally on a large scale, were a regular feature of the Russian occupation of Kazakhstan well into the twentieth century. In what ultimately proved an unequal contest, the Kazakhs did enjoy some successes, as in 1870 when Lieutenant Colonel Rukin—in true Custerian fashion—led a group of about 50 Russian troops in a punitive expedition against several thousand Kazakhs and was annihilated (Pierce 1960: 53). Russian reprisals were predictably harsh and considerable blood was shed on both sides, but over time the Kazakhs were driven back or forced into submission.

The pacification of the Kazakhs led to radical changes in the use of the land, as new administrative codes

provided a legal basis for expropriating the Kazakhs and accommodating Russian colonists. Land was decreed a state property that government would allocate for the use of Kazakhs and others. Private property was recognized only where charters were granted to Cossack officers or to descendants of the Khans. The remaining land was considered common, although plots with structures belonged to individual households. Kazakhs could thus rent land to Russian settlers, but they could not sell it. (Stebelsky 1983: 157).

Surveys conducted by the Russian government in 1895 and 1905 revealed considerable arable land "surplus" to the requirements of the Kazakhs; such assessments emboldened the authorities to promote settlement through various acts and committees. Encouraged by land made available through changes in tenure, an increasing number of migrants moved into the steppe frontier, culminating in the huge influx in the last decade of the nineteenth and the early twentieth centuries; between 1906 and 1915 alone over one million peasants migrated to Kazakhstan (Stebelsky 1983: 159). Such a volume of migration inundated the indigenous Kazakhs in many areas of the steppe, and began the long process through which they eventually became a minority in their own ethnoterritory.

Russian colonialism in Central Asia

With the Kazakh steppe absorbed into the empire, the Russians in the second half of the nineteenth century turned their attention further south toward Central Asia, an area of long historical settlement, cultural advancement, and several golden ages, but by this time in a period of relative decline. Russian interest in Central Asia was of long standing but, owing to geography, at some remove; trade between Muscovy and Central Asia was ongoing and embassies were exchanged at least as far back as the fifteenth century. Peter the Great took an active interest in establishing links with the region, and dispatched an expedition to Khiva in 1717 which met with disaster after reaching that city. Although this beginning of the official Russian penetration of Central Asia was inauspicious, surveys and other knowledge gained would prove valuable in later years (Donnelly 1975: 210–12). Other contacts and a growth of trade between Russia and Central Asia through the eighteenth century were precursors to a more ambitious move against the region in the next century.

After another unsatisfactory campaign against Khiva in 1839, the Russians resolved to advance in a less dramatic but ultimately relentless fashion. In the early 1850s the tsarist government launched a major two-pronged movement against Central Asia; one axis of advance was from the west up the Syrdar'ya River and the other from the east along the foothills of the Tien Shan Range. The envelopment closed at Chimkent in 1864, and in the process included the founding of Vernyy (now Alma-Ata) in 1854 and the capture of Kzyl-Orda in 1853. With Kazakhstan now completely encircled, the Russians confronted the three indigenous states in the area, the emirate of Bukhara and the khanates of Khiva and Kokand. After the capture of Tashkent (1865), further campaigns resulted in the collapse of the local polities as Kokand was defeated (1866) and annexed (1876); Khiva was taken (1873); and Bukhara was reduced to vassalage (1868) and partially annexed (Pierce 1960: 22–37). The final piece in the Central Asian puzzle fell into place with the Russian capture of the area which had become known as Transcaspia, the ethnic homeland of the Turkmen people. Despite fierce resistance to the Russian invaders, Transcaspia was overrun by 1885 and the regional borders rounded out by international agreement with the British by 1895 (Pierce 1960: 37–42).

Unlike the Kazakh steppe to the north, Central Asia proper was not to be settled by large numbers of Russians, except for the foothill region known then as Semirech'ye, where Cossack and peasant found excellent arable land (Stebelsky 1983: 163). Rather, the Russian presence was mercantile or military; the region was ruled by Russian military government, the most noteworthy head of which was General K. P. Von Kaufman, who held sway in the territory from 1867 until his death in 1881 and who had tremendous influence on the development of the colonial administration. Outside of Semirech'ye, Russians and other outsiders restricted themselves mainly to the cities, where they could be protected by military garrisons and where they lived apart from the indigenous inhabitants.

After establishing political control over Central Asia, the Russian colonial authorities embarked on a major effort to restructure agriculture in the region, especially to promote the growing of cotton. The American Civil War had disrupted cotton supplies for Russia's huge textile industry, so an increase in Central Asian production would reduce the empire's dependency on imports (Matley 1967: 274–7). Some of this increase resulted from expansion of irrigated land, but a significant amount was attributable to a displacement of traditional food crops. Together with the gains realized from expansion of grain cultivation by Russian colonists in Semirech'ye, by the time of the 1917 Revolution, Central Asia had become a major exporter of industrial crops and foodstuffs to other areas of the empire. The economic integration of the region into the imperial economy was facilitated by the construction of several major railroad lines, which tied into the transportation network of Siberia and European Russia. These railroads also allowed for the movement of troops and material within Central Asia in the event of local disturbances (Pierce 1960: 45). Mining activities in Central Asia were well underway by the late nineteenth century, financed and directed by Russians and other outsiders. Thus, in rather short order the Central Asian colony was providing raw materials to the industries of the Russian metropolis (Pierce 1960: 190–5).

Curiously, the Russian colonial administration embarked on an extraordinarily progressive and far-sighted land reform program in Central Asia. By turning land over to those who worked it and giving them hereditary claims, a move "more liberal than anything the government had dared to undertake in Russia," and by severely restricting the right of Russians to buy land in the region (outside of Semirech'ye), Von Kaufman avoided clashing directly with the local population and may have neutralized resentment to outside rule (Pierce 1960: 149–50). A relatively enlightened tax policy and a judicious regard for water rights were other hallmarks of Von Kaufman's rule.

Although Von Kaufman deserves some credit for his wisdom in dealing with the Central Asians in such matters, resistance to rule by outsiders continued, and occasionally flared into violence. It is certainly the case that internecine conflict occupied Central Asia even after Russian dominance was established, but the outsiders were in numerous instances the target of attack (Manz 1987: 267–8, 281). Such hostility was still evident in the Basmachi rebellion during the early Soviet period (Olcott 1981: 352–69).

THE RUSSIAN EXPERIENCE IN CENTRAL ASIA IN COMPARATIVE PERSPECTIVE

As we have seen, the expansion of the Russian Empire into Central Asia involved all of the elements of the frontier and colonial models derived from the growth of European hegemonism in North America, Australia and New Zealand, Africa, and elsewhere. In the early period, trade and diplomacy brought the Russians into contact with the nomadic peoples of the Eurasian steppe, a contact which eventually escalated into conflict as the lure of agricultural land and the promise of more lucrative trade with the sedentary societies of Central Asia pulled the Russians ever southward. Territory was annexed in turn and land was expropriated as the frontier advanced, and alien settlers poured in to take advantage of agricultural opportunities. The Russian acquisition of Central Asia proper was a logical progression from the pacification of the Kazakh steppe, urged on by strategic geopolitics, a sense of manifest destiny, and the call to glory for military commanders.

However, the colonial system established in Central Asia differed significantly from the frontier model imposed on Kazakhstan. First, Central Asia was already inhabited by sedentary cultivators with a system of landownership in place, and accordingly there was relatively little surplus land available in that region without expanding the costly irrigation network. The number of migrants who could be accommodated in Central Asia without completely disrupting the status quo was, therefore, relatively small compared with Kazakhstan. By the time of the 1926 Soviet census, over 2.2 million Russians and other Slavs lived in Kazakhstan, whereas only about 600,000 were enumerated in Central Asia (Lewis et al. 1975: 290). Second, the motives for and the method of annexation of the two regions were substantially different. In the case of Kazakhstan, Russian expansion was probably seen as more of a "natural" extension of the agricultural frontier, a process begun centuries earlier far to the west. With regard to Central Asia, conquest followed a more deliberate plan, in part because the logistics of mounting military compaigns across vast expanses of inhospitable terrain required a more organized effort. Also, the Russian imperialists of the late nineteenth century were no doubt in touch with kindred spirits elsewhere, some of whom (such as the British in South Asia) posed a threat to Russian interests; all of this lent an aura of legitimacy to the pursuit of empire. Thus, Fieldhouse, in his survey of European colonialism, saw Russian Central Asia as "a typical colonial society, autocratically governed by aliens, with a growing settler population, a vast cultural and linguistic gap between Central Asians and immigrants, and a dependent primary economy" (1967: 339).

What sets Central Asia apart from the mainstream of imperialist history is, of course, the fact that the flow of the nineteenth century—which brought about 85 percent of the Earth's territory under European control—did not ebb in his region in the twentieth century as it did in almost every other part of the world. Consequently, Central Asia today is a multiethnic region in which indigenous and intruder groups interact within a larger, ethnoterritorial state, the Soviet Union. The colonial heritage continues to manifest itself to this day, largely because the ethnic stratification system established in tsarist times has bent, but has not broken. As will be evident in the chapters that follow, every facet of modern Soviet Central Asia is to some degree influenced by the legacy of the Russian imperial past.

REFERENCES

Allworth, Edward 1967. Encounter. In Edward Allworth (ed.), *Central Asia: A Century of Russian Rule*, New York: Columbia University Press.

Baumann, Robert, F. 1987. Subject nationalities in the military service of Imperial Russia: the case of the Bashkirs. *Slavic Review* 46: 489–502.

Bowman, Isaiah 1931. *The Pioneer Fringe*. Special Publication No. 13. New York: American Geographical Association.

Connell, Evan S. 1984. *Son of the Morning Star: Custer and the Little Bighorn*. New York: Harper & Row.

Demko, George J. 1969. *The Russian Colonization of Kazakhstan, 1896–1916*. Bloomington, Ind.: Indiana University Publications.

Donnelly, Alton S. 1968. *The Russian Conquest of Bashkiria, 1552–1740*. New Haven, Conn.: Yale University Press.

Donnelly, Alton S. 1975. Peter the Great and Central Asia. *Canadian Slavonic Papers* 17: 202–17.

Eidt, Robert C. 1971. *Pioneer Settlement in Northeast Argentina*. Madison, Wis.: University of Wisconsin Press.

Elkins, Stanley M. and McKitrick, Eric L. 1954. A meaning for Turner's frontier. *Political Science Quarterly* 69: 321–53.

Enloe, Cynthia H. 1980. *Ethnic Soldiers: State Security in Divided Societies*. Athens, Ga: University of Georgia Press.

Fieldhouse, D. K. 1967. *The Colonial Empires: A Comparative Survey from the Eighteenth Century*. New York: Delacorte Press.

Gibson, James R. 1969. *Feeding the Russian Fur Trade*. Madison, Wis.: University of Wisconsin Press.

Horne, Alistair 1977. *A Savage War of Peace: Algeria 1954–1962*. New York: Penguin.

Hughes, Robert 1987. *The Fatal Shore*. New York: Knopf.

Lewis, Robert A., Rowland, Richard H. and Clem Ralph S. 1975. Modernization, population change and nationality in Soviet Central Asia and Kazakhstan. *Canadian Slavonic Papers* XVII: 286–301.

Lewis, Robert A., Rowland, Richard H., and Clem, Ralph S. 1976. *Nationality and Population Change in Russia and the USSR*. New York: Praeger.

Lobanov-Rostovsky, A. 1957. Russian expansion in the Far East in the light of the Turner hypothesis. In Walker D. Wyman and Clifton B. Kroeber (eds), *The Frontier in Perspective*, pp. 79–94. Madison, Wis.: University of Wisconsin Press.

Manz, Beatrice Forbes 1987. Central Asian uprisings in the nineteenth century: Ferghana under the Russians. *Russian Review* 46: 267–81.

Matley, Ian Murray 1967. Agricultural development. In Edward Allworth (ed.), *Central Asia: A Century of Russian Rule*, pp. 266–308. New York: Columbia University Press.

Meinig, D. W. 1969. A macrogeography of Western imperialism: some morphologies of moving frontiers of political control. In Fay Gale and Graham H. Lawton (eds), *Settlement and Encounter*, pp. 213–40. Melbourne: Oxford University Press.

Meinig, D. W. 1971. *Southwest, Three Peoples in Geographical Change, 1600–1970*. New York: Oxford University Press.

Mikesell Marvin W. 1960. Comparative studies in frontier history. *Annals of the Association of American Geographers* 50: 62–74.

Miller, David Harry and Savage, William W. Jr 1977. Ethnic stereotypes and the frontier: a comparative study of Roman and American experience. In David Harry Miller and Jerome O. Steffen (eds), *The Frontier: Comparative Studies*, pp. 109–37. Norman, Okla.: University of Oklahoma Press.

Morris, Donald R. 1965. *The Washing of the Spears: The Rise and Fall of the Zulu Nation*. New York: Simon & Schuster.

Olcott, Martha Brill 1981. The Basmachi or freemen's revolt in Turkestan 1918–24. *Soviet Studies* 33: 352–69.

Olcott, Martha Brill 1987. *The Kazakhs*. Stanford, Calif.: Hoover Institution Press.

Pierce, Richard A. 1960. *Russian Central Asia, 1867–1917: A Study in Colonial Rule*. Berkeley, Calif.: University of California Press.

Pollock, Norman 1980. Contacts between settlers and native peoples. In Anthony Lemon and Norman Pollock (eds), *Studies in Oversea Settlement and Population*, pp. 81–101. London: Longman.

Price, A. Grenfell 1950. *White Settlers and Native Peoples*. Cambridge: Cambridge University Press.

Robinson, Geroid T 1932. *Rural Russia under the Old Regime*. New York: Macmillan.

Shaw, Denis J. B. 1983. Southern frontiers of Muscovy, 1550–1700. In James H. Bater and R. A. French (eds), *Studies in Russian Historical Geography*, pp. 117–42. New York: Academic Press.

Shibutani, Tamotsu and Kwan, Kian M. 1965. *Ethnic Stratification: A Comparative Approach*. New York: Macmillan.

Stebelsky, I. 1983. The Frontier in Central Asia. In James H. Bater and R. A. French (eds), *Studies in Russian Historical Geography*, pp. 143–73. New York: Academic Press.

Thompson, Leonard and Lamar, Howard 1981. Comparative frontier history. In Howard Lamar and Leonard Thompson (eds), *The Frontier in History*, pp. 3–13. New Haven, Conn.: Yale University Press.

Turner, Frederick Jackson 1920. *The Frontier in American History*. New York: Holt, Rinehart & Winston.

Webb, Walter Prescott 1951. *The Great Frontier*. Austin, Tex.: University of Texas Press.

Chapter 3

The Political Geography of Soviet Central Asia: Integrating the Central Asian Frontier

Lee Schwartz

INTRODUCTION

The study of political geography is primarily concerned with the relationships between man's political institutions and his surrounding environment. In order to understand the operational aspects of such a relationship, it is important to understand the origins as well as the functions of political-geographic characteristics such as boundaries (both internal and international), administrative areas, and features of national and local autonomy. The consequences of the system of complex interactions defining any politically-organized region are reflected in decision making related to particular spatial concerns, including issues of administrative organization, political representation, investment, and foreign policy. Soviet Central Asia, which is often viewed today as a distinct geographic region, is nonetheless an area of rich spatial diversity. How and why this region does or does not operate as a functional political whole will be a major concern of this chapter.

Soviet Central Asia (the Turkmen, Uzbek, Kirgiz, and Tadzhik Soviet Socialist Republics) is a particularly interesting area of the world to look at from the perspective of political geography, given its intermediary position in the so-called "pivot" of Eurasia, between east and west, north and south (Lattimore 1950: ix). Located in the historic crossroads between Europe and Asia, and characterized physically by its vast flat and easily traversed terrain, its history has been one more of being subject to conquest from outside than of political consolidation from within. Over the centuries this region has come under the authority of dynastic rulers from Alexander the Great to Ghengis Khan to Tamerlane, and has been subject to the influence of peoples as diverse as the Persians, Huns, Arabs, Turks, Mongols, and, in the most recent century and a half, the Russians.

History has often been considered a bedfellow of geography, and the study of the political geography of any particular region necessarily involves the understanding of the historical influences which have forged the context of present-day political institutions and structures. This does not, however, necessitate becoming excessively bogged down in descriptive detail, but instead requires the selective isolation of primary historical influences on the political organization of Central Asia today. This chapter will therefore focus first, and briefly, on the essential aspects of the historical political geography of what is now Soviet Central Asia—the early formation of boundaries and regions, the nature of the system of national autonomy and federalism, and the competing interests which have operated over time. The historical time period most pertinent to this chapter lasts from the late nineteenth century through the early years of Bolshevik rule. These are the years when the political authority of first the Russian Empire and then the Soviet state was firmly established and explicitly defined in the broad geographic region of Central Asia.

The scale of the investigation is of importance to any geographic study. Various schools of geographic thought have evolved over the years, many of which have depended to a large extent on the scale of analysis. Three levels of generalization which are readily categorized in most cases are those at the macro (small), meso (middle), and micro (large) scales of analysis. It is important to keep in mind that none of these scales is mutually exclusive; that is, each contributes to a greater understanding of political-geographic relationships, depending on what one is trying to portray. For example, a study of population density at the macro scale of analysis might be conducted on a global or national basis. A similar investigation would reveal strikingly different patterns when conducted at the meso and micro levels, utilizing increasingly larger scales of analysis, such as states, cities, counties, or census tracts.

After a brief review of the historical political geography of what is today Soviet Central Asia, this chapter will investigate selected aspects of all three levels of analysis. Soviet Central Asia first will be considered as an integral whole (at the macro scale), in terms of both its relationship to the entire Union, as well as its strategic or geopolitical importance as a geographic fulcrum between Europe and Asia. The next section will focus on the republic, or meso-level, scale in order to investigate the federal nature of the Soviet state, as well as certain nationality aspects of political participation. Last, a larger (micro) scale of analysis will be used to trace the changing political and territorial-administrative organization within the republic of Soviet Central Asia over the years.

It is not the goal of this chapter to cover all of the various factors contributing to the political organization of space in Soviet Central Asia. Rather, I will be highlighting several issues and concerns which contribute to an understanding of both change and stability throughout the region. Each scale of analysis used will reveal different aspects of the integrating process related to the region and republics of Central Asia.

POLITICAL INTEGRATION OF THE CENTRAL ASIAN PERIPHERY

Within the context of political geography, one way of viewing the prospects for consolidating the Central Asian periphery is from the perspective of political integration. A political system is considered highly integrated to the extent that there exists a pool of commonly accepted norms and institutions throughout the various political communities of a state (Ake 1967: 3). In the Soviet Union, a hierarchical duplication of governmental structure and party organs at both the federal and local levels has produced a high degree of *institutional* integration throughout the country. According to the functionalist school of integration theory, public legitimacy should gradually develop as people become aware of the tangible benefits of membership in a community or, in the Soviet case, a federation.

In Central Asia, such "tangible benefits" may be construed as the relative proportion of resources allocated from the center to the Kirgiz, Tadzhik, Turkmen, and Uzbek republics. In a Soviet system which has been geared towards a "merging of socialist nations," regional equalization has produced a degree of convergence in measured levels of economic investment, social opportunity and mobilization, and political administration and representation. Because other chapters in this book (see Liebowitz, this volume) deal with developments related to factors of socioeconomic equalization, the concern here will be with integration that is political in nature.

A question which is interesting to investigate further in a discussion of the political geography of Soviet Central Asia is whether the equalization process has resulted in a situation whereby the political autonomy of the different units, at various scales, has functioned in a way to make them a more functional part of a larger aggregate. When members of a group obey and conform to the commands and injunctions of political institutions, this is generally construed as support (DeVree 1972: 109). In this sense, integration is concerned with, and in fact may depend upon, the actions of a loyal political community.

Political community, therefore, is a condition in which specific groups and individuals show more loyalty to their central political institutions than to any other political authority, in a specific period of time and in a definable

geographic space. ... A population may be said to be loyal to a set of
symbols and institutions when it habitually and predictably over long
periods obeys the injunctions of their authority and turns to them for the
satisfaction of important expectations. (Haas 1958: 5)

In the Soviet Union, political institutions such as the Communist
Party and state and local legislatures are structures which have the
potential to maintain republic support for the federation. A high degree
of interaction between the center and periphery can also contribute to
the political consolidation of a community of peoples. On the other
hand, critics of traditional integration theory contend that, with
modernization and equalization, there are tendencies which operate
against loyalty toward the center: the rise of national awareness, the
inertia of bureaucracies which tend to usurp tangible benefits, and the
problems which arise from the need to appeal to parochial rights to
maintain support.

A primary concern of political geography is that of the political
regionalization of space. When political subdivisions are also based
on national composition, the scheme of administrative structure and
authority clearly has a heightened and more salient potential for serving
as a force of regional integration or disintegration. Integration, therefore, in the context which applies herein, refers to the level and degree
of interaction between and within the various political units of Soviet
Central Asia, and between these units and the rest of the Union.

HISTORICAL POLITICAL GEOGRAPHY OF SOVIET CENTRAL ASIA

After centuries of incursions by various foreign powers, the Uzbeks
emerged, beginning in the sixteenth century, as the first dominant local
power in Western Turkestan,[1] eventually conquering much of the
former domain of Tamerlane centered at Samarkand. To the west, the
Turkmens gradually consolidated their presence in the Kara Kum Desert
between the Atrek and Amudar'ya rivers, and the ethnically Persian
Tadzhiks were settled in the Pamir Mountains to the east. To the north,
the Russians were slowly making incursions into the Kazakh steppe,
where they had established a series of frontier outpost fortifications
thrusting down toward the Central Asian oases.[2]

By the beginning of the nineteenth century, the region surrounding
these oases had become loosely organized under the administration of
the three khanates of Khiva, Bukhara, and Kokand.[3] Even when the

entire region had come under the control of the Russian Empire by the second half of the nineteenth century, the tsarist administration paid little attention to the specifics of the political subdivisions in Central Asia. The three Uzbek-dominated khanates, along with the Turkmen region of Transcaspian Turkestan, remained the basis for the territorial-administrative structure of much of Central Asia up until the years following the Bolshevik Revolution. By the end of the nineteenth century the khanate of Kokand had been incorporated into the governor-generalship of Turkestan, while the khanates of Bukhara and Khiva were made protectorates under the tsar in 1868 and 1876, respectively. While this meant nominal sovereignty in terms of maintaining their own army and judiciary, Khiva and Bukhara had, in effect, become vassal states of the Russian Empire. The annexation of the Pamir region to Central Asia in 1895 marked the end of Russian territorial acquisitions in the region (see Wheeler 1964).

Although the territory of Central Asia was the last major geographic region to come under the control of the Russian Empire, its subjugation occurred relatively rapidly. The completion of a railway network greatly facilitated Russian access to the heart of Central Asia. By 1888 the Trans-Caspian Railroad had been extended to Samarkand and the Orenburg–Tashkent Railroad was completed in 1906 (Matley 1967: 327). Less than 50 years passed between the time Russian authority was firmly established throughout the Kazakh steppes and the time when the final thrust to the Afghan and Persian borders was completed. Of interest for the historical context of this period is the fact that the American Civil War had caused a severe disruption in the supply of cotton to Europe, and there was a heightened interest in gaining access to the fertile lands of the Fergana Valley. Furthermore, the time of the most aggressive Russian activity directed successfully toward the southern Asia frontier corresponded with a series of military setbacks suffered in the west and east, beginning with the Crimean War in 1854 and ending with defeat in the Russo-Japanese War of 1905.

The term "frontier" as it applies to this region of the Soviet Union has two different meanings. One describes the settlement frontier of Slavic expansion through the south Russian steppes and toward the heart of Kazakhstan. This region was rather sparsely populated and would eventually become dominated numerically by the Russian and other Slavic settlers who moved in to farm the land. This former frontier is still very much evident in the patterns of settlement, land use, and investment allocations. In a true sense, in terms of topography, land use, population, and socioeconomic measures, the northern two-thirds of Kazakhstan has more similarities to the Russian republic than to those of Central Asia.[4]

The other way in which the term "frontier" applies in Central Asia is to the external boundary of the Soviet Union, at the so-called pivot of

Eurasia. This frontier is an example of a colonial frontier, where the limit of aggressive tsarist expansion was demarcated before the area in between the former frontier and current boundary became subjugated. This area, containing what are today the four republics of Soviet Central Asia, has remained relatively unscathed by Russian cultural influences, even though it has been politically and economically incorporated within the Soviet Union. The southern frontier of Soviet Central Asia traverses a zone of physiographic and cultural similarities, and in this sense is less distinct than the political, economic, and ideological delimitation marked by the actual boundary. This zone, reaching towards what is today Turkey, Iran, and Afghanistan, also played the classic frontier role of a buffer between tsarist and British expansion in the nineteenth century.

Internally, the region of Central Asia eventually was subdivided into oblasts, which were further divided into uyezds and smaller districts called "uchastki." Within the framework of this structure, territorial reorganizations took place in both 1882 and 1898 (McIntyre 1974: 58). By 1922 the former khanates of Khiva and Bukhara had become the Khorezmian and Bukharan "People's Soviet Republics," and in 1923 and 1924 they became the Soviet Socialist Republics of Khorezm and Bukhara. In 1924 the administrative boundaries of Soviet Central Asia were delimited based on nationality criteria and by the end of the year the entire region had been divided into the four autonomous jurisdictions of the Turkmen, Uzbek, Tadzhik, and Kirgiz Soviet Socialist Republics (SSRs).[5]

The federal system of union and autonomous republics at the primary and secondary nationality levels, and oblasts (regions) and rayons (districts) at the purely administrative tertiary sublevel, defines the structure of politically organized space in Soviet Central Asia today (Map 3.1). It is interesting to note that while the administrative and economic subdivisions within Central Asia have been in a state of near constant flux over the years, the boundaries of the republics and autonomous regions have remained particularly inviolable. There have been numerous internal administrative changes made—such as the upgrading in autonomous status, the shifting of jurisdictional authority, the creation of subunits, and some minor territorial adjustments and transfers—but the basic external boundaries of the four Central Asian republics, and two autonomous regions, have remained largely unchanged. (The exception is the Uzbek SSR, which gained a significant amount of territory from the Kazakh SSR in 1963, part of which was transferred back to the Kazakh SSR in 1971.) This is not to say that there were not numerous problems encountered when trying to divide this vast region based on a population whose distribution was largely uncharted, characterized by nomadic migrations and a conspicuous lack of national self-awareness. The actual boundaries were for the most part

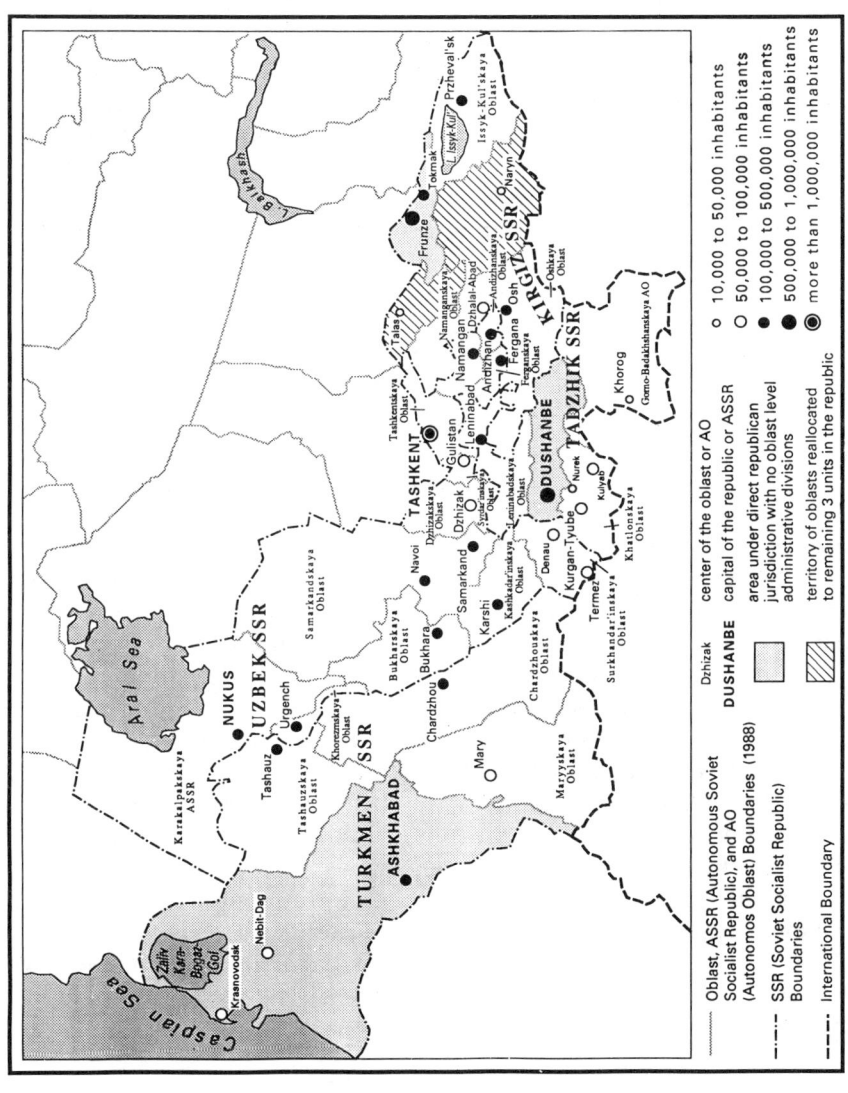

Map 3.1 Political-administrative structure of Soviet Central Asia, 1988.

delimited by local republic party officials and based on historic political divisions. A good deal of dispute over the establishment of these republics has revolved around the contention that the boundaries which ostensibly divided the indigenous peoples were artificial creations (see, for example, Bennigsen and Lemercier-Quelquejay 1967: 130–7). The fact remains that over the years these borders have taken on a permanence in the way they describe settlement patterns, divide nations and ethnic groups, determine investment, and encompass government and party institutions as the secondary and tertiary levels of administration.

The *inter*national boundaries of Soviet Central Asia resulted from centuries of conflict and dispute, and are based on a different historical context than are the internal divisions of Central Asia. By the middle of the nineteenth century the boundaries with the adjoining countries of Persia, Afghanistan, and China were still far from firmly established. The major events consolidating this frontier were: the 1881 truce establishing the Russian frontier with Persia after the battle of Gok Tepe; the 1887 and 1895 demarcation of the Afghan-Russian boundary by joint commissions consisting of Russian and British representatives; and the 1860 Sino-Russian Treaty which signified China's acquiescence to the Russian annexations along the Sinkiang-Uygur frontier. These international borders were determined through bilateral treaty negotiations which often were initiated due to the threat or outbreak of armed conflict. One result of this process of boundary delimitation was the separation of a number of ethnic groups between two or more countries. These included the Turkmens, the Uzbeks, the Tadzhiks, the Uygurs, and the Kazakhs. Such a situation clearly has the potential to serve as a disruptive, centrifugal force for the future of regional political stability in the Soviet Union.

Now that the historical political geography of the region has been provided, the remainder of this chapter will focus on selected issues in the political geography of Soviet Central Asia over the past several decades (the stated time frame of this book is the post-World War II period). The organizational framework will be based on utilizing the scale that is most appropriate for the analysis of a particular topic, be it the regional (macro), republic (meso), or intra-republic (micro) level. Whenever possible, for purposes of consistency, census years (1959, 1970, 1979, 1989) will be used as benchmark dates for the compilation of data; for the decade of the 1980s, an effort will be made to provide as current information as is available.

CENTRAL ASIA AS A UNIFIED REGION: CONSOLIDATING THE FRONTIER

The initial, macro level of analysis in this context views the four republics of Soviet Central Asia as comprising an integrated whole, and examines the role such a region plays in two ways. One investigates the influence of this region in terms of domestic policy, and the other looks at the foreign policy implications, or international concerns, of Soviet Central Asia.

In the latter sense, one can view the political geography of Soviet Central Asia from the realm of geopolitics, or "applied political geography." This disciplinary offshoot of the field of political geography provides a valuable perspective for analyzing spatial information for use in the application of state policy, both domestic as well as international. In a geopolitical sense, the *orientation* of the region is seen as most important. Do the political-geographic characteristics of the four Central Asian republics taken together cause the region as a whole to function as a unifying or divisive force for Soviet interests, both internal and international? This approach is particularly appropriate today in light of concerns expressed over the strategic importance of this region's rapidly growing population, distinct ethnic and religious heritage, and location in proximity to recent wars and revolution in neighboring Afghanistan and Iran.

Domestic geopolitical issues in Soviet Central Asia

There are three main domestic issues the Soviet leadership is faced with concerning Central Asia today. They are:

(a) concern over the disproportionate share of Soviet population growth caused by the high rates of natural increase among the Muslim populations of Central Asia;
(b) decisions for regional investment strategies based on priorities of intensive vs extensive growth and the desire for a more cost-effective return on investment; and
(c) the prospects for an Islamic revival or, at the least, a rising political opposition based on a sense of a common religious heritage.

While none of these concerns is specifically political in nature, each has the potential to become a politically divisive force in a region which by many measures is a distinctive entity. Specific problems caused by the disequilibrium of population growth and regional investment will be covered in other chapters. Suffice it to say that issues such as the nationality composition of the military and discussions over the possibility of river diversion projects have caused the Central Asian republic to be considered very much as a distinct regional political whole.

The third domestic issue I have identified in this geographic region—the prospects for political solidarity based on a sense of a common religious heritage—deserves some elaboration, as it is directly related both to the region's unity as well as to its potential for divisiveness. It has become clear that the Turkic-Muslim populations of Central Asia have resisted integration with the dominant Slavic segment of Soviet society. The reasons for this resistance are complex and interrelated, but have nevertheless resulted in a society which has become stratified in many ways based on a common religious affiliation.

It would appear that the sector of society which is most resistant to integration is that which is also most suceptible to potentially problematic regional movements based on a sense of solidarity which is rooted primarily in the countryside and usually associated with an Islamic religious tradition. With more and more indigenous Central Asians entering political and economic elites, the potential for a further dichotomization of society is great. There is a growing indigenous intelligentsia which could provide the political leadership for a movement for political consolidation based on a common regional, if not religious, consciousness.

Throughout Central Asia there have been numerous protests and demonstrations which have led to confrontation between members of the indigenous Central Asian nations and Russians, some of which may be related to religious or ethnic rights, some to economic or political competition, and some to a rebellion against Soviet authority. Most notable of these have been altercations between Kazakhs and Russians in Alma-Ata. On the other hand, several incidences of violent conflict between two ethnic groups, both of which have been Muslim, bring into serious question whether Islam can properly by construed as a potent integrating force throughout Central Asia. Examples of this latter type of conflict include the riots between Uzbeks and Meskhetian Turks in the Fergana Valley, and cross-border clashes between Tadzhiks and Kirgiz. Such incidents point to the increasing importance of the republics as national homelands, which would tend to dissipate movements for the consolidation of the entire Islamic population of the region.

Whether or not the bases for these local disputes are religious or national in origin, or based on core–periphery or equalization relations, the very existence of Islam as a potentially powerful unifying political force throughout each of the republics is often considered a factor in the prospects for consolidating this region as an integral whole. At this point in time, however, evidence would appear to point against religion as an effective unifying force among the peoples of the various republics of Soviet Central Asia (Kaiser, this volume).

Early fears of Islamic or pan-Turkic solidarity were allayed by dividing the region into distinct republics and encouraging their development and relations with the Soviet center. Now that this administrative distinction has taken on a sense of permanence, the

Soviet leadership has moved toward a program of rationalizing economic investment, which could contribute to a further disintegration of this region from the center. It seems that, despite the stated goals of Marxism-Leninism, there has been a desire in recent years *not* to integrate the republics of Central Asia as a whole functional region of the Soviet Union, whether for reasons of economic efficiency, ethnic tensions, or the implications for political control.[6]

The international geopolitical situation—relative location

The last item to consider regarding the view of Soviet Central Asia as a regional whole is the geostrategic value of this area, due not only to its location in the "pivot" of Asia, but also due to its proximity to the political dynamism of the Islamic renewal movements in the contiguous countries of Afghanistan and Iran. During the years of Soviet intervention in Afghanistan, the USSR media wrote against the view, advanced primarily in the West, that the decision to engage in military activities was based in part upon the fear of a Soviet Islamic revival.

There are signs, however, that since the military invasion of Afghanistan, the atmosphere of official friendship and cooperation between Soviet authorities and the Soviet Turkic-Muslim leadership has been strained. These include a surge in anti-Islamic publications, a crackdown on Sufi religious activities, an officially expressed concern about the increased availability of information about Islamic activities abroad, and the early withdrawal of Central Asian troops from service in Afghanistan.

While there does not appear to be strong evidence for Islamic religious fundamentalist activism in the Soviet Union (unlike the Muslims of the Shiite Iranian fundamentalist movement, the Soviet Central Asian Muslims are predominantly Sunni), there is concern over the fact that ethnic brethren of the Uzbek, Turkmen, and Tadzhik peoples, specifically, are located just across the Soviet border in the northern provinces of Iran and Afghanistan. Many of these are direct descendants of those who fled Soviet Central Asia during the Civil War, Basmachi rebellion, and in the 1930s due to collectivization and religious persecution.

On the other hand is the view that the continuing civil struggle has the potential to separate further the cross-border peripheral nations if it can succeed in bonding Afghanistan's various different opposition groups together. While the implications for Central Asia of a decade of active Soviet involvement in Afghanistan are not yet clear, Soviet policy appears to have been clearly governed more by security and strategic concerns than by ethnic or religious ones. Still, as Table 3.1 reveals, the international boundaries of the Central Asian republics transgress the traditional settlement patterns of a number of rather numerous nations.

Table 3.1 includes the extraterritorial populations of the four major Central Asian nations as well as Kazakhs (in the People's Republic of China) and Azerbaydzhani (in Iran). These latter two are included due to their large numbers and relative proximity to the boundaries of Soviet Central Asia. In addition to the peoples listed in Table 3.1, the countries encircling Soviet Central Asia contain a large number of Uygurs and Kurds.[7] While neither of these latter groups are indigenous to Soviet Central Asia, their significant presence on both sides of the border creates the potential for the expression of irredentist desires for unifying the historic homeland.

Table 3.1   Indigenous Soviet Central Asian nationalities living in countries bordering the "southern tier," 1975–83.

	1978	1983
People's Republic of China		
Uzbeks	14,000	13,000[1]
Tadzhiks	20,000	26,500
Kirgiz	120,000	115,000[1]
Kazakhs	840,000	930,000
Iran		
Turkmen	550,000	650,000
Azerbaydzhani	5,800,000	6,200,000
Tadzhiks	40,000	44,000
Afghanistan		
Uzbeks	1,300,000	1,500,000
Tadzhiks	3,000,000	3,500,000
Kirgiz	10,000	5–8,000[2]
Turkmen	300,000	350,000
Turkey		
Turkmen	100,000	120,000

[1] The numbers here decrease from 1978 due to inaccurate estimates made prior to the 1983 PRC census.
[2] This approximation is due to the fact that a large number of Kirgiz peoples were relocated to Turkey in 1982, following the Soviet invasion of Afghanistan.
Sources: Bruk 1981: 377, 441, 459, 469, 526, 1986: 419, 438, 523; Bennigsen and Wimbush 1986: 36.

There has also been raised the question of economic motives of the Soviet Union in Afghanistan (natural gas) as well as the contention that the Soviet invasion in Afghanistan represented an initial step in the revival of a centuries-old quest for a warm-water port on the Persian Gulf. This brings up the broader issue of viewing Soviet Central Asia as part of a larger geopolitical region referred to sometimes as "Inner Asia" which includes Afghanistan, Sinkiang Province of the PRC, Mongolia, and Tibet.[8] This is the area referred to earlier as the geographical "pivot" of Asia, and the Soviet Union appears to have the desire to maintain a strategic advantage in this region for defensive purposes if nothing else. This has an impact on the tone of Sino-Soviet relations concerning "Inner Asia" as well. Indeed, the argument is easily made that this region falls clearly within the geopolitical sphere of influence of the Soviet Union.

There emerges a final strategic value which has been attributed to Soviet Central Asia, that is, as a model for development in Islamic and other Third World countries in general. By any standard, the economic development that has occurred among the peoples of Central Asia has been remarkable, particularly when compared to the kindred national groups just across the Soviet border. The Soviets have over the years conducted programs of exchanges with representatives from under-developed countries to showcase Central Asia as an example of what Soviet-style economic development could do in *their* countries. While the assessment of this program's success in Islamic countries has been marginal, it is said to have been partially effective with some lesser developed countries. In the realm of international affairs, therefore, the continued development of Central Asia as an integral part of the Soviet state is a desirable policy to pursue.

THE NATURE OF SOVIET FEDERALISM: THE POLITICAL STRUCTURE

The next logical level of analysis of the regional political geography of Soviet Central Asia is at the union and autonomous republic level. At this (meso) level it is important to recognize that, while there are certain underlying features common throughout the republics of Soviet Central Asia, there are important differences as well. While most of these similarities and differences are usually depicted as cultural, religious, and linguistic in nature, there are political characteristics which also should be brought to light.

The political geography of any region in the Soviet Union can be analyzed based on the nature of Soviet federalism. The modern federal structure of the Soviet state was established on the principle of national autonomy and this concept can be analyzed further from two perspectives: that of political autonomy, and that of jurisdictional autonomy. The question is debatable as to just how nominal such an autonomous distinction is; one should hesitate to conclude that autonomy is in name only. There have in the past been occasions, particularly those concerning matters of language, education, and territory, where the mechanisms of the system of national autonomy have made significant contributions to policy formation and decision making in the Soviet Union (Bloembergen

1967: 28). The perestroyka-related reform policies of regional "khozraschet" (self-financing) and the creation of spheres of local autonomy have remained primarily economic in nature, and the calls for political secession or realignment that have proliferated in the Baltics and Transcaucasus have not been as rife in Central Asia. Still, if the rate of reform continues apace, the future will doubtless bring an increase in the degree of national autonomy and local control exercised by the individual republics throughout the USSR, including Central Asia.

Political autonomy

The first perspective on the nature of the federal system referred to above—political autonomy—is more or less constitutionally determined (Articles 70–2) and relatively consistent throughout the Soviet Union. The jurisdictional aspect, on the other hand, is more esoteric, has implicit geographic implications, and is important if one is to understand the more subtle distinctions among the various republics of Soviet Central Asia.

In terms of political autonomy, each union republic within the Soviet federation (with the unique exception of the Russian Republic) has its own party and government structures parallel to those of the union-wide organizations. These include a Communist Party central committee and politburo, a judiciary system, a council of ministers, local as well as union-wide soviets (councils) of government, and a constitution which mirrors the Soviet constitution but takes into account the special national, historic, and economic features of the republic. The Soviet constitution (USSR 1978: Article 72) also gives to each union republic the right to secede, which is one reason why there exists no union republic without an external (that is, international) border. Due to the fact that the republics are based on a specific national population, the legislatures are unicameral, requiring no separate Soviet of Nationalities. Instead, where significant concentrations of national minorities exist, they are delimited into lower levels of autonomous jurisdictions—the autonomous republics, oblasts, and okrugs.

Constitutionally, therefore, the Soviet Union allows for a great deal of autonomy in decision-making, but because the system of planning and allocation of resources remains highly centralized, most decisions at the republic level depend on directives from Moscow. Nevertheless, the right to separate political representation at the union-wide level is maintained by the constituent republics, which each send 11 representatives to the Council of Nationalities in the newly constituted bicameral Soviet legislature.[9] In addition, there has traditionally been a representative from each union republic with the special status as a vice-president of the presidium of the Supreme Soviet of the USSR.

The structure of the Soviet federal system has resulted in few differences among the individual republics in terms of regional political authority; a great deal of institutional standardization exists within and throughout the republics of the Soviet Union. At the same time, regional politics have proven more durable than originally expected (see Bahry 1987). There has been a recognition of the difficulty involved in achieving efficient central control over a polity that is organized into 15 union republics which include over 150 oblasts and more than 40,000 local government councils, or soviets (Khorev 1981: 147–8, 222–3). One reason for a lack of greater autonomy in terms of political authority lies in the need for more integration ("sochetaniye") between branch, sectoral, and territorial needs within the economy. Despite numerous attempts over the last several decades at regional reorganization (from Sovnarkhozy to Industrial Management Regions to Territorial Production Complexes), the relationship between central planning and regional political authority remains ambiguous, and highly inefficient (Dellenbrant, 1986: 62–84). Thus, even as local authority on many matters specific to the republics of Central Asia has increased, there remains a high amount of centralization in the allocation and execution of such authority.

Various researchers trying to determine the Central Asian hierarchy of power from a nationality perspective have investigated changing rates of Communist Party membership by nationality within each union republic (see Burg 1984: 40–62; Karklins 1986: 77–100). Table 3.2 shows the changing pattern of the absolute and relative shares of total republic Communist Party membership held by both the titular Central Asian groups and the Russians and other nontitular nations since 1950. (When figures were not available for the years listed in the table heading, the year from which the totals were provided is noted.) This table reveals the fact that, in general, the titular groups have made proportional gains in Communist Party membership within their own republics over the past several decades, relative to the total membership registered by other groups within these republics. The latest (1990) figures do not yet indicate that substantial numbers of indigenes have resigned from the communist party organizations of the four Central Asian republics, although the share of the Russian party members has dropped significantly.

The implications of increased numbers of the titular Central Asian nationals having joined the ranks of the Communist Party are unclear. Some say it was related to the educational advancement of the indigenous nationalities. This is supported by the fact that the group with the highest rate of education in Central Asia, the Uzbeks, is the most politically active, both within the Uzbek SSR relative to its Russian population, as well as at the all-union level.

Another means of analysis, however, points to the fact that the

Table 3.2 Changing rates of Communist Party membership in Soviet Central Asia: percentages indigenous and Russian of total Republic memberships

Republic[1]	1950 No.	%	1960 No.	%	1970 No.	%	1979 No.	%	1984 No.	%	1989[@] No.	%
Uzbek	132,336	100.0	202,865	100.0	412,321	100.0	518,350[#]	100.0	644,095[$]	100.0	664,520	100.0
Uzbeks	57,901	43.8	102,663	50.6	226,357	54.9	306,324	59.1	393,027	61.0	436,179	65.6
Russians	35,811	27.1	46,514	22.9	84,849	20.6	N.A.	N.A.	N.A.	N.A.	90,706	13.7
Others	38,624	29.1	53,688	26.5	101,115	24.5	212,026	40.9	251,068	39.0	137,635	20.7
Kirgiz	44,790	100.0	61,646	100.0	103,028	100.0	126,402[+]	100.0	139,085	100.0	154,650	100.0
Kirgiz	14,567	32.5	22,159	35.9	38,881	37.7	55,928	44.3	65,425	47.0	77,538	50.1
Russians	17,369	38.8	23,306	37.8	38,847	37.7	N.A.	N.A.	N.A.	N.A.	43,337	28.0
Others	12,854	28.7	16,181	26.3	25,300	24.6	70,474	55.7	73,660	53.0	33,775	21.9
Tadzhik	31,234	100.0	47,920	100.0	84,236	100.0	103,641	100.0	118,546	100.0	126,881	100.0
Tadzhiks	12,401	39.7	21,579	45.0	40,346	47.9	51,484	49.7	61,005	51.5	68,841	54.2
Russians	9,173	24.4	11,927	24.9	18,593	22.1	N.A.	N.A.	N.A.	N.A.	19,912	15.7
Others	9,660	35.9	14,414	30.1	25,297	30.0	52,157	50.3	57,541	48.5	38,128	30.1
Turkmen	33,463	100.0	45,152	100.0	66,390	100.0	93,556[+]	100.0	105,891	100.0	115,008	100.0
Turkmen	14,865	44.4	22,918	50.8	37,515	55.3	56,788	60.7	66,711	63.0	76,565	66.6
Russians	10,270	30.7	11,906	26.4	14,761	23.2	N.A.	N.A.	N.A.	N.A.	16,278	14.1
Others	8,328	24.9	10,328	22.8	14,114	21.5	36,768	39.3	39,180	37.0	22,165	19.3

Explanation of Symbols: # = 1978 + = 1981 $ = 1981 s = 1985 @ = 1985 @ = 1 January 1990 N.A. = Not Available
Sources: Institut Istorii Partii, *Kommunisticheskaya Partiya Ubekistana v Tsifrakh (1924–1977)* (Tashkent: Uzbekistan, 1979), pp. 217, 279; Institut Istorii Partii, *Kommunisticheskaya Partiya Turkmenistana v Tsifrakh (1924–1974)* (Ashkhabad: Turkmenistan, 1975), p. 210; Institut Istorii Partii, *Kommunisticheskaya partiya Turkmenistana 60 let (1924–1984)* (Ashkhabad: Turkmenistan, 1984), pp. 182, 184. Institut Istorii Partii, *Kommunisticheskaya partiya Tadzhikistana v Tsifrakh za 60 let (1924–1984 gg.)* (Dushanbe: Izdatel'stvo "Irfon", 1984), pp. 54, 79. Institut Istori Partii, *Kommunisticheskaya partiya Kirgizii v Tsifrakh, 1918–1984* (Frunze: Kirgiziya, 1984), pp. 58, 90, 93. Institut Istori Partii, *Kommunisticheskaya Partiya Kirgizii* (Frunze: Kirgizstan, 1971), p. 11; Burg 1984, pp. 56–57; *Sovetskiy Entsiklopedicheskiy Slovar'*. Moscow: "Sovetskaya Entsiklopediya", 1986, p. 1379; *Pravda Vostoka* (5 June 1990), p. 5; *Izvestiya TsK KPSS*. (No. 5, 1990): 60–63.
[1] Figures in this chart represent the nationality compositions of the full and candidate members of the communist party organizations of the four Central Asian republics.

gains made actually lag behind the relative increases in total population made by the titular groups in comparison with the rest of their republics' populations. Table 3.3 reveals that, in the period 1960–89, the percentage gained in the Communist Party memberships of the titular Central Asian groups, compared to the percentage increase in their actual republic populations, was markedly less than that of the nontitular peoples within each republic.[10]

It can be concluded from the figures in Table 3.3 that, while the *total* nontitular Communist Party membership has had its relative share of each republic's total party membership decrease over the years, their rates of participation *per capita* reflect a less precipitous decline relative to the titular groups. The most recent intercensal period (1979–1989) reveals a marked decrease in the ratio of communist party membership to population increase in the category of "others", reflecting in part the increase in the rate of out-migration of Russians from Central Asia during these years.

Recent (1989) figures reveal that the percentage of each republic's population which belongs to the Communist Party is by far the lowest in all of the USSR—3.4, 3.6, 2.5, and 3.3 percent for, respectively, the Uzbek, Kirgiz, Tadzhik, and Turkmen republics (USSR 1989, pp. 3–9; *Izvestiya Tsk KPSS* 1990, pp. 60–63). These rates are well below the USSR total figure of 6.8 percent. In considering the large deviation from the all-union average, one should take into account the fact that the rapid rate of natural increase over the past several decades has resulted in a very young age structure within the republics of Central Asia. To give a more accurate comparison, the above figures should be adjusted only for the population eligible for Communist Party membership (age 18 and above). While such an adjustment still results in the Communist Party membership rates within the Central Asian republics being substantially below the all-union rate, the numbers are not as widely divergent. The adjusted figures are: total USSR, 10.1 per cent CPSU membership; Uzbek SSR, 7.6 percent; Tadzhik SSR, 6.2 percent; Kirgiz SSR, 7.3 percent; and Turkmen SSR, 7.6 percent (United States 1987; USSR 1988: 49, 61, 81, 85, 91).[11]

Lastly, there is the view which holds that positions of effective party power are held by non-Central Asians and that the local nationals in positions of visibility have often been either figureheads or scapegoats. Members of the nomenklatura, those in positions of power, have been more heavily drawn from the non-Central Asian peoples (primarily Russian and Ukrainian) population since World War II. Slavs continue to occupy a disproportionate number of the key positions within the party apparatus, the government structure, the security and armed forces, and key economic enterprises (Radio Liberty/Radio Free Europe 1986: 1–12). There also exists the practice of "doubling" important Turkic-Muslim functionaries with Slavic counterparts. This includes the

Table 3.3 Percentage increase in Communist Party membership for indigenous and other nationalities living in Central Asia, 1959–1989: Compared to the total Republic population increases of each group

Republic and nationality	Percentage increase in population				Percentage increase in CP membership				Ratio of percent increase in CP membership to percent increase in population			
	1959–70	1970–79	1979–89	1959–89	1959*–70	1970–79#	1979–89+	1959–89	1959–70	1970–79	1979–89	1959–89
Uzbek SSR	45.6	30.4	28.7	144.4	103.2	25.7	28.2	227.6	+126.3	−15.5	− 1.7	+57.6
Uzbeks	53.3	36.8	33.6	180.3	120.5	35.3	42.4	324.9	+126.0	− 4.1	+26.2	+80.2
Others	32.8	18.3	17.9	85.3	85.6	14.0	7.7	127.9	+161.0	−23.5	−57.0	+49.9
Kirgiz SSR	42.0	20.1	20.9	106.1	67.1	22.7	22.3	150.9	+ 59.8	+12.9	+ 6.7	+42.2
Kirgiz	53.5	31.3	32.1	166.3	75.5	43.8	38.6	249.9	+ 41.1	+39.9	+20.2	+50.3
Others	34.1	11.4	10.6	65.1	62.5	9.9	9.4	95.3	+ 83.3	−13.2	−11.3	+46.4
Tadzhik SSR	46.5	31.3	33.7	157.1	75.8	23.0	22.4	164.8	+ 63.0	−26.5	−33.5	+ 4.9
Tadzhiks	55.1	37.2	41.6	201.4	87.0	27.6	33.7	219.0	+ 57.9	−25.8	−19.0	+ 8.7
Others	36.7	23.6	22.4	106.8	66.6	18.8	11.3	120.3	+ 81.5	−20.3	−50.4	+12.6
Turkmen SSR	42.4	28.1	27.0	131.6	47.0	40.9	22.9	154.7	+ 10.8	+45.6	−15.2	+17.6
Turkmen	53.4	33.5	33.4	173.3	63.7	51.4	34.8	234.1	+ 19.3	+53.4	+ 4.2	+35.1
Others	25.2	17.6	13.2	66.7	29.9	27.3	4.6	72.9	+ 18.7	+55.1	−63.2	+ 9.3

Sources: See Sources, Table 2; also USSR 1962, Tsentral'noye Statistitheskoye Upravleniye SSSR, *Itogi Vsesoyuznoy Perepisi Naseleniya 1959 Goda, Svodniy Tom*, pp. 206–208; USSR 1973, Tsentral'noye Statisticheskoye Upravleniye SSSR, *Itogi Vsesoyuznoy Perepisi Naseleniya 1970 Goda*, Vol. 4, pp. 202, 284, 295, 306; USSR 1984b, Tsentral'noye Statisticheskoye Upravleniye SSSR, *Chislennost'i Sostav Naseleniya SSSR*, pp. 110, 130, 132, 134; USSR 1988, *Naseleniye SSSR 1987. Statisticheskiy Sbornik*, pp. 49, 61, 81, 85, 91; USSR 1989, Goskomstat, *Natsional'nyy Sostav Naseleniya*, Moscow: Information-Publication Center, pp. 64, 87, 90, 94.
Additional Source: Izvestiya Tsk KPSS (No. 5, 1990), pp. 60–63.
* The 1959 CP membership totals were those listed on 1 January 1960.
# Figures for Uzbeks were calculated based on 1978 data; those for Kirgiz and Turkmen were based on 1981 data.
+ Figures for 1989 CP membership are the numbers listed for 1 January 1990.

key position of party second secretaries (where the party secretary is a Central Asian national), the head of department of organizational party work and of administrative organs, and heads of security sections (Rywkin 1982: 127–8). The percentage of indigenous nationals occupying leading party positions in the hierarchy of the Central Asian republics has been below that of the other republics in the USSR (Hodnett 1978: 101–3).

A study which documents over 1,000 cases of personnel transfers of oblast party officials between republics within the same economic region reveals very little geographical mobility from republic to republic within Central Asia (Miller 1983: 78–9). Much more mobility is evident between the republics of Central Asia and party organs at the all-union center. This would seem to suggest that a high degree of centralization in the Communist Party hierarchy is being maintained, with no measurable strategy for combating regionalism and nationalism by dispersing staff. It should be noted that the turnover rate of Communist Party and government officials in the four Central Asian republics during the past three years (1986–9) of the Gorbachev administration has been the most tumultuous of any region in the USSR. The top party leadership in each republic has been largely replaced in response to a CPSU Central Committee mandate to eradicate the corruption and inefficiencies that had become endemic throughout the region. Included in these purges have been the first secretaries of the Communist Party organizations in all but the Tadzhik Republic.[12]

At any scale, access to power is an essential aspect of political geography when looking at the variety of inputs which contribute to political systems of decision making. Positions of Communist Party power, from the trade unions to the central committee, are not dominated by the indigenous nationalities in Central Asia to the extent that their population composition would indicate. The same can be said of the all-Union Communist Party and government organizations— there were, after all, no Central Asian members on the CPSU Politburo until the recent reorganization which made first secretaries of union republics members of the Politburo.

Jurisdictional national autonomy

The second category of this republic-level analysis of the Soviet federal structure is *jurisdictional national autonomy*, which refers to the territorial extent of the whole system of features based on the unique national character of the indigenous population. This refers to the institutional confines of national autonomy, that is, the system of education, publishing, the press, language, and literature, all of which have taken on an official and unique national character. The geographic focus of these jurisdictional boundaries of national institutions has been formalized as a national homeland, a specific territorial unit which not only is legitimized politically, but also confines a whole system of cultural attributes.

At the time the original boundaries of the republics of Central Asia were delimited, the distinctions among the various peoples and tribes living throughout the region of Western (or Russian) Turkestan were at times quite unclear. There were sedentary and nomadic peoples speaking the same languages, Persian and Turkic types living together in the eastern mountains, and a number of various tribes broadly designated as either Uzbek or Turkmen. The uncertainty of the Russian authorities at the time of the federal compromise is evident by the fact that the Kirgiz republic was at first designated as the Kara-Kirgiz AO, while what is now the Kazakh Republic originally bore the title of Kirgiz ASSR.

Many accounts consider the boundaries separating the Central Asian republics as largely artificial, resulting in the separation of indigenous peoples in Central Asia who had already begun the process of assimilation. Regardless of the degree of artificiality to these boundaries, they have remained largely unchanged to this day.[13] The permanence of these boundaries, describing cultural institutions specific to a given nation, provided the opportunity for groups to develop and focus their sentiments toward a particular territorial jurisdiction and its related national designation. In a 1 July 1989 speech on nationality problems that was televised union-wide, First Secretary Gorbachev reaffirmed the importance of maintaining the sanctity of the federal boundaries. "And we must [take] . . . the most resolute measures, in accordance with the law and the people's vital interests, against those who provoke interethnic clashes and call for borders to be redrawn and national minorities to be expelled." (Gorbachev 1989: 1).

The degree to which regional distinctions are manifest among the Uzbeks, Karakalpaks, Turkmen, Tadzhiks, Kirgiz, and various other minority nationalities is therefore directly related to the specific nature of Soviet jurisdictional national autonomy in Central Asia. There has been a general retreat in rhetoric from the 1961 party program adopted at the Twenty-Second Party Congress which referred to the republic borders as "increasingly losing their significance" (Gleason 1987: 1). The political expression of national distinction is seen in the acceptance of, and even the insistence on, the republic boundaries as defining certain rights which are national in character (*Current Digest of the Soviet Press* 1988b: 11–15).[14] In Central Asia, the most vociferous protests of late have involved minorities living within, and thus subject to the jurisdiction of, the republic boundaries. Violent clashes broke out in early June 1989 between members of a minority nationality, the Meskhetian Turks, and the local Uzbek population. What began as a local incident in a market in the city of Fergana escalated into a republic-wide uprising resulting in over 100 dead and thousands of refugees. Central to this dispute was the desire to preserve the ethnic "purity" of the homeland. Involved also were related issues of unemployment, discrimination, and the right for the Meskhetian Turks to be repatriated to their homeland, from which they were expelled during World War II. The emotional association with a particular political jurisdiction as a national homeland thus remains an important force throughout the Soviet federation, at the republic as well as the inter- and intra-republic levels.

LOCAL POLITICAL GEOGRAPHY: ADMINISTRATIVE REORGANIZATION AND POLITICAL REPRESENTATION

At the micro-level scale of analysis, the political geography of a region focuses largely on questions of administrative organization and change, political representation, public welfare, the allocation of goods and services, and other local concerns. Except for the subject of administrative reorganization and political representation, such studies have typically involved a level and consistency of information not readily available when dealing with the Soviet Union. Furthermore, the influence of public opinion, lobbies, and political action groups has not been as evident in the Soviet Union as in the West, although with the new initiatives represented by the glasnost' reforms, the potential for further democratization is a very real possibility. Already in the Soviet Union there are several nongovernmental environmental organizations which are active, and quite influential, at the local level of decision making. In Soviet Central Asia, concerns over the depletion of water resources in particular have mobilized the population and the political leadership to call for actions such as the southward diversion of Siberian rivers in order to rejuvenate the Aral Sea (Sinnot, this volume).

The micro-scale level of analysis in the Soviet Union involves making use of data collected below the republic level, primarily by oblasts and rayons, or by various sectors of the economy. At this scale, questions of regional differentiation focus in large part on the issues of investment, employment, and urban-rural differences. Some of the chapters which follow will deal with these issues. The focus here is largely on the local territorial subdivisions—how space is politically allocated within the republic of Central Asia (Map 3.1). Tables 3.4 through 3.8 reveal the history of jurisdictional formation and change which has occurred primarily at the oblast level of administration in Soviet Central Asia. With the exception of the transfer of the Karakalpakskaya ASSR from Kazakhstan, and other relatively minor

Table 3.4   Major political subdivisions, dates of formation, and changes:
Central Asian republics, 1988.

Administrative unit	Date of original formation	Date abolished	Date reestablished	Date abolished
Uzbek SSR	1924	n.a.[1]	n.a.	n.a.
Karakalpakskaya ASSR[2]	1932	n.a.	n.a.	n.a.
Andizhanskaya Oblast	1941	n.a.	n.a.	n.a.
Bukharskaya Oblast	1938	n.a.	n.a.	n.a.
Dzhizakskaya Oblast[2]	1973	1988	1989	n.a.
Kashkadar'inskaya Oblast	1943	1960	1964	n.a.
Navoiyskaya Oblast	1982	1988	n.a.	n.a.
Namanganskaya Oblast	1941	1960	1967	n.a.
Samarkandskaya Oblast	1938	n.a.	n.a.	n.a.
Surkhandar'inskaya Oblast	1941	n.a.	n.a.	n.a.
Syrdar'inskaya Oblast	1963	n.a.	n.a.	n.a.
Tashkentskaya Oblast	1938	n.a.	n.a.	n.a.
Ferganskaya Oblast	1938	n.a.	n.a.	n.a.
Khorezmskaya Oblast	1938	n.a.	n.a.	n.a.
Kirgiz SSR[4]	1924	n.a.	n.a.	n.a.
Issyk-Kul'skaya Oblast	1939	1959	1970	n.a.
Narynskaya Oblast[5]	1939	1962	1970	1988
Oshskaya Oblast	1939	n.a.	n.a.	n.a.
Talasskaya Oblast[6]	1944	1956	1980	1988
Tadzhik SSR[7]	1924	n.a.	n.a.	n.a.
Gorno-Badakhshan AO	1925	n.a.	n.a.	n.a.
Kulyabskaya Oblast[8]	1939	1955	1973	1988
Kurgan-Tyubinskaya Oblast[9]	1939	1947	1977	1988
Leninabadskaya Oblast	1939	1962	1970	n.a.
Khatlonskaya Oblast	1988	n.a.	n.a.	n.a.
Turkmen SSR	1924	n.a.	n.a.	n.a.
Ashkhabadskaya Oblast[10]	1939	1959	1973	1988
Krasnovodskaya Oblast[11]	1939	1955	1973	1988
Maryyskaya Oblast	1939	1963	1970	n.a.
Tashauzskaya Oblast	1939	1963	1970	n.a.
Chardzhouskaya Oblast	1939	1963	1970	n.a.

1 Not applicable

2 Karakalpakskaya ASSR was originally formed as part of the Kazakh SSR in 1932. In 1936 it was transferred to the Uzbek SSR.

3 Dzhizakskaya Oblast was in 1988 reestablished and returned to its previous borders ("Decree issued on organization of Uzbek oblasts" FBIS-SOV-90-046, 8 March 1990.)

4 The Kirgiz SSR was constituted originally as the Kara-Kirgiz ASSR in 1924, and administratively fell within the boundaries of the RSFSR. The Kara-Kirgiz ASSR was renamed the Kirgiz ASSR in 1926, and was separated from the Russian federation and elevated to union republic status in 1936.

5 Originally formed as Tyan'-Shanskaya Oblast in 1939; abolished in 1962, reestablished as Narynskaya Oblast in 1970. In October 1988, Ak-Talinskiy, At-Bashinskiy, Dzhumgalskiy, Kochkorskiy, and Tyan'-Shanskiy rayons were incorporated into Issyk-Kul'skaya Oblast. Toguz-Torouskiy Rayon was incorporated into Oshskaya Oblast.

6 An oblast of this name, but with different borders, existed between 1944 and 1956, when it was incorporated into Frunzenskaya Oblast, which was itself abolished in 1959. In October 1988, Kirovskiy, Leninpol'skiy, and Talasskiy rayons and the city of Talass were given direct republican jurisdiction. Toktogul'skiy Rayon and the city of Kara-Kul' were incorporated into Oshskaya Oblast.

7 Originally created as an ASSR within the Uzbek Republic. Elevated to union republic status in 1929.

8 In September 1988, rayons of the Kulyabskaya Oblast were incorporated into Khatlonskaya Oblast.

9 In September 1988, rayons of the Kurgan-Tyubinskaya Oblast were incorporated into Khatlonskaya Oblast.

10 In August 1988, rayons of the Ashkhabad Oblast were transferred under direct republican jurisdiction.

11 In 1988, rayons of the Krasnovodskaya Oblast were transferred under direct republican jurisdiction.

Sources: USSR 1959, 1960, 1970, 1971, 1979, 1980, 1986, 1987; *Geograficheskiy Entsiklopedicheskiy Slovar'* 1989: 582–6; FBIS 1990; also Shabad 1956: 505; Craumer 1984.

Table 3.5 **Changes in the numbers of political subdivisions: union republics of Soviet Central Asia.**

Republic	1959	1970	1979	1986	1991
Uzbek SSR					
Oblasts	9	10	11	12	10
Rayons (districts)	114	112	148	155	149
Autonomous republics[1]	1	1	1	1	1
Cities	33	43	93	123	124
Rayons in cities	5	9	17	17	17
Urban settlements	66	83	95	95	104
Village soviets (councils)	803	909	1,061	1,207	1,245
Kirgiz SSR					
Oblasts	2	1	3	4	2
Rayons (districts)	37	32	38	40	40
Cities	14	15	18	21	21
Rayons in cities	0	3	4	4	4
Urban settlements	30	35	31	28	29
Village soviets	37	156	372	388	390
Tadzhik SSR					
Oblasts	1	0	3	3	2
Rayons	33	40	43	45	45
Autonomous regions[2]	1	1	1	1	1
Cities	14	17	18	18	19
Rayons in cities	3	3	4	4	4
Urban settlements	31	43	49	49	48
Village soviets	214	272	313	322	330
Turkmen SSR					
Oblasts	3	0	5	5	3
Rayons	39	34	44	44	41
Cities	13	15	15	16	16
Rayons in cities	2	2	3	3	3
Urban settlements	61	68	74	74	74
Village soviets	225	226	240	266	276

[1] Karakalpakskaya ASSR.
[2] Gorno-Badakhshanskaya Autonomous Oblast.
Sources: USSR 1960, 1971, 1980, 1987; see also USSR 1959, 1970, 1979, 1986.
[1] Totals listed are for 1 January 1990.

adjustments over the years in the Kazakh–Uzbek boundary, the external boundaries of the republics have remained sacrosanct. And at the oblast level of organization, there were only three jurisdictions existing in 1988, all of them in the Uzbek SSR, which were not in existence prior to the end of World War II, despite the intervening abolishment and reconstruction of several oblast territories (see Table 3.4).

While internal oblast boundaries throughout most of the Soviet Union have been fairly stable in the past two decades, the oblast boundaries within the republics of Central Asia have witnessed relatively numerous, albeit minor, adjustments in recent years. Tables 3.4 through

Table 3.6 Changes in the areas of Central Asian jurisdictions, 1959–90.

Republics and oblasts	Area (000 sq km)[1]			
	1959	1970	1979	1987
Uzbek SSR	409.4	449.6	447.4	447.4
Karakalpakskaya AO	156.1	165.6	165.6	164.9
Andizhanskaya Oblast	4.2	4.3	4.2	4.2
Bukharskaya Oblast	122.9	143.2	143.2	39.4
Dzhizakskaya Oblast	n.a.	n.a.	20.3	20.5
Kashkadar'inskaya Obl.	29.2	28.4	28.4	28.4
Navoiyskaya Oblast	n.a.	n.a.	n.a.	110.8
Namanganskaya Oblast	7.0	7.8	7.9	7.9
Samarkandskaya Oblast	37.8	29.2	24.5	16.4
Surkhandar'inskaya Obl.	20.1	20.8	20.8	20.8
Syrdar'inskaya Oblast	n.a.	23.1	5.3	5.1
Tashkentskaya Oblast	20.5	15.6	15.6	15.6
Ferganskaya Oblast	7.1	7.1	7.1	7.1
Khorezmskaya Oblast	4.5	4.5	4.5	6.3
Kirgiz SSR	198.5	198.5	198.5	198.5
Issyk-Kul'skaya Oblast	n.a.	n.a.	43.5	43.5
Narynskaya Oblast	50.6	n.a.	50.2	51.1
Oshskaya Oblast	44.8	73.9	73.9	65.6
Talasskaya Oblast	n.a.	n.a.	n.a.	19.6
Tadzhik SSR	142.5	143.1	143.1	143.1
Gorno-Badakhshanskaya AO	63.7	63.7	63.7	63.7
Kulyabskaya Oblast	n.a.	n.a.	12.9	12.0
Kurgan-Tyubinskaya Obl.	n.a.	n.a.	11.7	12.6
Leninabadskaya Oblast	25.6	n.a.	26.1	26.1
Turkmen SSR	488.0	488.1	488.1	488.1
Ashkhabadskaya Oblast	n.a.	n.a.	95.4	95.4
Krasnovodskaya Oblast	n.a.	n.a.	138.5	138.5
Maryyskaya Oblast	118.6	n.a.	86.8	86.8
Tashauzskaya Oblast	75.4	n.a.	73.6	73.6
Chardzhouskaya Oblast	93.5	n.a.	93.8	93.8

Sources: see sources, Table 3.4; also USSR Goskomstat 1990, pp 11–12. Details of the most recent oblast area changes were not yet available at the time this was written.
[1] The totals of the oblasts do not always sum to the total area of the republic due to the fact that in several instances there are territories which fall directly under the administration of the republic.

3.7 indicate the extent of some of these changes over the past 30 years, the reasons for which have varied from consolidating the administration of economic activity, to reviving historical configurations, to managing irrigation systems (Shabad 1982: 468).

The 1989 preliminary census results indicate that a further reorganization of Central Asian oblasts has occurred since 1988. These latest changes have resulted in a consolidation of administrative jurisdictions within the republics' boundaries. This was done either by the absorption of small oblasts by larger contiguous ones, by extending the territorial jurisdictions of capital city oblasts, or by creating additional numbers of "republic-subordinate rayons." Table 3.8 lists the recent changes which have occurred at the oblast level. In both the Uzbek and Tadzhik republics, the least populated oblasts were eliminated and absorbed by neighboring units. In Uzbekistan, Dzhizakskaya and Navoiskaya Oblasts were absorbed by neighboring administrative territories. The Tadzhik SSR combined two oblasts to create a new one (Khatlonskaya Oblast) and most of the territory of the two smallest oblasts of the Turkmen Republic was placed under the direct administrative control of the republic. In the Kirgiz republic, two new oblasts were created, Dzhlai-Abad and Chu, out of parts of the territories of Narynskaya and Talasskaya Oblasts (*Pravda* 11 December 1989, p. 2).

In addition to the elimination of some oblasts and the creation of new ones listed in Table 3.8, numerous changes have continued to occur below the oblast level of jurisdiction. The most substantial of these have occurred within the Uzbek SSR, which has undergone substantial administrative and territorial reorganization at the rayon as well as the oblast level. The most recent territorial transfers affecting such boundaries occurred in May of 1989. A decree issued by the Uzbek SSR Supreme Soviet presidium resolved:

1 To grant the petitions of the executive committees of Samarkand and Bukhara oblast soviets of people's deputies to transfer a part of the territory of Samarkand Oblast, consisting of Kanimekhskiy, Kyzyltepinskiy, Navoiyskiy, Tamdynskiy, and Uchukuduskiy rayons, including the cities of Uchkuduk, Navoi, and Zarafsham, to Bukhara Oblast.

2 To recognize the deputies of the Samarkand Oblast Soviet, who were elected from voting districts situated within the territory being transferred, as deputies of Bukhara Oblast. (Joint Public Research Service 1989b: 22)

The adjustments between Samarkand and Bukhara oblasts were in part due to the fact that Bukhara Oblast had lost a significant portion of its former territory in an earlier (1982) transfer (see Table 3.7).[15]

The attachment to national boundaries expressed at even the lowest

Table 3.7 Internal political-administrative reorganization in Soviet Central Asia: major oblast area transfers and changes, 1959–89.

Uzbek SSR (1924)[1]	1963	Over 40,000 sq km transferred from Kazakh SSR
	1971	Part of Syrdar'inskaya Oblast ceded back to Kazakh SSR in 1971
Karakalpakskaya ASSR (1932)	1960	Gained nearly 10,000 sq km in area, much of which (5,400 sq km) was from Bukharskaya Oblast
	1982	Part of ASSR ceded to Khorezmskaya O.
Bukharskaya Oblast (1938)	1963	Gained over 27,000 sq km, mostly from Kazakh SSR
	1982	Land ceded to Novoiyskaya and Khorezmskaya Oblasts
	1989	Gained land from Samarkandskaya Oblast
Dzhizakskaya Oblast (1973)	1973	Formed from parts of Surkhandar'inskaya and Samarkandskaya Oblasts
	1979	Small portion (200 sq km) of territory transferred from Syrdar'inskaya Oblast
	1988	Eliminated and merged with Syrdar'inskaya and Samarkandskaya oblasts
	1989	Returned to pre-1988 borders
Kashkadar'inskaya (1943)	1960	Abolished and merged with Surkhandar'-inskaya Oblast
	1964	Reestablished
Navoiyskaya Oblast (1982)	1982	Formed from parts of Bukharskaya and Samarkandskaya oblasts
	1988	Abolished and merged with Syrdar'inskaya and Samarkandskaya Oblast
Namanganskaya Oblast (1941)	1960	Abolished and merged with Andizhanskaya and Ferganskaya oblasts
	1967	Reestablished
Samarkandskaya Oblast (1938)	1973	Portion ceded to Dzhizakskaya Oblast
	1982	Portion ceded to Navoiyskaya Oblast
	1988	Land transferred back from Dzhizakskaya and Navoiyskaya oblasts
Syrdar'inskaya Oblast (1963)	1963	Formed from parts of Tashkentskaya, Samarkandskaya, and Chimkentskaya (Kazakh SSR) oblasts
	1979	Lost small portion (200 sq km) of its territory to Dzhizakhskaya Oblast
	1988	Gained parts of Dzhizakskaya and Navoiyskaya oblasts
	1989	Lost the above part of Dzhizakskaya Oblast
Khorezmskaya Oblast (1938)	1982	Gained territory from Bukharskaya Oblast and Karakalpakskaya ASSR
Kirgiz SSR (1924)		
Dzhalal-Abadskaya Oblast (1939)	1958	Abolished

Table 3.7 Continued.

Issyk-Kul'skaya Oblast (1939)	1959	Abolished
	1970	Reestablished
	1988	Gained territory from elimination of Narynskaya and Talasskaya oblasts.
Frunzenskaya Oblast (1939)	1959	Abolished
Narynskaya Oblast (1939)	1962	Abolished (orig. named Tyan'-Shanskaya)
	1970	Reformed as Narynskaya Oblast
	1988	Abolished
Oshskaya Oblast (1939)	1960	Gained territory from former republic-administered rayons
	1980	Lost 15,000 sq km (to 59,500 sq km), mostly to neighboring Talasskaya Oblast
	1984	Chatkalskiy Rayon transferred from Talasskaya Oblast
	1988	Gained territory from Narynskaya and Talasskaya oblast
Talasskaya Oblast (1944)	1980	Recreated partly from territory of Oshskaya Oblast, partly from former republic-administered rayons; 1980 area was 25,700 sq km
	1984	Territory of Chatkalskiy Rayon (6,100 sq km) transferred to Oshkskaya Oblast
	1988	Abolished
Tadzhik SSR (1924)		
Kulyabskaya Oblast (1939)	1955	Abolished; reestablished in 1973
	1979	Territory (900 sq km) transferred to neighboring Kurgan-Tyubinskaya Oblast
	1988	Abolished
Kurgan-Tyubinskaya (1939)	1955	Abolished; re-established in 1977
	1979	Territory (900 sq km) transferred from neighboring Kulyabskaya Oblast
	1988	Abolished
Leninabadskaya Oblast (1939)	1955	Abolished
	1970	Reestablished
Khatlonskaya Oblast	1988	Created by the consolidation of Kurgan-Tyubinskaya and Kulyabskaya oblasts
Turkmen SSR (1924)		
Ashkhabadskaya Oblast (1939)	1959	Abolished (May)
	1973	Recreated from republic-administered rayons plus part of Maryyskaya Oblast
	1988	Abolished
Krasnovodskaya Oblast (1939)	1947	Abolished
	1952	Recreated
	1955	Abolished
	1973	Recreated
	1988	Abolished
Maryyskaya Oblast (1939)	1963	Abolished; reformed in December 1970

Table 3.7   Continued.

Tashauzskaya Oblast (1939)	1963	Abolished; reformed in December 1970
Chardzhouskaya Oblast (1939)	1963	Abolished; reformed in December 1970

[1] Dates in parentheses indicate the years the units were originally formed.
Sources: see sources, Table 3.3; for more complete information on the specific nature of the administrative changes made, consult *Vedomosti Verkhovnogo Soveta SSSR*, which is the monthly journal of the Supreme Soviet of the USSR.

Table 3.8   Recent territorial consolidation: Central Asian oblasts, 1989–1990.

Republic	Oblasts eliminated	Oblasts gaining territory	New oblasts created
Uzbek SSR	Dzhizakskaya Navoiyskaya	Samarkandskaya Syrdar'inskaya Bukharskaya	
Kirgiz SSR	Narynskaya Talasskaya	Issyk-Kul'skaya Oshskaya Republic-subordinate rayons	Dzhalal-Abadskaya Chuskaya
Tadzhik SSR	Kulyabskaya Kurgan-Tyubinskaya		Khatlonskaya
Turkmen SSR	Ashkhabadskaya Krasnovodskaya	Republic-subordinate rayons	

Sources: Joint Public Research Service 1989a: 20–1; also USSR 1988: 28–9.

of geographic scales is manifest by a territorial battle waged between two small villages located across from one another along the Tadzhik–Kirgiz border. The ongoing dispute over land and water resources between Kirgiziya's Leninabadskaya Oblast and Oshskaya Oblast in the Tadzhik SSR is rooted in the distant past, prior to the demarcation of the present-day republic boundaries. In April 1989, Tadzhik villagers from Oktyabar were prohibited by the Kirgiz village of Smarakandyk from taking sheep and cows to pasture in the mountains on the Kirgiz side of the border. In response, residents of Oktyabar used rocks to block the Matchoi Canal, which runs through their territory and provides drinking and irrigation water for Smarakandyk. The dispute simmered for three months before the canal was cleared following a series of high-level negotiations involving both Tadzhik Communist Party leader Makhamov and the chairman of the Kirgiz Council of Ministers, Dzhumgulov (Foreign Broadcast Information Service 1989c: 75). More recently (June 1990), a dispute over land allocation provided the spark for violent clashes between Kirgiz and Uzbek inhabitants of the Muslim holy city of Osh.

Electoral geography

The March 1989 elections for the newly constituted Congress of People's Deputies are of concern at both the federal and the republic level. Each Central Asian republic reelected both its first and second party secretaries, and of all the regions in the USSR, the republics of Central Asia were the most successful in reelecting their regional party secretaries as well. These results, combined with the fact that nearly 90 percent of the elected delegates were Communist Party members, indicates that access to political power remains largely based on the party hierarchy.

In terms of political representation, one must keep in mind that, within the Soviet federation, representation bears no constitutional relationship with the nationality upon which the republic is based. At the union-wide level, delegates from Central Asia have generally been equitably represented in the Soviet of Nationalities, one of two chambers comprising the Supreme Soviet. In the second chamber, the Soviet of the Union, the figures have in recent decades indicated a trend towards a proportionate *under*-representation of Central Asians nationals from within their own republic (McIntyre 1974: 96; Vanneman 1977: 73; FBIS 1989a: 29–44).[16] Of those delegates elected to the newly constituted Supreme Soviet, the composition of the Soviet of the Union (ostensibly determined based on population distribution) is curiously skewed in favor of the Slavic republics at the expense of Central Asia. The delegates from the RSFSR, Belorussia, and Ukraine comprise 2.5 percentage points more of the delegates than their combined populations would warrant; this entire representative surplus is made up by a corresponding deficit in the four republics of Central Asia, despite their significantly smaller populations (FBIS 1989a: 29–44).

The indigenous Central Asian nationalities have also not been significantly increasing their representation in their republic Supreme Soviets in recent years (*Current Digest of the Soviet Press* 1988b: 12). For example, after making substantial gains in their percentage composition of deputies to the Uzbek Supreme Soviet between 1951 and 1959 (63.8 to 73.2 percent), the Uzbeks began to lose representation relative to Russians living in Uzbekistan (Akademiya Nauk Uzbekskoy SSR 1970: 79). This trend continued even as Russian out-migration from Central Asia intensified and high rates of natural increase gave the Uzbeks a growing majority within their republic.

In May 1989, two months after numerous incumbents were defeated in the elections for the new Congress of Peoples Deputies, Gorbachev announced that local elections, scheduled for the fall of 1989, would be postponed until the spring of 1990. The reason, ostensibly, was to give local government officials more time to prepare better for reelection. At a Supreme Soviet session held on 25 July 1989, however, several deputies put forth the persuasive argument that new elections to party organizations at all levels could help restore trust in local authorities. Gorbachev responded to these arguments by reversing his earlier stance and announcing that local elections should be held as soon as possible in order to reevaluate past policies (*Washington Post*, 25 July 1989: A1). These results reveal regional differences in electoral geography, changes in the local nomenklatura, and issues of cultural, religious, and ecological concern. Preliminary figures have shown that the local party authorities in the four republics of Central Asia have maintained their power at a much higher rate than in local elections in the other Soviet republics.

INTEGRATING THE CENTRAL ASIAN FRONTIER: DIFFERENT PERSPECTIVES

The three levels of analysis of the political geography of Soviet Central Asia reveal different perspectives with regard to the relative political integration of the region. Taken as a whole, the major features of the political geography of Soviet Central Asia seem predominantly disintegrative. Factors such as common religion, language, ethnicity, and a shared history operate to distinguish this region from the rest of the USSR. Yet because Central Asia is a strategically located region, it remains in the vital interests of the Soviet Union to orient its republics toward the center as much as possible.

At the republic level of analysis, the political geography of each unit is institutionally and structurally identical, and consistent with all other republics as well. Such a symmetry lends itself to a sense of community and is conducive to building loyalty toward central authorities. At this level, official support for an imposed political structure is balanced by an increased identification with the individual republics as national homelands (Kaiser, this volume). At the same time, active political participation remains primarily limited to Communist Party members, who continue to comprise a small percentage of each republic's eligible population.

Lastly, administrative and territorial reorganization at the oblast level and below is often seen as related to the degree of regional centralization being encouraged from above. At the local level there are also disintegrative tendencies present throughout Central Asia, manifest as either anti-Soviet or anti-Russian sentiments, interrepublic conflict, and local animosity between various different ethnic groups.

While new elections have the potential to mobilize the population and legitimize the system and structures of political authority, increased democratization has also led to a rise in demands for regional and political autonomy. Clearly, both integrating and disintegrating political forces are operative throughout the republics of Soviet Central Asia, and the scale of analysis tends to determine which tendency is dominant in each particular instance.

NOTES

1. The region of Turkestan between the Caspian Sea and the Tien Shan Mountains is referred to as Western Turkestan to distinguish it from Eastern Turkestan, comprising the Chinese province of Sinkiang. For a general source detailing the history of this region's organization see Bartol'd (1927).

2. This gradually expanding series of fortifications defining the Russian-Kazakh frontier contributes to the justification of considering Kazakhstan, particularly what was the northern steppe province, as distinct from the four Central Asian republics in terms of its political geography.

3. The khanate of Kokand was actually established by the Begs of the Fergana Valley. These three khanates were hardly at peace with one another, and the region was constantly in turmoil with local battles.

4. Functionally, the southernmost oblasts of Kazakhstan—particularly Kzyl-Ordinskaya, Chimkentskaya, Dzhambulskaya, and Alma-Atinskaya oblasts—share many common characteristics with the four republics of Soviet Central Asia.

5. The Kirgiz and the Tadzhik SSRs were actually first established as autonomous republics within the Russian and Uzbek union republics, respectively. The Kirgiz SSR which emerged in 1936 was originally constituted in 1924 as the Kara-Kirgiz Autonomous Oblast, was reconstituted as the Kirgiz AO in 1925, and was transformed into the Kirgiz ASSR in 1927. The Gorno-Badakhshan Autonomous Oblast was formed in 1925 out of the Pamir district of the Tadzhik ASSR, which was declared a union republic in 1929. Another autonomous republic, the Karakalpak ASSR, formed originally as an autonomous oblast within Kazakhstan in 1924, was transferred to the Uzbek SSR in 1936.

6. Jerry Hough, for one, has suggested that the multinational character of the USSR contributes to the maintenance of control of the periphery by the center (Hough 1988: 42–3).

7. There are nearly 12 million Uygurs in the PRC, and the 1979 Soviet census listed 29,817 in the Kirgiz SSR (147,943 in Kazakhstan). There are over 6 million Kurds living in Turkey, an estimated 2 million in Iran, and some 10,000 in Afghanistaan. The 1979 total Soviet Kurdish population was 115,858; 9,544 were found in the Kirgiz SSR.

8. There is a good deal of discrepancy over the precise definition of this term. In fact, Soviet terminology refers to its four republics as Middle Asia ("Srednyaya Aziya"), and the broader region as Central Asia ("Tsentral'naya Aziya").

9. Prior to the recent election reforms, each Soviet Socialist Republic (SSR)

sent 32 representatives to the Soviet of Nationalities. Accordingly, the Karakalpak ASSR now sends only 5 representatives (formerly 11) and the Gorno-Badakhshan Autonomous Oblast sends 2 (formerly 5) to the Soviet of Nationalities. Many of the structures of both government and party at these latter two levels approximate those of the union republics. The authority of the autonomous oblasts is more strictly limited, however, primarily to cultural affairs.

10. It should be noted that the category "others" in Table 3.2 also includes other Central Asian indigenous nations. Only in the Tadzhik SSR does their portion of the nontitular population comprise greater than 10 percent of the republic's total population. In addition, because Communist Party membership is an age-specific measure, and the indigenous Central Asian nations have high birth rates, the figures for the indigenous groups' population change would likely be lower (making their percentage increase compared to the party membership slightly higher) if the percentages were based on the ethnic group populations of each republic above the age of 18. Unfortunately, at the time of this writing, such ethnic breakdowns of the population were not available.

11. These figures were calculated based on the reported 1987 age structure, 20 and above. These figures, adjusted for age-specific mortality rates, will approximate the 1985 18-and-above populations.

12. M. Gapurov was replaced by S. A. Niyazov in the Turkmen Republic; T. U. Usubaliyev was replaced by A. M. Masaliyev as first secretary of the Kirgiz party organization; and R. N. Nishanov replaced G. P. Razumovskiy in Uzbekistan. M. K. Makhkamov remains the first secretary of the Tadzhik Communist Party central committee.

13. According to Article 79 of the 1977 Soviet constitution (USSR 1978), autonomous republics have the right to change their own political boundaries. (Note the changes indicated in Table 3.5).

14. Such expressions become vocalized most evidently when rights are infringed upon or taken away, i.e. the demonstrations in Alma Ata following the replacement of the Communist Party first secretary Kunayev (a Kazakh) with Kolbin, an ethnic Russian (who, with his election as chairman of the USSR People's Control Committee, was recently replaced by a Kazakh national, N. A. Nazarbayev—see Foreign Broadcast Information Service 1989b: 84). Attempts to change aspects of the constitution, federation, and language and educational policies have met with fierce resistance when applied selectively to nationality groups. This has been evident most recently in the Transcaucasus, where Armenians are calling for the annexation of Nagorno-Karabakh Autonomous Oblast from the Azerbaydzhan SSR and Abkhazi have expressed the desire to secede from the Georgian Republic.

15. For details of recent changes in the Uzbek SSR administrative and territorial divisions, see *Pravda Vostoka* (1988).

16. Until the 1989 election reform, the composition of the Soviet of the Union was comprised of one delegate per 300,000 population. The 1989 elections were based on one delegate for approximately 375,000 persons.

REFERENCES AND BIBLIOGRAPHY

Akademiya Nauk Uzbekskoy SSR 1970. *Sovety Deputatov trudyashchikhsya Uzbekskoy SSR v Tsifrakh (1925–1969)*. Tashkent: Izdatel'stvo "Fan" Uzbekskoy SSR.

Ake, Claude 1967. *A Theory of Political Integration*. Homewood, Ill.: Dorsey Press.

Akiner, Shirin 1983. *Islamic Peoples of the Soviet Union*. London: Kegan Paul.

Allworth, Edward (ed.) 1967. *Central Asia: A Century of Russian Rule*. New York: Columbia University Press.

Allworth, Edward 1968. The "nationality" idea in czarist Central Asia. In Erich Goldhagen (ed.), *Ethnic Minorities in the Soviet Union*, pp. 229–50. New York: Praeger.

Arkhipov, Konstantin 1923. Types of Soviet autonomy. *Vlast' Sovetov* 10: 9–14.

Bacon, Elizabeth E. 1966. *Central Asians under Russian Rule*. Ithaca, NY: Cornell University Press.

Bahry, Donna 1987. *Outside Moscow*. New York: Columbia University Press.

Bartol'd, V. V. 1927. *Istoriya Kul'turnoy Zhizni Turkestana*. Leningrad: Akademiya Nauk.

Bennigsen, Alexandre 1971. Islamic or local consciousness among Soviet nationalities? In Edward Allworth (ed.), *Soviet Nationality Problems*, pp. 168–82. New York: Columbia University Press.

Bennigsen, Alexandre and Lemercier-Quelquejay, Chantal 1967. *Islam in the Soviet Union*. New York: Praeger.

Bennigsen, Alexandre and Wimbush, S. Enders 1986. *Muslims of the Soviet Empire*. Blooomington, Ind.: Indiana University Press.

Bloembergen, Samuel 1967. The union republics: how much autonomy? *Problems of Communism* 16: 27–35.

Bruk, S. I. 1981. *Naseleniye Mira*. Moscow: Izdatel'stvo "Nauka."

Bruk, S. I. 1986. *Naseleniye Mira*. Moscow: Isdatel'stvo "Nauka."

Burg, Steven L. 1984. Central Asian political participation and Soviet political development. In Roi 1984, pp. 40–62.

Carrere d'Encausse, Helene 1979. *Decline of an Empire*, trans. Martin Sokolinsky and Henry A. La Large. New York: Newsweek Books.

Chew, Allen F. 1970. *An Atlas of Russian History*. New Haven, Conn.: Yale University Press.

Coates, W. P. and Coates, Zelda 1969. *Soviets in Central Asia*. New York: Greenwood Press.

Craumer, Peter R. 1984. Areas of secondary and tertiary administrative units of the USSR, 1949–1983. Unpublished monograph.

Current Digest of the Soviet Press 1988a. Nishanov picked as new head of Uzbek party. XL, 3: 4.

Current Digest of the Soviet Press 1988b. Ethnic assimilation debated by scholars. XL, 9: 11–15.

Czaplicka, M. A. 1918. *The Turks of Central Asia*. London: Curzon Press.

Dellenbrant, Jan A. 1986. *The Soviet Regional Dilemma*. Armonk, NY: M. E. Sharpe.

Deutsch, Karl W. 1964. Communication theory and political integration. In

Philip E. Jacob and James V. Toscano (eds), *The Integration of Political Communities*, pp. 75–97. Philadelphia, Pa: Lippincott.

De Vree, Johan K. 1972. *Political Integration: The Formation of Theory and its Problems*. The Hague, Mouton.

Foreign Broadcast Information Service 1989a. *Pravda* lists new USSR Supreme Soviet members. *FBIS Daily Report, Soviet Union*, 5 June: 29–44.

Foreign Broadcast Information Services 1989b. Plenums held. *FBIS Daily Report, Soviet Union*, 7 July: 84. Translated from *Pravda*, 23 June 1989, 2nd edn: 2.

Foreign Broadcast Information Service 1989c. Reporting details: time to clear away the rocks. *FBIS Daily Report, Soviet Union*, 27 July: 75. (From *Trud*, 21 July 1989: 4.)

Foreign Broadcast Information Service. 1990. Decree issued on organization of Uzbek oblasts. *FBIS Daily Report, Soviet Union*, 8 March. From *Pravda Vostoka*, 17 February 1990: 1.

Friedgut, Theodore H. 1979. *Political Participation in the USSR*. Princeton, NJ: Princeton University Press.

Geograficheskiy Entsiklopedicheskiy Slovar'. 1989. Moscow: "Sovetskaya Entsiklopediya."

Gleason, Gregory 1987. Soviet federalism: anatomy of a debate. Talk given to the Harriman Institute Seminar on Soviet Republics and Regional Issues, Columbia University, 10 April.

Gleason, Gregory 1990. *Federalism and Nationalism: The Struggle for Republican Rights in the USSR*. Boulder, Colo. Westview Press.

Gorbachev, M. S. 1989. Vystupleniye general'nogo Sekretarya TsK KPSS, predsedatelya verkhovnogo soveta SSSR M. S. Gorbacheva po tsentral'nomy televideniyu. (Speech on Central Television by the General Secretary of the Central Committee of the KPSS, President of the Supreme Soviet of the USSR, M. S. Gorbachev.) *Pravda*, 2 July: 1.

Haas, Ernst B. 1958. *The Unity of Europe—Political, Social, and Economical Forces, 1950–1957*. London: Stevens.

Harasymiw, Bohdan 1984. *Political Elite Recruitment in the Soviet Union*. London: Macmillan Press.

Hodnett, Grey 1978. *Leadership in the Soviet National Republics: A Quantitative Study of Recruitment Policy*. Ontario: Mosaic Press.

Hough, Jerry F. 1988. *Opening Up the Soviet Economy*. Washington, DC: The Brookings Institution.

Institut Istorii Partii 1971. *Kommunisticheskaya Partiya Kirgizii*. Frunze: Kirgizstan. Izvestiya TsK KPPS 1990. No. 5, pp. 60–3.

Institut Istorii Partii 1975. *Kommunisticheskaya Partiya Turkmenistana v Tsifrakh (1924–1974)*. Ashkhabad: Turkmenistan.

Institut Istorii Partii 1979. *Kommunisticheskaya Partiya Uzbekistana v Tsifrakh (1924–1977)*. Tashkent: Uzbekistan.

Institut Istorii Partii 1984a. *Kommunisticheskaya Partiya Turkmenistana 60 Let (1924–1984)*. Ashkhabad: Turkmenistan.

Institut Istorii Partii 1984b. *Kommunisticheskaya Partiya Tadzhikistana v Tsifrakh za 60 let (1924–1984 gg.)*. Dushanbe: Izdatel'stvo "Irfon."

Institut Istorii Partii 1984c. *Kommunisticheskaya Partiya Kirgizii v Tsifrakh, 1918–1984*. Frunze: Kirgiziya.

Joint Public Research Service 1989a. Preliminary census results published.

USSR Report. 19 May: 17–21. (Translated from Moscow *Pravda*, 29 April 1989, 2nd edn.)

Joint Public Research Service 1989b. UzSSR: Samarkand territory transferred to Bukhara Oblast. USSR Report. 27 July: 22. (Translated from *Pravda Vostoka*, 17 May 1989: 1.)

Kaganskiy, V. L. 1982. Geographical boundaries: contradictions and paradoxes. *Soviet Geography: Review and Translation* 1: 39–48.

Karklins, Rasma 1986. *Ethnic Relations in the USSR: The Perspective from Below.* Boston, Mass.: Allen & Unwin.

Khorev, B. S. 1981. *Territorial'naya Organizatsiya Obshchestva* Moscow: "Mysl.''

Kozlov, V. I. 1982. *Natsional'nosti SSSR.* Moscow: "Finansy i Statistika.''

Kushner, P. I. 1951. *Etnicheskiye Territorii i Etnicheskiye Granitsy.* Moscow: Izdatel'stvo Akademii Nauk SSSR.

Lattimore, Owen 1950. *Pivot of Asia.* Boston, Mass.: Little, Brown.

Lattimore, Owen 1962. The new political geography of Inner Asia. In Owen Lattimore (ed.), *Studies in Frontier History. Collected Papers, 1928–1958,* pp. 165–79. London: Oxford University Press.

Liebowitz, Ronald D. 1985. Spatial and ethnic dimensions of Soviet regional investment: 1956–1975. PhD dissertation, Columbia University.

Matley, Ian Murray 1967. Industrialization. In Edward Allworth (ed.), *Central Asia: A Century of Russian Rule,* pp. 324–53. New York: Columbia University Press.

McIntyre, T. V. 1974. Soviet Central Asia: its geopolitical role in contemporary Asian affairs. PhD dissertation, The American University.

Miller, John 1983. Nomenklatura: check on localism?. In T. H. Rigby and Bohdan Harasymiw (eds), *Leadership Selection and Patron Client Relations in the USSR and Yugoslavia,* pp. 62–97. London: Allen & Unwin.

Morrison, J. A. 1938. The evolution of the territorial-administrative system of the USSR. *American Quarterly on the Soviet Union* 1 (November): 25–46.

Morrison, Minion K. C. 1982. *Ethnicity and Political Integration.* Syracuse, NY: Maxwell School of Citizenship and Public Affairs.

Parker, W. H. 1969. *An Historical Geography of Russia.* Chicago: Aldine.

Pierce, Richard A. 1960. *Russian Central Asia, 1867–1917.* Berkeley, Calif.: University of California Press.

Pipes, Richard 1954. *The Formation of the Soviet Union: Communism and Nationalism, 1917–1923.* Cambridge, Mass.: Harvard University Press.

Pravda 1989 11 December: 2.

Pravda Vostoka 1988. 3 September: 3.

Pravda Vostoka 1990. 5 June: 5.

Radio Liberty/Radio Free Europe 1986. First secretaries of the krai and oblast' party committees. *Radio Liberty Research Bulletin* RL 256/86, 7 July: 1–12, and supplement 1/86.

Rakowska-Harmstone, Teresa 1970. *Russia and Nationalism in Central Asia.* Baltimore, Md: Johns Hopkins University Press.

Ro'i, Yaacov (ed.) 1984. *The USSR and the Muslim World.* Boston, Mass.: Allen & Unwin.

Rumer, Boris Z. 1989. *Soviet Central Asia: ''A Tragic Experiment.''* Boston, Mass.: Unwin Hyman.

Rywkin, Michael 1982. *Moscow's Muslim Challenge*. Armonk, NY: M. E. Sharpe.

Rywkin, Michael 1983. *Russia in Central Asia*. New York: Collier.

Shabad, Theodore 1951. *Geography of the USSR: A Regional Survey*. New York: Columbia University Press.

Shabad, Theodore 1956. The administrative-territorial patterns of the Soviet Union. In W. Gordon East and A. E. Moodie (eds), *The Changing World*, pp. 365–84. Yonkers-on-Hudson, NY: World Book Company.

Shabad, Theodore 1982. News notes. *Soviet Geography: Review and Translation* 23 (June): 468.

Sherstobitov, V. P. (ed.) 1979. *Natsional'nyye Otnosheniya v SSSR na Sovremennom Etape. Na Materyalakh Respublik Sredney Azii i Kazakhstana*. Moscow: "Nauka."

Sovetskiy Entsiklopedicheskiy Slovar' 1986. Moscow: "Sovetskaya Entsiklopediya."

Tuzmuhamedov, R. 1973. *How the National Question Was Solved in Soviet Central Asia*, Moscow: Progress Publishers.

USSR Goskomstat 1989. *Natsional'nyy Sostav Naseleniya*. Moscow: Information-Publication Center.

USSR Goskomstat 1990. *Demografisheskiy Yezhegodnik 1990*. Moscow: "Finansy i statiska."

USSR Goskomstat 1990a. *SSR v Tsifrakh v 1989 Godu*. Moscow: "Finansy i Statistika."

USSR 1928. Izdaniye Tsentral'nogo Izdatel'stogo Komiteta SSSR. *Atlas Soyuza Sovyetskikh Sotsialisticheskikh Respublik*. Moscow.

USSR 1959. Prezidium Verkhovnogo Soveta. *SSSR: Administrativno-territorial'noye Deleniye Soyuznykh Respublikh na 1959 Goda*. Moscow: Izdatel'stvo "Izvestiya sovetov Deputatov Trudyashchikhsya SSSR."

USSR 1960. *Narodnoye khozyaystvo SSSR v 1959 Godu*. Moscow: Gosstatizdat.

USSR 1962. Tsentral'noye Statisticheskoye Upravleniye SSSR. *Itogi vsesoyuznoy Perepisi Naseleniya 1959 Goda*. 16 vols Moscow: Gosstatizdat.

USSR 1970. Prezidium Verkhovnogo Soveta. *SSSR: Administrativno-territorial'noye Deleniye Soyuznykh Respublikh, na 1970 Goda*. Moscow: Izdatel'stvo "Isvestiya Sovetov Deputatov Trudyashchikhsya SSSR."

USSR 1971. *Narodnoye Khozyaystvo SSSR v 1970 G*. Moscow: Izdatel'stvo: "Statistika."

USSR 1973. Tsentral'noye Statisticheskoye Upravleniye SSSR. *Itogi Vsesoyuznoy Perepisi Naseleniya 1970 Goda*. 7 vols Moscow: "Statistika."

USSR 1978. *Konstitutsiya (osnovnoy zakon) soyuza sovetskikh Sotsialisticheskikh respublik*. Moscow: Izdatel'stvo Politicheskoy Literatury.

USSR 1979. Prezidium Verkhovnogo Soveta. *SSSR: Administrativno-territorial'noye Deleniye Soyuznykh Respublikh, na 1970 Goda*. Moscow: Izdatel'stvo "Izvestiya Sovyetov Deputatov Trudyashchikhsya SSSR."

USSR 1980. *Narodnoye Khozyaystvo SSSR v 1979 G*. Moscow: Izdatel'stvo: "Statistika."

USSR 1983. Tsentral'noye Statisticheskoye Upravleniye SSSR. *Naseleniye SSSR. Spravochnik*. Moscow: Izdatel'stvo Politicheskoy Literatury.

USSR 1984a. *Narodnoye Khozyaystvo SSSR v 1985 G*. Moscow: "Finansy i Statistika."

USSR 1984b. Tsentral'noye Statisticheskoye Upravleniye SSSR. *Chislennost' i Sostav Naseleniya SSSR*. Moscow: "Finansy i Statistika."

USSR 1986. Prezidium Verkhovnogo Soveta. *SSSR: Administrativno-territorial'noye Deleniye Soyuznykh Respublikh, na 1986 Goda.* Moscow: Izdatel'stvo "Isvestiya Sovetov Deputatov Trudyashchikhsya SSSR."

USSR 1987. *Narodnoye Khozyaystvo SSSR v 1986 G.* Moscow: "Finansy i Statistika."

USSR 1988. *Naseleniye SSSR 1987. Statisticheskiy Sbornik.* Moscow: "Finansy i Statistika."

United States, Central Intelligence Agency 1959. *Atlas of Soviet Territorial Administrative maps.* 2 vols. Washington DC: CIA.

United States 1987. *The USSR: A Spatial Perspective.* Washington, DC: Department of State, Office of the Geographer; and Center for International Research, Bureau of Census.

Vanneman, Peter 1977. *The Supreme Soviet: Politics and the Legislative Process in the Soviet Political System.* Durham, NC: Duke University Press.

Wheeler, Geoffrey 1964. *The Modern History of Soviet Central Asia.* New York: Praeger.

Chapter 4

The Physical Geography of Soviet Central Asia and the Aral Sea Problem

Peter Sinnott

INTRODUCTION

This chapter is divided into two parts. The first provides a basic overview of the physical geography of Soviet Central Asia; the second provides a focus on a major environmental problem in the USSR: the Aral Sea.

Historically, the words "Central Asia" have evoked an area of diverse and uncertain dimension. Even using Soviet terminology there has been some confusion. "Central Asia" ("Tsentral'naya Aziya") is commonly used to portray an area that incorporates portions of western China along with the Soviet republics of Kazakhstan, Kirgiziya, Uzbekistan, Tadzhikistan, and Turkmeniya. "Middle Asia" ("Srednyaya Aziya") is used primarily in reference to an area encompassing the Soviet republics of Kirgiziya, Uzbekistan, Tadzhikistan, and Turkmeniya, which is also consolidated as an economic region. It is also sometimes used, however, to include southern Kazakhstan, which in terms of cultural lifestyle as well as physical geography is strongly linked to the Central Asian Economic Region to its south. This chapter includes some of the southern desert region of Kazakhstan for unity in the physical region of Soviet Central Asia (Map 4.1).

Central Asia is an arid land dominated by deserts for most of its main expanse and mountains on the east and south. The Caspian Sea forms its western border. Its northern border is characterized by the physical transition zone of plateau and semidesert in southern Kazakhstan. Life in Central Asia, as in most of the world, is strongly linked to climate. Settlement is chiefly in oases and along the two main rivers which dissect the desert plain—the Amudar'ya and Syrdar'ya, as well as on a transitional piedmont zone between the mountains and the desert plain.

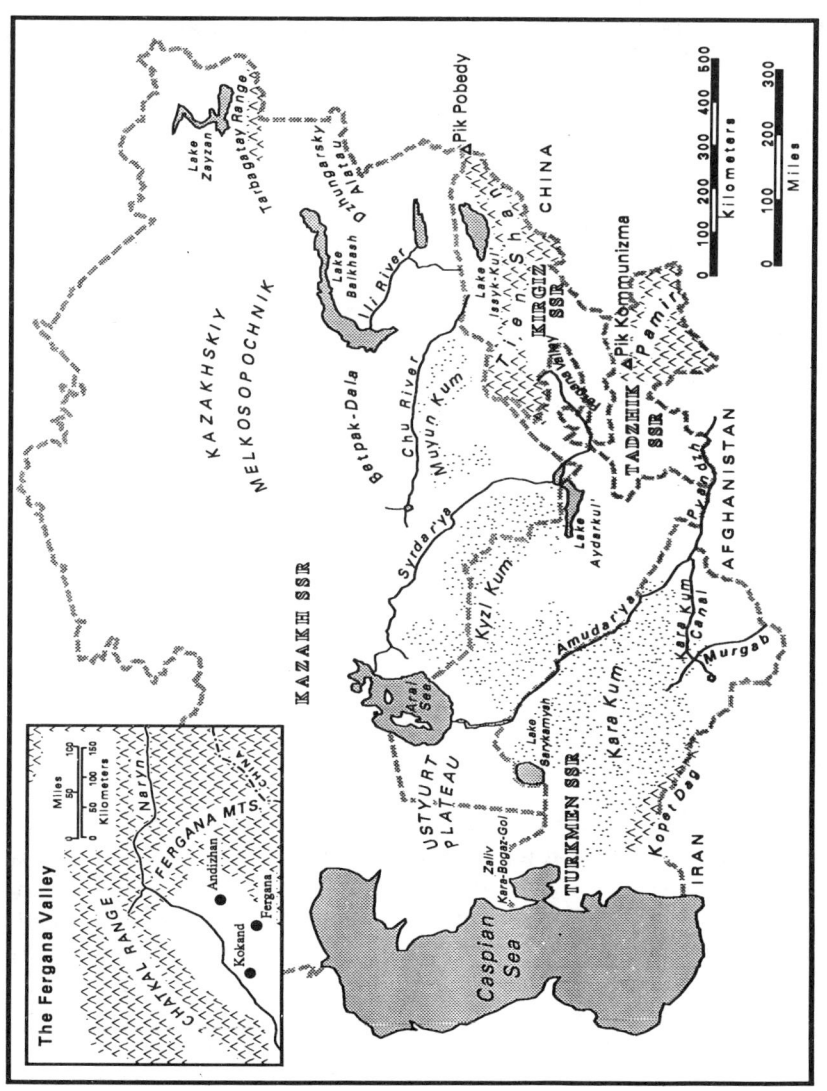

Map 4.1 Physical geography.

CLIMATE

Atmospheric circulation is primarily affected by the continental position and the division of Central Asia into a terrain of desert and mountains. Central Asia lies at the periphery of two important pressure systems which help to explain the continentality of the climate. In the winter, pressure systems in Central Asia are dominated by the Asiatic High which originates over northwestern Mongolia. Lydolph notes that the pattern of pressure divergence that forms over northern Kazakhstan results in a high-pressure ridge where winds diverge north toward Siberia and the Arctic and south toward Central Asia (Lydolph 1977: 151). This brings a northeasterly flow of air into Central Asia in winter and a decreasing barometric pressure north to south with isobars trending east–west. As a result of the transfer of air from a more northerly part of the continent to a more southerly, these prevailing northeastern winds are very cold. In fact, the winter is surprisingly cold relative to the latitude. In summer the other major pressure system, an eastern extension of the Azores High which brings moisture to southern Europe, has lost almost all of its moisture by the time it has crossed the Caucasus Mountains and reached Central Asia. This tropical air mass brings a northwesterly flow of air to Central Asia. Air from the north reaches the region as well. Borisov found continental temperate and continental tropical air to account for 90 percent of the air masses in Central Asia (Borisov 1965: 22).

Seasonal climate variation

Winters in Central Asia are short but cold. Temperatures increase latitudinally north to south, with January normally the coldest month (Map 4.2). Typically, the Aral Sea along with the lower course of the Syrdar'ya is frozen for four or five months, as is Lake Balkhash occasionally. Lake Issyk-Kul' ("Hot Lake" in Turkic), high in the ranges of the Tien Shan, however, does not freeze over due to locally mild winters, salinity, and high water volume (Suslov 1961: 458). The northern half of the desert lowland has a more severe winter, with monthly average temperatures below freezing from November to March. Snowfall is possible within the same period, but is of negligible cover, varying from 20 to 50 days in most desert areas. Depth and duration decrease from north to south. According to Borisov, Kazalinsk on the lower Syrdar'ya might typically have 70 days of snow cover, while Bayram-Ali near Mary in Turkimeniya might have only four days (Borisov 1965: 181). The southern half of the desert lowland has only two months that are truly wintery: January and

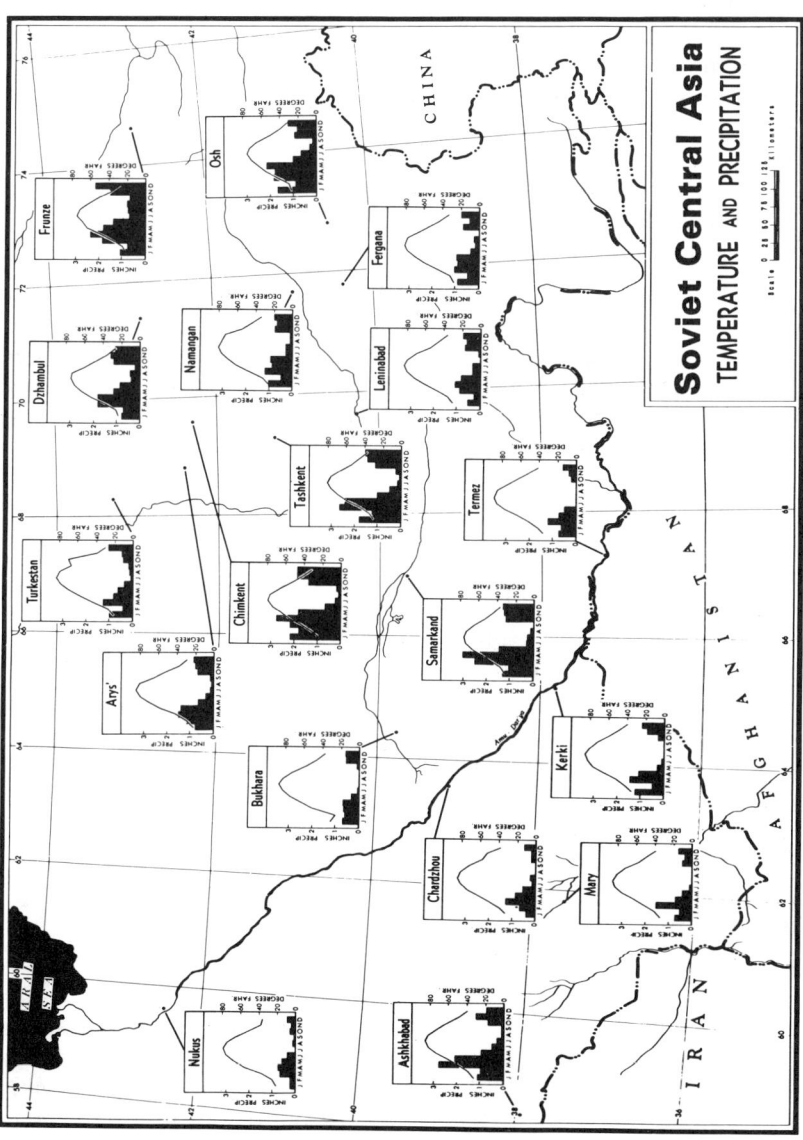

Map 4.2   Temperature and precipitation

February. Yet, even in those months there are days that are well above freezing.

Spring is very short, lasting only one month: April in the north, March in the south. Winter snow disappears as temperatures rise quickly. Within this short season the majority of the year's precipitation falls in short torrential bursts.

The summer is long and dry. It begins in May after the spring rains and lasts through September. The sky becomes clear. In the northern parts of the desert, cloud cover averages 17 percent in August, while the south has only 4 percent (Suslov 1961: 439). The lack of cloud cover contributes to the extraordinary heat for this latitude as well as to the large temperature fluctuation, since the day heat quickly escapes at night. Average daily temperatures are above 15°C and can reach daytime maxima of 75.4°C (Kazalinsk) and 70.8°C (Bayram-Ali) (Suslov 1961: 438). This climate is excellent for the growing of cotton and other irrigated crops. The growing season is as long as 204 to 288 days. Precipitation is practically nonexistent in summer, and the phenomenon of "dry rain" is encountered in the Golodnaya Steppe ("Hungry Steppe") and the sandy deserts as the extreme heat causes the rain to evaporate before hitting the ground. The lack of humidity also contributes to the dryness, with summer low-humidity averages of 50 percent for Kazalinsk and 30 percent for Bayram-Ali (Suslov 1961: 438).

The transition from a hot summer to a cool autumn begins in the desert plains toward the end of September or early October (western parts). The weather is characterized by stable, dry, and sunny days accompanied by increasingly cool nights. In autumn the rate of temperature decline is slower than its rise in spring, as the thermal depression and cyclonic activity return. There is increasing cloudiness and, occasionally, precipitation or even snow as frosts begin to occur toward the end of October.

MOUNTAINS

Central Asia's mountains were one of the last large blank spots on the map as the nineteenth century was drawing to a close. Large expeditions, both independent and government sponsored, were mounted from imperial Russia and western Europe as a means of increasing both geographical knowledge and political pressure. Still, by the time of Soviet power some blank areas still remained. Only during the 1930s and 1940s did several Soviet expeditions complete the basic mapping and classification of the landforms, flora, and fauna of Central Asia.

With respect to physical geography, Central Asia begins in northeastern Kazakhstan between the Tarbagatay Range and the Dzhungarskiy Alatau. This is considered the traditional border for several reasons. This transitional area forms a break between the mountains associated with Inner Asia and those of Central Asia and is called the Dzhungarian Gate. This barren, narrow, constantly windy gorge, a graben formation, served as the portal for nomadic pastoral breeders and caravans as well as conquerors. It was also a cultural divide, since further north the mountains gave way to Siberian forests that supported hunter-gatherers, while to the south sedentary oasis-based cultures thrived.

The present-day Central Asian mountain systems began as a large syncline in a great inland sea. With the gradual retreat of the sea and increasing lateral pressure during the lower and upper Paleozoic era, the northern and central ranges of the Tien Shan formed. By the Neocene period the mountains had undergone denudation and become a near peneplain. This was followed by Alpine folding which created the southern Pamir-Alay systems and uplifted the Tien Shan. The continuing process of the Indian subcontinent's collision toward Asia has made this region of folded mountains one of the most seismically active areas in the world, with more than a thousand local earthquakes recorded each year.

The Tien Shan mountain system (from the Chinese, "Heavenly Mountains") is an extensive, geologically complex group of high mountain ranges. They are an old mountain system that underwent rejuvenation and have thus acquired many characteristics associated with young mountains due to uplifting. They extend west to east for 2,450 kilometers through both the Soviet Union and China. Approximately half of this mountain system is situated in the Soviet Union. It stretches in long arcs from the northern edge of the Fergana basin up to the escarpment of the Kirgiz Range overlooking the Muyun Desert and the central Tien Shan's massive concentric series of glacial peaks. Among these peaks are Khan Tengri Peak, (6,995 meters) and Pik Pobedy ("Victory Peak," 7,439 meters). This latter mountain is the second highest peak in the USSR and is located near the point where Kazakhstan, Kirgiziya, and China meet. They present a landscape of high, closely grouped ranges surrounded by smaller ranges on high, broad plateaus dissected by young rivers and streams which gradually give way to lower, more widely spaced ranges. Between the ranges are mountain basins which are generally flat and characterized by steppe or desert conditions. The largest and most economically important, the Fergana Basin, is one of the most densely settled areas in Central Asia. Other significant basins are the Ili Depression just north of the Tien Shan and Lake Issyk-Kul'.

There are many large glaciers in the high catchment zones of the

Tien Shan and Pamir-Alay systems. The Tien Shan alone has 10,200 square kilometers of glaciers. Vegetation in the high mountains is sparse and typically alpine as meadow and meadow steppe predominate. The landscape of green alpine meadows and snows gives a mottled appearance at a distance. Hence, the Ala-tau (from Turkic, "mottled") appendage to many of the mountain names. Below the high mountains vegetation is strongly characterized by zonation, with distinct horizontal belts of vegetation at different altitudes. The lower foothills are typically semidesert, with higher foothills giving way to a broad-leaved forest belt from 600 to 1,800 meters, followed by a coniferous zone that begins at 1,700 to 1,800 meters, which thins out into a subalpine zone that might stretch to 3,000 meters, where alpine meadows predominate (Knystautus 1987: 142).

The Pamir-Alay mountain system begins on the southern flank of the Fergana basin in the Trans-Alay Range and reaches China and Afghanistan where the countries touch near the "Pamir Knot." The Pyandzh River forms much of the southern border with Afghanistan's Hindu Kush. The Pamirs are the highest mountain system in the Soviet Union. They contain the highest peak (Pik Kommunisma, 7,495 meters), as well as numerous glaciers which are an important water source for the desert lowlands below. The largest is the 77 kilometer long Fedchenko Glacier in the Akademiya Nauk Range. This glacier is named for a Russian explorer of the Pamirs in the nineteenth century. The Pamir is a large massif of high upland of contrasting relief. The eastern Pamir is characterized by a high upland with broad, flat, treeless valleys that are surrounded by flattish peaks whose height relative to the valley floor is small—2,000 to 3,000 meters. This area is the driest in Central Asia; it is even drier than the desert lowland below, due to the altitude of much of the area being above the clouds, as well as the blocking of precipitation by the Tien Shan, on the one hand, and the Hindu Kush, on the other. The influence of the glaciers also retards erosional processes. In contrast, the western Pamirs are strongly dissected by river erosion; slopes are steeper, and peaks are more pronounced and capped by snow and glaciers. Vegetation is very sparse, as valleys and depressions tend to be filled with rock debris. Settlement is mainly along the wider lower reaches of the Pyandzh River. The climate is continental there, with mean temperatures of 18°C to 22°C in summer and −8°C in winter (Ginzburg 1986: 399).

The Kopet Dag Mountains in southern Turkmeniya are detached from the Central Asian mountain systems and are properly identified as a part of the Near Eastern Highland Zone (Knystautus 1987: 220). They are an extension of the Iranian desert tableland and run astride the Soviet–Iran border, rising to heights of more than 2,700 meters. They lie in a subtropical zone and contain no permanent snowline or meadows. They are deeply dissected, and the landscape is semidesert and mountain-steppe, with vegetation limited to drought-resistant plants and ephemerals that come to life with spring rains. Light juniper forest is found from 1,000 to 2,500 meters. The foothills contain rich wind-swept loess soil, while the base of the mountains is occupied by a strip of clayey desert.

Piedmont plains and loess foothills form a transitional area between the desert lowlands and mountains. They are found in an irregular belt adjacent to the mountains' base that stretches from the Kopet Dag to the Pamir and the Zayli-Alatau Range in the Tien Shan, as well as flanking the Fergana Valley. Lewis divides this zone into two parts (Lewis 1966: 472). The first contains well-defined and dissected loess-covered foothills and low uplands as well as mountain spurs. The second, lower part of the piedmont consists of a gradually sloping alluvial plain that eventually merges with the desert lowlands. Both are traditional areas of settlement.

Typical of the lower piedmont zone is the Golodnaya Steppe (known as Mirza Chyl in Uzbek). This large plain lies at elevations of 250 to 300 meters and is situated just below the Fergana Valley between the Dzhizak and Syrdar'ya rivers. Its sierozem soils could not be irrigated until a large-scale Soviet settlement and irrigation development plan was implemented after World War II. It produces cotton with the help of irrigation as well as vegetables which are raised mainly in greenhouses.

The Fergana Valley is an important cultural, population, and economic center for Central Asia. It stretches 300 kilometers and reaches 170 kilometers in width. Its center is mainly desert and desert-steppe, but its flanks contain many loess-rich foothills and mountain slopes through which mountain streams carry sediment to the alluvial fans below. So great is the irrigation usage for agriculture that these mountains streams are nearly consumed within the valley. Agriculture is principally cotton and vinoculture oriented.

DESERTS

More than three-quarters of the territory in Central Asia is desert lowland, which varies greatly in configuration, as sandy, stony, salt, and clay deserts are found. The greatest of these deserts is the Kara Kum ("Black Sands" in Turkic), which stretches more than 350,000 square kilometers. It lies between the mountains and the Amudar'ya. On the west it touches the pre-Caspian lowland and the desolate Ust-Urt Plateau, and on the east it dies on the foothills of the Pamir-Alay Mountains. The origins of this desert are primarily alluvial, but aeolian erosion contributes as well. The Kara Kum is divided into two regions, the Central Kara Kum and the Zaunguz Plateau. Sandy desert and related landforms predominate in this arid expanse. Ridge sand—sand dunes held in place by psammophilic (sand-loving) vegetation and stretching longitudinally in the direction of the wind—is the most common form in the Kara Kum. Suslov points out that the predominance of such a landform stretched in a nearly meridional direction facilitated the movement of the caravans (Suslov 1961: 455). Crescent-shaped and mobile sand dunes, barchans, are also encountered in the Kara Kum. They form when sand-fixing vegetation is absent or destroyed. They are most common on the left bank of the Amudar'ya. Between the sand ridges one commonly finds "takyr" claypans (Korzhenevskiy 1960: 57). They are formed by spring rains which create lakes in the depressions between sand ridges. The intense heat causes the lakes to quickly evaporate and form clayey alluvium bottoms which resist water penetration. They dry out and crack in polygonal shapes. They harden as the summer progresses, providing a surface that sand rarely fixes upon. Another desert relief form, which is typical to the Zaunguz Plateau, is the salt flat, called "sors," which appears as a white salty layer on the soil.

Vegetation cover is primarily ephemeral drought-resistant plants. Unique to Central Asian sandy deserts is a psammophyte semibrushwood, the saxaul tree. Its leaves are very narrow so that the amount of water evaporated through them is quite small. It has excellent sand-fixing capabilities, and its wood is extraordinarily dense, yielding an excellent charcoal.

The Kyzyl Kum Desert ("Red Sands" in Turkic) lies between the Amudar'ya and Syrdar'ya rivers. It stretches from the Aral Sea to the foothills of the Tien Shan. The Kyzyl Kum, like the Kara Kum, is a sandy desert. Ridge sand and sandy hills occur, but mobile barchans do not. Except for the pre-Aral lowland, the landscape of the Kyzyl Kum offers higher elevations and more complex relief than the Kara Kum. Remnant tablelands, monadocks, with elevations over 500 meters can be found toward its center.

Between the Caspian and the Aral seas lies the arid Ustyurt plateau in northwestern Central Asia. It is a flat plain with occasional small hills containing salt lakes, salt marshes, and sand. Rain is practically nonexistent, and there is no running water on the plateau. Elevations in the central plateau run 200 to 220 meters, with heights of 320 to 340 meters along the southwest corner where they encounter the relatively small Mangyishlak Mountains with peaks around 550 meters.

WATER RESOURCES

Central Asia is a closed basin with no open sea or ocean outlet, and its hydrologic cycle is thus strongly affected by climatic features. The very limited precipitation on the lowland plain, coupled with the high temperatures, low humidity, and high degree of solar radiation, provides for a high degree of potential evapotranspiration. Water resources are thus mainly surface waters that are formed in the mountains and spent on the plains of Central Asia. Meltwater from permanent snowfields is the main source of water for the rivers, with glaciers and seasonal snowfields contributing as well. The result is a fairly reliable river flow that deviates only slightly year to year.

The most important rivers are the Amudar'ya and Syrdar'ya. The Amudar'ya is the largest river in Central Asia, stretching some 2,540 kilometers in length. It begins in the Hindu Kush and is fed by the meltwater from snow and glaciers. This produces two close but distinct flood seasons: spring (April and May) for snow thawing, and summer (June and July) for glacial melting, both of which are very favorable periods for irrigation. Because of the gradual slope of the plain the river has no real valley in its middle course, and because of the fine texture of the alluvium the river's course often changes quickly. During high water the channel has a width of 500 to 1,500 meters (Suslov 1961: 471). In the lower course, which is more heavily irrigated, artificial banks define much of the channel; and approximately 400 kilometers from its depository, the Aral Sea, the river braids and for the last 150 kilometers become a delta. The alluvium, rich in phosphates, lime, and potassium, is excellent for agriculture and needs little, if any, fertilizer. The Amudar'ya once carried more sediment than any other river in the world, with as much as 210 million tons carried to its delta annually up until 1960 (*Great Soviet Encyclopedia* 1973, vol. II: 363). More recently, it is estimated that the delta region near Nukus receives 128 million tons of sediment annually (Kes' and Klyukanova 1990: 604).

When the Syrdar'ya begins in the region of permanent snow and glaciers in the central Tien Shan it is known as the Naryn River. Later, merging with the Karadar'ya it becomes the Syrdar'ya and flows through the Fergana Valley and the lowland plain to the Aral Sea. Its flood season is June, and its overall water discharge and sediment transported to the Aral Sea are much less than from the Amudar'ya. It has been estimated that the Syrdar'ya annually transported 38 million tons of sediment to the central Asian plains (Kes' and Klyukanova 1990: 604).

As a result of a December 1978 call for the drafting of measures to improve water use in the economy of Central Asia and southern Kazakhstan, a breakdown of surface water resources was published. It measured total surface water resources in the Aral Sea basin at 127.5 cubic kilometers for the average year (Kes' et al. 1982: 414). On average the same survey estimated that over three-quarters of the surface water resources are used by irrigation channels, 96.7 cubic kilometers. In some dry years, however, the percentage was greater—88.4 percent, for instance, in 1972 for both rivers.

IRRIGATION AGRICULTURE AND REGIONAL DEVELOPMENT

While the great irrigation growth of recent years has been an economic boon for agriculture, the possibility of it having a far-reaching negative effect on the environment has only very recently become an issue.

During the Soviet period, agricultural needs have been met more through extensive growth than intensive growth. This has allowed the introduction of new lands to mask ecosystem damage that has at times been wrought by an agricultural system tuned to production goals with a short time frame in mind, usually no more than five years. Also, typically, there has been a centrally imposed rigidity to crop structure since the Stalin era that has resulted in various agricultural areas, such as Central Asia, being associated with monocultures even though earlier they presented a more diversified agricultural mix destined principally for local needs.

Soviet Central Asia is still a predominantly rural region, and its economy has been so linked to the cotton crop that there are aspects of a "plantation economy" in its monoculture. The success of the regional economy is still centered around irrigation, agriculture and cotton production. Southern Kazakhstan and the Central Asian republics produce over 90 percent of the Soviet Union's cotton crop. This skewed development continues to be defended by some from a socialist inter-nationalism stance: "Placed upon us was we could say an exceptional historic mission: to guarantee cotton independence for our country and the entire socialist community" (Yusupov and Ziyadullayev 1988: 4). In 1913 the irrigated area of the Aral Sea basin stood at close to 2 million hectares. Real development planning was initiated in the Soviet period and expansion was at first gradual. By 1950, 4.7 million hectares of agricultural land was irrigated in the Aral basin, primarily for cotton, by 1960, 5.1 million hectares. Agricultural irrigation increased dramatically only after the May 1966 CPSU Central Committee plenum decision enacted a policy of irrigation expansion. At present about 7 million hectares of agricultural land are under irrigation in southern Kazakhstan and Central Asia, an increase of more than one-third since 1965. Until recently the irrigation of an additional million hectares in the Aral Sea basin was planned by the end of the century (Voropayev et al. 1988: 9).

RIVER DIVERSION SCHEMES

A key component in the promotion of further irrigation expansion in the Aral Sea basin was the promise of additional water sources from Siberia to supplement the increasingly strained Amudar'ya and Syrdar'ya rivers. During the Brezhnev era two water diversion plans had been drawn up in great detail and were close to implementation. The first, the Northern Rivers Project, would have diverted water from rivers and lakes in the northern part of European Russia to the Volga basin. The second and much larger project, the Siberian River Diversion Project, was to provide a 50 percent increase in irrigated lands in the Aral Sea basin by the turn of the century. Some critics opposed the plan as dangerously premature with respect to understanding its effects on both the ecosystem of the Ob' River basin and the Arctic ice supply. Further criticism appeared in *Pravda* on the eve of the Twenty-Seventh Party Congress (Aganbegyan et al. 1986). Abel Aganbegyan and other academicians criticized the projects' economic and scientific underpinnings: "Above all, it is desirable to abandon the costly work to redistribute water resources between basins. The idea for redistribution arose from the impression that crop farming could continue to grow in extensive terms, which would inevitably result in increased requirements." (Aganbegyan *et al.* 1986:3). Furthermore, their view that there "exist enormous reserves for reducing water consumption, above all in agriculture," brought projections of a worsening water deficit by the ministries and institutes involved under close scrutiny (Sinnott 1988: 3).

A joint resolution of the CPSU Central Committee and the USSR Council of Ministers, published in August 1986, seemingly ended speculation about the future of the proposed diversion of part of the flow of northern and Siberian rivers to the southern regions of the country (*Pravda* 20 August 1986). To compensate for this, the resolution sought a 15–20 percent reduction in water consumption, with investment in the modernization of existing irrigation systems.

These various water diversion schemes had been championed by those favoring agricultural expansion, especially by many prominent party members in the Central Asian republics who saw it as a way of further expanding the irrigated land areas. Their cancellation and the public discussion of the possible economic and ecological consequences had the projects been approved left those championing further agricultural expansion without the addition of Siberian water. Without it the question of continued agricultural irrigation expansion has to be balanced with the immediate cost: further decline in the size of the Aral Sea.

THE ARAL SEA

The Aral Sea, once the second largest inland water body in the Soviet Union and the fourth largest in the world, continues to recede. A saline lake, its water volume and area have been nearly halved since 1961 and its shores have receded dramatically along its shallow eastern banks. In some places it has retreated more than 120 kilometers. Since late 1989 the northern part of the Aral has separated from the larger southern part due to dessication. Sea level has accordingly dropped sharply as well. The present sea level of approximately 39 meters is 13 meters less than in 1961. This past decade the annual loss has been nearly 90 centimeters a year (Oreshkin 1988). Some of the effects of this dramatic drop have been a rising mineralization level in the water, with a related near extinction of fish in the Aral Sea, and wind storms containing dust and salt from the dried-up sea bed known locally as "the dry tears of the Aral" (Oreshkin 1988). The dried-up sea bottom now covers an area of more than 27,000 square kilometers, and salt and dust storms arising from its surface now carry some 75 million tons annually. They can move in belts as broad as 40 kilometers and inflict damage on soil and crops thousands of kilometers away.

The dimensions of the current ecological crisis in the Aral Sea basin continue to broaden. The most immediate area of impact, the area adjoining the Aral Sea, shows significant deterioration in a broad range of health indicators such as waterborne diseases and infant mortality. These ongoing problems led the USSR Health Minister, Evgeniy Chazov, in the summer of 1988 to send in several medical teams on a long-term basis to help in some of the immediate problems (Brown 1988). These appear to be connected to the recent dramatic change in the surrounding Aral ecosystem (Bohr 1988). The roles of the Aral and its arteries, the Amudar'ya and Syrdar'ya rivers, in the ecosystem as well as their interrelationship with future economic development are being intensely discussed.

The controversy has come to center on both the economic and environmental impact of the receding sea level of the Aral Sea. The sea level of the Aral is dependent on local precipitation and the inflow of its tributaries. It is receding primarily due to water withdrawal from the two rivers for irrigation purposes (including a diversion to the Kara Kum Canal through Turkmeniya), industrial water needs, hydroelectric dams, and other sociocultural needs as well. But irrigation development is clearly the leading cause. One estimate places irrigation usage in the Aral Sea basin at more than 110 cubic kilometers of water (Micklin 1986: 167). The increased irrigation usage of the river flow has strongly reduced the Aral's inflow to a mere trickle (Table 4.1).

Until 1960 the inflow of water to the Aral Sea had a yearly average of 56 cubic kilometers (Epstein 1987: 144; Kotlyakov et al. 1987: 24). The combined flow of the Amudar'ya and Syrdar'ya averaged 42.9 cubic kilometers for 1960–71, and 16.1 for 1971–80, but for 1981–4 the flow was lowered to 4.2 cubic kilometers (Akramov and Rafikov 1986: 168). The Syrdar'ya flow has more or less ceased since 1978, and the Amudar'ya

Table 4.1   Inflow into the Aral Sea (in cubic kilometers).

Years	Amudar'ya (at Temirbay)	Syrdar'ya (at Kazalinsk)	Total
1959	40.0	18.3	58.3
1960	37.8	21.0	58.8
1961	29.2	0.0	29.2
1962	29.1	5.7	34.9
1963	29.9	10.6	40.5
1964	36.5	14.9	51.4
1965	25.2	4.6	29.9
1966	33.1	9.5	42.6
1967	28.6	8.6	37.3
1968	28.9	7.2	36.1
1969	55.1	17.5	72.6
1970	28.7	9.8	38.6
1971	15.3	8.1	23.5
1972	15.5	6.9	22.4
1973	33.4	8.9	42.3
1974	6.2	1.9	8.1
1975	10.0	0.6	10.6
1976	10.3	0.5	10.8
1977	7.2	0.4	7.7
1978	18.9	0.0	18.9
1980	8.3	0.0	8.3
1984	7.9	0.0	7.9
1987	n.a.	n.a.	10.5
1988	n.a.	n.a.	23.6
1989	n.a.	n.a.	4.3

Sources: Rafikov 1986: 11; Hydrometeorological Service, Uzbek SSR; interview with Viktor Dukhovny, SANIIRI, 16 July 1990.

flow has been reduced to 1 to 5 cubic kilometers per year. The restoration of some flow for 1987 and 1988, as presented in Table 4.1, is due more to extraordinary snow melt than to significantly less water use. The amount of water taken from the Syrdar'ya has increased from 22 cubic kilometers in 1950 to 43 cubic kilometers in 1980, of which 6 cubic kilometers was returned (Akramov and Rafikov 1986: 167). Because water use has exceeded streamflow since 1970, the Syrdar'ya basin is dependent on the supplemental use of reservoir water and drainage return flow which is frequently polluted with pesticides from the fields (Izrael' 1988a).

To understand the economic importance of irrigation to the economy better one has to examine the statistics in light of natural phenomena such as drought. Rafikov, an Uzbek scientist who has been involved with the study of conditions in the Aral Sea since the 1960s, noted that, though there was drought condition in 1970, and while the Syrdar'ya's average yearly streamflow was reduced, the volume of water marked for irrigation remained steady, and "during 1974–75 a strong drought occurred which reduced the average flow of the

Syrdar'ya by 38 percent . . . at the same time the volume of water which was diverted from the river increased 1.5–2 times" (Akramov and Rafikov 1986: 167). For 1970, 71.5 percent of the Amudar'ya and Syrdar'ya combined streamflow was diverted for irrigation; in 1971, 83 percent; in 1972, 88.4 pernent; and in 1973, 71 percent (Micklin 1986: 346).

Before the contemporary irrigation era, the Amudar'ya delta was characterized by the Russian geographer L. S. Berg in 1903 as the "Aral Sea landscape" (Rafikov 1983: 550). It contained reeds interspersed with small lakes and bogs which formed a hydromorphological complex. The lakes and bogs formerly comprising an area of 300,000 hectares in western and central areas of the delta have been replaced by thick salt marshes and rushes which have become solonchak-type soil due to the drop in the subterranean water table and rising mineralization and salt accumulation in the soil (Micklin 1983; Akramov and Rafikov 1986). This complex also contains fringing forest, known locally as "tugay," which formerly occupied an area of 760,000 hectares. The area is now reduced to less than 100,000 hectares as the reed-tugay vegetation is giving way to tamarisk, saltwort, and other salt family vegetation associated with an arid environment. This area of pasture whose vegetation formerly produced 6–16 centners of pasture and hay per hectare now yields only 0.5–3 centners (Akramov and Rafikov 1986: 170).

The regulation of the Amudar'ya's flow by irrigation as well as the Takhitosh hydroelectric station since 1974 has changed the delta's ecosystem and geographical zones. The restricted flow of the river has resulted in lowering the amount of alluvial nutrients and micro-organisms introduced into the delta soils (Kamalov 1987). Also, due to reduced streamflow, the river's water storage has been reduced, as water now tends to sink below. Subterranean waters have dropped by as much as 6–8 meters (Akramov and Rafikov 1986: 170), resulting in an accumulation of salts at the surface and in the root layer of the soil (Rafikov 1983: 347), which promotes the formation of various forms of solonchak soil. Salt accumulations throughout the Amudar'ya delta have been tremendous in recent years, as noted in a recent study:

In hard, smooth and closed depressions, solonchak, which is to a strong degree salinized, and pasture-*takyr* soils are scattered. In the soils the salt content reaches 30 percent and at 1.5–2 meters depth, salt horizons of a 30–40 centimeter thickness are encountered. In them the amount of salt in the upper layers as compared with the lower is very strong. On the Aral Sea's southeast shore in inumerable solonchak structures, there is a very large amount of salt existing (in every acre up to a meter thick). The chemical composition of the salts is nitrate-chloride and sulfate-nitrate. (Akramov and Rafikov 1986: 170)

As a result, the soils in the Amu delta are increasingly evolving from those associated with a hydromorphological or polyhydromorphological complex to those associated with an arid environment. Sand hills and barchans now inhabit the old shorelines of the Aral to a breadth of 10–12 kilometers (Rafikov 1986: 11). There is increasing deflation whereby wind erosion is sweeping loose salts throughout the Aral Sea basin, threatening the fertility of much of Central Asia. "Every year here 15 to 75 million tons of dust are raised into the atmosphere. For every hectare of land in the Aral region on average up to 520 kilograms of sand and salt are blown away" (Izrael' 1988a). This also affects living conditions in some of the towns adjoining the former shoreline. In Kungirat, "dust and salt have covered the streets and houses like a hoar frost as the walls of buildings have become white from the blown salt" (Baranov 1988: 3).

All this is in sharp contrast to the confident prediction of the former president of the Turkmeniya Academy of Sciences, A. Babayev, who wrote:

I belong to that group of scientists which considers the drying out of the Aral much more advantageous than its preservation. Firstly, from its area we receive excellent, fertile land. According to preliminary calculations it might yield 1.5 million tons of cotton a year. Cultivation of just this culture pays for the current existence of the Aral Sea with its fishing, shipping and other industries. Secondly, it is the conviction of many scholars . . . that the disappearance of the sea would not influence the landscape of the region. (Kamalov 1987: 24)

Similarly, the comment of the first deputy minister of the Ministry of Land Reclamation and Water Management (Minvodkhoz) that "the Aral should die beautifully," is indicative of an attitude held by many involved in the management of irrigation agriculture and water resources (Shermukhamedov et al. 1988: 4)—namely, that any expansion in agricultural product justifies the means of obtaining it. Many of those associated with the management of agriculture and water resources have sought to emphasize the economic significance of the irrigation-agriculture-centered economy while belittling the ecological impact of the Aral's demise:

For the past 26 years, since the beginning of the decline in sea level, the volume of production in the agrarian sector for the region has increased by a billion rubles per year—but in total perspective, including organs associated with the development of the water management complex—more than 20 billion rubles per year. . . . General damage to the national economy and the natural environment of the Aral basin and the delta is estimated at 100 million rubles. (Lapkin et al. 1988: 8)

Furthermore, in the opinion of several authors affiliated with the water management industry, maintaining even the current shrunken sea level of the Aral would require a yearly inflow of 30–40 cubic kilometers a year (Epstein 1987; Lapkin et al. 1988): a sum which they all warned is impossible without inflicting damage on the economy. Moreover, in their prescription for solving the conflict between agricultural irrigation expansion and its ecological consequences for the Aral Sea basin, they cautioned that water economizing alone would be insufficient to surmount the "constantly increasing deficit from the growing demand of the region until the arrival of Siberian water" (Lapkin et al. 1988: 8).

Other authors have been more pessimistic in regard to the economic costs associated with the decline of the sea. Kamalov, an Uzbek geographer, writing in *Ekonomicheskaya Gazeta*, has retorted:

> At the end of the 1970s the total of losses in the lower Amudar'ya due to the lowered profitability in fishing and agriculture . . . was estimated to be more than 700 million rubles for Uzbekistan alone. By the mid-1980s the sum of losses for the sea and adjoining areas including the Syrdar'ya delta reached 1.5–2 billion rubles a year—as much as the additional production promised by those supporters of the Aral Sea's liquidation. (Kamalov 1987: 24)

Another estimate by the Scientific Research Institute of Geography of the USSR Academy of Sciences placed "losses in natural production in connection with the drying of the Aral equal to 7 billion rubles" (Kamalov and Kabulov 1988: 2).

The growing water deficit symbolized by the Aral's desiccation and the diminishment of inflow from the Amudar'ya and Syrdar'ya to a merely nominal level has led to fundamental questioning of the role and course for the irrigation monoculture orientation. Kotlyakov characterized the water deficit as "more a symptom than a cause of the difficult ecological confusion in the region" (Kotlyakov 1988: 3). He concluded:

> The Aral Sea basin ecological crisis is a complex territorial problem. It should in no account be taken as just a water deficit. Blame needs to be placed not just on the designers and builders but on the entire mechanism which is not interesting land users in correct water use.

In an interview with *Kommunist*, academicians Laskorin and Tikhonov expanded on the economic argument presented earlier by Aganbegyan and others in *Pravda* that the water shortages in Central Asia are merely symptomatic of an agricultural system based on extensive growth. Laskorin maintained that "the theory of a water deficit" had been invented in the depths of the interested departments and in the institutions serving them (Laskorin and Tikhonov 1988: 90).

On one level there was significant criticism in the press. On another level there was an organizational opposition provided by the Committee for Saving the Aral. It was formed in the summer of 1986 by the Uzbekistan Writers' Union under the leadership of Pirmat Shermukhamedov. The Committee for Saving the Aral not only kept the ecological issue of the Aral Sea basin alive through literary activity, it also put together expeditions which traveled throughout Central Asia organizing and reporting on the Aral ecological situation, and it organized a public bank fund for the collection of donations in both Uzbekistan and Kazakhstan for the restoration of the Aral. In many ways its level of organizational activity and opposition to the entrenched cotton and water management interests rivaled those of a political party.

CHOICES FOR THE ARAL SEA

Perhaps the highest level of public discussion occurred in the summer of 1988, when in Aral'sk, Kazakhstan, a former Aral Sea shipping and fishing port, an "Aral meeting" was held.

> The work of the "Aral meeting" began by the entrance to a hotel where, on a broad square before a facade, gathered hundreds of inhabitants of the city and nearby settlements. They came here to talk of their plight and to pose an increasingly unified question: What will become of the sea and of us? (Bektepov 1988: 2).

Indeed, it provided a public forum where the general population, writers, scientists, and representatives of the party and government could not only actively debate the present conditions of the Aral region, but plan for its future development touching upon the roles of water and irrigation in society.

Much of the debate centered around a proposal called the "Kazakhstan project" (Bektepov 1988). This project envisages the division of the Aral into a greater and lesser sea whereby the smaller sea in the northeastern part of the Aral would stand at around 7 cubic kilometers with redeveloped fishing. While the division of the sea into larger and smaller sections is already indirectly achieved by the massive drop in volume, some warned that its promotion would form a broad, constantly growing isthmus which will unite the desert of Western Kazakhstan with the Turanian lowland. Thus, the natural water shield which has from time immemorial defended Kyzyl-Orda Oblast from salt storms which come from the west will be destroyed. This means that the unfavorable ecological conditions in the Northeast Aral Sea basin will intensify. (Bektepov 1988: 2)

A similar proposal for the division of the Aral into a lesser and greater sea had been put forward by Voropayev of the Institute of Water Problems as a means to further limit irrigation expansion earlier that same year (Voropayev et al. 1988). A local fisherman responded:

Not one fisherman I know is a supporter of the dismemberment of the sea. Who might support such a project, please? Bold transformers. ... By dividing off a small sea ... we deprive fish of a natural spawning area. ... Could this lead to the death of the entire ecological system of the Aral? (Bektepov 1988: 2)

Another proposal attempted a three-stage solution to the Aral's decline. The first stage called for an inflow from the Amudar'ya and Syrdar'ya of 25–30 cubic kilometers yearly to prevent further desiccation of the sea. This would be achieved by establishing a "sharp cutback in lands sown to cotton and rice, substituting for them less water-intensive crops" (Bektepov 1988: 2), accompanied by the construction of water treatment plants. The second stage would involve the reconstruction of the existing land reclamation system and the diversion to the sea of many man-made reservoirs of collector-drainage water like Lake Sarykamysh. The third stage envisions a large-scale redistribution of the country's water resources. This includes the possibility of pumping water from the Caspian Sea by means of a 500 kilometer long canal as well as a revival of support for a reexamination of the feasibility of bringing in water from Siberia.

The feeling of weakness in the decision-making process by the local inhabitants was underscored by the then Kyzyl-Orda obkom party first secretary Auelbekov:

Unfortunately, one must note that for the last three years many voices everywhere spoke of the misfortune of the Aral's plight, but no one came to us from Minvodkhoz. For two years we demonstrate the necessity of rayon statistics—we have an income per capita two times lower than average in the country. ... We speak of the fate of the Aral, that it is a catastrophe. However, no one helps the Aral people who are struck by the disaster. (Bektepov 1988: 2)

Indeed, *Pravda* reported that "the workers of Minvodkhoz USSR and other powerful central departments whose hands have helped to create a crisis situation . . . amicably ignored the '[Aral] meeting,' preferring not to be confronted with the fruit of their activities and the indignation of their opponents" (Oreshkin 1988: 3).

In 1988, following a summer noted by heated public debate as well as demonstrations in response to the ecological crisis related to the Aral's decline, a resolution was adopted by the CPSU Central Committee and the USSR Council of Ministers for the "necessity of realizing radical measures for the restoration of the destroyed ecological balance in the Aral region and the preservation of the Aral Sea" (*Pravda*, 2 September 1988).

A broader explanation of the specifics of this resolution was provided by R. N. Nishanov, then first secretary of the Communist Party in Uzbekistan (Nishanov 1988). Basically the resolution seeks to maintain agricultural production at least at current levels while gradually modernizing the utilization of water resources in Central Asia. This involves the reconstruction of much of the existing irrigation network as well as recognizing the necessity of constructing numerous sewage treatment centers and water purification plants. The problem of irrigation drainage water being discharged into the river is to be solved by a pipeline along the Amudar'ya and Syrdar'ya rivers for the release of these contaminated waters into the Aral Sea. Yet, as outlined by Nishanov, the Aral Sea, "due to its own peculiar circumstances, evidently cannot manage to return to its former state. But its life can be saved on a smaller scale." The parameters outlined for a guaranteed inflow to the Aral promise continued desiccation for some time. The projected minimum by 1990 is 8.7 cubic kilometers per year; by 1995, up to 11 cubic kilometers; by 2000, 15–17 cubic kilometers; and by 2005, 20–21 cubic kilometers including collector drainage water. With "favorable weather conditions the Aral might receive a yearly inflow of 20–30 cubic kilometers and by 2005, from 30–40 cubic kilometers" (Nishanov 1988: 1).

Thus, under "favorable weather conditions" only by 2005 would the Aral Sea receive an inflow which some analysts cited as the minimum inflow needed to sustain its current shrunken state. Receiving such a minimum inflow only by the year 2005 means continued diminishment through the 1990s.

Nishanov's lengthy report was candid in describing the current situation:

> Today, the fact remains that errors connected with water use and cotton pesticides have seriously complicated the situation especially in the Karakalpak ASSR. There is a very high level of infectious diseases and child mortality. It is a very, very difficult, if not extraordinary, ecological

condition that remains. Natural wasting and salinization of the soil, and deep degradation of the ecological system and the flora and fauna world, have occurred. The quality of drinking water has worsened intolerably, and sewage has not ceased flowing into water reservoirs. The majority of cities and settlements lack a centralized system of water supply and sewers. Significant areas of land sown to cotton have become contaminated by diseases and pestilence among the plant life. (Nishanov 1988: 1)

Nishanov reported that the resolution commissioned Gosplan USSR to select "Aralvodstroy" (a new organization of Minvodkhoz) to realize "state plans of economic and social development for the USSR and necessary capital work and material resources." The selection of Minvodkhoz in such a crucial organizational role certainly serves to promote the established authorities despite their being much besieged by critics of the ecological situation.

In late June 1990 the party and government leaders of the republics of Central Asia and Kazakhstan met in Alma-Ata and reached agreement for cooperation on a broad range of economic, scientific-technical, and cultural issues on a regional basis that views each republic as equal and sovereign. To implement this regional cooperation a coordinating council with permanent staff is to be established in Alma-Ata. Besides this agreement, a "Declaration of the Leaders of the Republics of Central Asia and Kazakhstan" addressed the acute problem of the region—the Aral ecological catastrophe. (*Pravda Vostoka*, 24 June 1990). The declaration specifies that an interrepublic commission will be set up in order to unite the local effort for the restoration of the Aral Sea basin as well as a fund for the people of the Aral region. They charged that the "growing [water] deficit and the pollution of water resources place the population of the region on the brink of an ecological catastrophe which demands bringing in the means and resources of our region as well as the country as a whole." They asked Gorbachev to "declare the Aral Sea region a zone of national tragedy" and create a national ecological fund with United Nations specialists enlisted for the resolution of this problem. Furthermore, they linked resolution of the Aral Sea problem to reviving an appeal for diverting "part of the flow of Siberian rivers as one of the principal means of saving the Aral and ensuring food for the population" (*Pravda Vostoka*, 24 June 1990).

CONCLUSION

Soviet Central Asia is enclosed by mountains on its east and south, by desert and arid steppe along most of its northern expanse, and by the Caspian on the west. Historically these physical features have served to isolate much of Central Asia from the outside world. Settlement, despite the Soviet government's efforts to broaden it through the irrigation of marginal and peripheral land, is still mainly in the oases that adjoin the Amudar'ya and Syrdar'ya and throughout the piedmont zone between the mountains and the desert plain. In such an environment man's relationship with nature is always precarious and greatly dependent on water resources. In Soviet Central Asia the limits of expansion are clearly being felt as the consequences of relying on unabated irrigation expansion for the cotton monoculture has led to a real breakdown in the ecosystem. The crisis has been brought about through long-term neglect of the rivers' role in supporting the ecological equilibrium of the Aral Sea basin and the domestic needs of the population of the delta regions in order to satisfy the immediate economic needs of short-term planning.

In the era of Gorbachev and perestroyka there has been a decided turning away from the grandiose and expensive state projects of the Brezhnev and earlier eras. The consequence is that not only is the Siberian water diversion project canceled but several other water projects throughout the Soviet Union as well. As a result the emphasis is on economic justification and self-reliance for project approval. The cotton interests and the water management institutions not only cannot count on an ever-expanding irrigation base, but also are encountering through the Committee to Save the Aral an organized opposition which wants to curtail their role in water use.

As the regional economy undergoes the radical restructuring being introduced throughout the Soviet Union at this time, there is a great deal of uncertainty about what effect a more market-oriented economy would have on crop structuring. Even if there is a decided curtailment in cotton production, some ask: "will switching many fields from a monoculture crop like cotton to crop rotation for food production, as well as increasing the area of personal plots, really result in less water demand?" (Epstein and Eingori 1990: 2). The question remains whether the water requirements of this rapidly growing population can be satisfied for the long term without the revival of the Siberian River Diversion Project.

As evidenced by the resolution of the CPSU Central Committee and USSR Council of Ministers, the Aral Sea basin is viewed as an ecological disaster. The situation has been decades in the making, yet it requires massive immediate help to restore its ecological equilibrium. The solution presented by the resolution is one that depends on a centralized form of management under the very organs responsible for the regional planning and management which led to the present situation. Whether they can affect the problem and truly influence a more rational use of water resources remains to be seen.

REFERENCES AND BIBLIOGRAPHY

Resolution, CC CPSU and USSR Council of Ministers, *Pravda*, 20 August 1986: 1.

Resolution, CC CPSU and USSR Council of Ministers, *Pravda*, 2 September 1988: 1; for complete text of resolution: *Kommunist Uzbekistana* 10 (1988).

Resolution, Sobraniya Partiyno-Khozyaystvennogo Aktiva Respubliki, *Pravda Vostoka*, 11 October 1988: 1.

Agreement, Economic, Scientific-Technical, and Cultural Cooperation, *Pravda Vostoka*, 24 June 1990: 1.

Declaration, Leaders of the Republics of Central Asia and Kazakhstan, *Pravda Vostoka*, 24 June 1990: 1.

Aganbegyan, A., Golitsyn, G., Tikhonov, V., Eneev, T. and Yanshin, A. 1986. Zemlya—glavnoe bogatsvo. *Pravda*, 12 February: 3.

Akramov, A. and Rafikov, A. 1986. Orol madad suraidi. *Shark Yulduzi* 6: 166–77.

Baranov, A. 1988. Orolning shuri. *Sovet Uzbekistoni*, 30 April: 3.

Bektepov, Bakhtizhan 1988. Kto rassypal sol'? *Kazakhstanskaya Pravda*, 5 July: 2.

Bohr, Annette 1988. Infant mortality in Central Asia. *Radio Liberty Research Bulletin* RL 352/88, 17 August.

Borisov, A. A. 1965. *Climates of the USSR*. Chicago: Aldine.

Brown, Bess 1988. Special medical teams combat infant mortality in Central Asia. *Radio Liberty Research Bulletin* RL 353/88, 17 August.

Epstein, L. 1987. Na chto potracheny sili Arala? *Zvezda Vostoka* 1: 143–8.

Epstein, L. and Eingori, F. 1990. Aral: emotsii i real'nost. *Pravda Vostoka*, 4 January: 2.

Ginzburg, N. S. 1986. A microgeography of settlement in the Pamir highlands. *Soviet Geography* XXVII, 6: 398–434.

Glazovskiy, N. F. 1991. Ideas on an escape from the Aral crisis. *Soviet Geography* XXXII, 2. 73–89.

Great Soviet Encyclopedia 1973. Amu Dar'ya. Vol. II.

Izrael', Yu. A. 1988a. Spasenie morya. *Pravda Vostoka*, 14 September: 1.

Izrael', Yu. A. 1988b. Aral i PriAral'e: trevoga vsenarodnaya. *Pravda Vostoka*, 12 October: 2.

Kamalov, B. 1987. Ob Arale v proshedshem vremeni? *Ekonomicheskaya Gazeta* 29 (July): 24.

Kamalov, S. and Kabulov, S. 1988. Kaplya v more. *Pravda Vostoka*, 8 September: 2.

Kes', A. S. and Klyukanova, I. A. 1990. Causes of the Aral Sea variation in the past. *Soviet Geography* XXXI, 9: 602–12.

Kes', A. S., Krenke, A. N., Minayeva, Ye. N., Tsigel'naya, I. D. and Reshetkina, N. M. 1982. The present state and future prospects of using local water resources in central and southern Kazakhstan. *Soviet Geography* XXIII, 6: 415–25.

Knystautus, Algirdas 1987. *The Natural History of the USSR*. New York:

Korzhenevskiy, N. L. 1960. Fiziko-geographicheskiye provintsii Azii. *Nauchny y Trud* (Tashkent), Novaya Seriya 183: 34–88.

Kotlyakov, Vladimir 1988. Mozhno li spasti more? *Pravda*, 14 April: 3.

Kotlyakov, Vladimir, Kuznetsov, N. T. and Kes', A. S. 1987. Itogi izucheniya antropogennogo opustynivaniya PriAral'ya i znacheniye ikh pri razrabotke perspektiv razvitiya KKASSR. *Obschestvennyye Nauki v Uzbekistane* 10: 23–7.

Lapkin, K., Lebedev, O. and Dukhovnyy, V. 1988. Vodosberezheniye—glavnyy rezerv Arala. *Pravda Vostoka*, 29 June: 8.

Laskorin, B. N. and Tikhonov, V. A. 1988. Novyye podkhody k resheniyu vodnykh problem strany. *Kommunist* 4: 90–100.

Lewis, Robert A. 1966. Early irrigation in west Turkestan. *Annals of the Association of American Geographers* 56, 3 (September): 467–91.

Lydolph, Paul 1977. *World Survey of Climatology*. Vol. 7: *Climates of the USSR*. Amsterdam:

Micklin, Philip 1986. The status of the Soviet Union's north–south water transfer projects before their abandonment in 1985–86. *Soviet Geography* XXVII, 5 (May): 287–329.

Micklin, Philip 1988. Desiccation of the Aral Sea: a water management disaster in the Soviet Union. *Science*, 2 September: 1170–6.

Nishanov, R. N. 1988. Aral: zabota vsekh i kazhdogo. *Pravda Vostoka*, 9 October: 1.

Oreshkin, D. 1988. Sine more u obryba. *Pravda*, 13 August: 3.

Rafikov, A. A. 1983. Environmental change in the southern Aral region in connection with the drop in the Aral Sea level. *Soviet Geography* XXIV, 5 (August): 344–53.

Rafikov, A. A. 1986. Kuplar xakida uillar. *Fan va turmush*, June: 10–12.

Rafikov, A. A. and Tetyukin, G. A. 1981. *Snizheniye Urovnya Aral'skogo Morya i Izmeneniye Prirodnykh Nizov'yev Amudar'yi*. Tashkent: Fan.

Shermukhamedov, Pirmat and Rasputin, Valentin 1988. Kto esli ne my? *Pravda Vostoka*, 20 February: 2.

Shermukhamedov, Pirmat, Kovalev, Yu and Mirzayev, S. 1988. Komu i zachem eto nuzhno? *Pravda Vostoka*, 7 July: 4.

Sinnott, Peter 1988. Water diversion Politics. *Radio Liberty Research Bulletin*, RL 374/88, 17 August.

Suslov, S. P. 1961. *Physical Geography of Asiatic Russia*. San Francisco: W. H. Freeman.

Voropayev, Grigoriy, Ismailov, G. Kh. and Bostandzhoglo, A. A. 1988. Intensifikatsiya ispol'zovaniya vodno-zemel'nikh resursov rek basseyna Aral'skogo Morya. *Obschestvennyye Nauki v Uzbekistane* 3: 5–14.

Yusupov, Erkin Yu. and Ziyadullayev, S. K. 1988. Sud'by Arala i PriAral'ya—v rukakh lyudey. *Pravda Vostoka*, 28 June: 4.

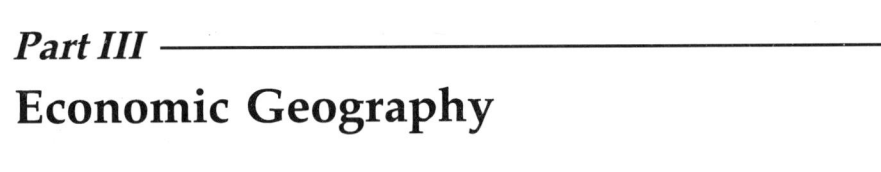

Part III

Economic Geography

Chapter 5

Soviet Geographical Imbalances and Soviet Central Asia

Ronald D. Liebowitz

INTRODUCTION

During the past two decades, a geographical imbalance among natural resources, labor resources, and industrial infrastructure has developed in the USSR and become more acute. Industrial resources in the USSR are largely east of the Ural Mountains; infrastructure and population are largely in the west, and the growth in the working-age population largely in the south. Consequently, the labor-rich Central Asian republics have taken on a more concrete economic and strategic importance than in the past. This chapter will focus on Soviet Central Asia and its economy in light of the worsening geographical imbalance afflicting the Soviet economy. It will outline the problem, provide an overview of the economic characteristics of the region that have had—or are likely to have—an impact on the national economy's geographical imbalance, and then highlight some of the constraints to solving the problem. A major emphasis of this chapter will be on capital investment flows in the region since the 1950s in absolute and relative terms.

THE GEOGRAPHICAL IMBALANCE

The increasing importance of Soviet Central Asia to the Soviet economy stems from a problem that is rather easy to identify, yet whose solutions are complicated because of the way they are likely to affect so many aspects of the multinational Soviet state: a decline in the growth rate of the national economy due in part to a worsening geographical imbalance among resources, labor, and infrastructure. The nonstructural reasons for this well-developed imbalance include the timing of Russian expansion beyond the European core, a number of irrational economic policies initiated since the late 1920s, and cultural or ethnic factors. Although these reasons for the geographical imbalance are statewide in scope, changes that have occurred primarily in the western, predominantly Slavic areas of the state, are most responsible for making Soviet Central Asia so important to the future economic vitality of the state. The unfavorable changes in the western regions are not new; the growth rate of industrial output declined in 7 of the 13 economic regions west of the Urals between 1960–5 and 1966–70, while all 13 regions experienced a similar decline over the following five years (Dienes 1983: 221).[1]

Although comparable regional data are not available for the 1976–85 period, there is no evidence of a reversal in this downward trend. By 1987, Soviet published data began to reveal the decline in the growth rates of industrial production for the entire national economy and for all of the 15 union republics (USSR 1987: 133–4).[2] For the national economy, average annual rates of industrial growth declined from 7.8 percent for the 1970–80 period to about 4.3 percent for 1980–8 (USSR 1989b: 339). Additionally, all 15 republics registered declines over the two time periods; Belorussia's average yearly increase declined from about 13 percent for the 1970–80 period to 7.0 percent for the 1980–8 period; Turkmen SSR's from 7.3 percent to 3.5 percent; and Estonia's from 7.4 percent to 3.3 percent. For industrial labor productivity, the rates of growth for the period 1980–8 declined from the 1970–80 rates in all 15 republics (USSR 1989b: 368). Approximate annual growth rates of industrial labor productivity for the 1970–80 period averaged between 2.9 percent in Tadzhikistan to 7.5 percent in Belorussia; but for the 1980–8 period these growth rates ranged between 1.9 percent in Turkmen SSR to 5.6 percent in Belorussia. For the national economy, the corresponding figures declined over the 18 years from an annual increase of 5.6 percent for the 1970–80 period to one of 4.0 percent for the 1980–8 period (USSR 1989b: 368).

The decline in performance of the western economic regions since the 1970s has been due to the increasing inefficiency of the existing industrial infrastructure, a declining pool of skilled labor, and the depletion of industrial resources that for so long were available in relatively close proximity to the centers of production and markets. The number of working-age people in the Russian Republic (RSFSR) is estimated to decline until 1995 (Baldwin 1980), and the non-Slavic south[3] is likely to account for almost three-quarters of the growth of the total working-age population in the 1980s (Lewis 1983). From 1979 to 1987, the Central Asian republics plus Kazakhstan, Georgia, Armenia, and Azerbaydzhan accounted for slightly less than half (46.1 percent) of the state's additions to the 20–59-year-old cohort while accounting for only

22 percent of the state's population (USSR 1988d: 48–93).[4] Thus, without significant in-migration from other areas, or a significant rise in labor productivity, there will be more pressure on the existing industrial infrastructure to compensate for declining labor inputs. The likelihood of an increase in capital productivity large enough to compensate for the expected decline in labor inputs seems unlikely even if one were to accept the productivity increases initially anticipated by the Soviet leadership as a result of perestroyka. Between 1960 and 1979 every republic except Turkmen SSR recorded declines in capital productivity,[5] and, as an example of the national economy's low levels of capital productivity, the growth in industrial output in the relatively advanced Russian Republic as early as the period 1968–75 could not match the growth in the republic's fixed capital (Dienes 1983: 220–6). Recent data reflect the decline in capital productivity since the 1970s in all branches of the national economy (USSR 1988c: 64).

The changing geography of Soviet industrial resources, especially energy resources, further accentuates capital and labor productivity problems in the western regions of the USSR. All the growth of oil and gas output, along with 90 percent of coal increases, will come from Siberia for the foreseeable future (Taaffe 1984). The spatial shift in energy resources is quite startling; as recently as 1970, 76 percent of the petroleum and 73 percent of Soviet natural gas were produced west of the Urals (Soviet Geography 1986: 252, 258). By 1988, the shares declined to 23 and 29 percent respectively (Soviet Geography 1989: 308, 318). Many of the nonenergy industrial resources face similar spatial shifts. The economic implications of the changing geography of resources include exorbitant expenditures for building new transportation networks, paying higher wages to attract and establish a viable work force to extract the resources, and establishing a social infrastructure for the workers to reduce high rates of labor turnover in labor-deficit regions. The costs for these new additions are great by any calculation; the fact that most of the newer resource-rich regions are located in difficult climatic and geologic areas makes the cost of development beyond what appears to be currently affordable.

The debate among Soviet economists and planners over whether to develop Siberian resources as such great costs or to use the capital for further resource exploration and development in European USSR continues (Dienes 1982, 1987; Schiffer 1989). The apparent advantage maintained by the pro-Siberia group through the 1970s appears to be declining due to the increasing costs of Siberian development. When one factors in the lost revenues caused by the recent decline in world prices for energy resources, the cost to develop pioneer regions makes the Siberian option less attractive than in the past for the capital-short Soviet economy. The emerging importance of Central Asia, then, stems primarily from its large and expanding reserve of labor resources when labor and capital shortages in the western parts of the country and Siberia are becoming more acute. The region is not nearly as well endowed with natural resources as Siberia. Natural gas is the leading resource of the region. The Turkmen SSR produces most of the natural gas currently produced in the region, but significant deposits are already developed in the Uzbek SSR (Gazli, Dzharkak, Achaksk, and Mubarek) and the Tadshik SSR (Nefteabad) as well, and there are at least three other major areas in Central Asia yet to be exploited (USSR 1989a: 235). These regions span a large area from the south Caspian region in the Turkmen SSR across to the Fergana region of Uzbekistan, and south into the Tadzhik SSR. During the 1980s, the Turkmen Republic became the region's largest natural gas and petroleum producer, as well as Central Asia's largest center for gas byproduct and petrochemical products. Although there is some nonferrous metallurgical extraction in the region (Angren and Almalyk, Uzbekistan; Altyn-Topkansk, Tadzhikistan), important deposits of mercury and other rare metals in southern Kirgiziya, and some coal, especially in the Kirgiz SSR (USSR 1989a: 248), Central Asia is not a resource-rich region.

CONSTRAINTS TO A SOLUTION

If the solution for declining growth rates in the Soviet economy entailed only eliminating what is now viewed as an overcentralized and often irrational economic mechanism, forecasts for the Soviet economy would be far less pessimistic given recent movements toward a decentralized and regulated market economy. There are, however, several nonsystemic constraints to solving the imbalance. These constraints include the USSR's ideology (and more importantly today its legacy), and the nationality based territorial political-administrative structure of the state, both of which become intertwined in any discussion of regional economic development in the Soviet Union.

Ideology as a constraint

Development strategy in the USSR for many decades was circumscribed to some extent by Soviet ideology, which prevented the implementation of many would-be economic policies if they were believed to promote inequality (Liebowitz 1987). The importance of post-World War II Soviet ideology, some argue, may be on the wane, as even President Gorbachev has begun to recognize the incompatibility of the egalitarian principles that for decades were cited in virtually all planning documents with the kind of economic growth that the USSR now needs. Nevertheless, the impact of the state's adherence to an egalitarian ideology for so many decades is apparent when one looks at the development patterns throughout the state, and should not be ignored because of what appears to be a move toward a market economy and an abandonment of socialist principles.

Soviet ideology, until perestroyka, contended that socialism could not be achieved until political, social, and economic equality were attained throughout a state's territory and enjoyed by its population. Given the great differences in the levels of socioeconomic development throughout the Russian Empire at the time of the Revolution, it is not surprising that many features of what were defined for so long as the guidelines for Soviet and socialist regional economic development were specifically concerned with solving regional inequality (Wagener 1973; Dyker 1983; Bahry 1987), and often appeared economically irrational. In addition to "paying" an economic price for pursuing equality over growth or efficiency, there was an unmeasurable yet real pressure of having had to adhere to the most "egalitarian" of ideologies. Thus, ideology was, and—because collective psyches change very slowly— continues to serve as, a major constraint to solving the geographical imbalance.[6]

The tenets of socialism, particularly until the second decade of the Brezhnev years, have led to what are by Western standards irrational investment decisions, stated most often to appease the peoples of relatively lesser developed regions (Dyker 1983). The results of such investment decisions were slower growth rates for the national economy, because the returns on capital and labor are so much lower in the lesser developed regions. Soviet planners have been ideologically constrained in developing rational economic plans (and will continue to be until Gorbachev's perestroyka produces a more efficient economic mechanism), because they have been expected to operate within social- ism's previously established "theoretical" yet economically irrational limits. Because the theoretical basis of Soviet regional economic develop- ment was for so many decades inconsistent and therefore open to conflicting interpretations, it is clear that a "socialist" solution to the economy's geographical imbalance would be very difficult given the ideological nature of many of the principles used to direct economic development.

The guidelines that have governed the location of productive forces in the USSR and acted as the theoretical basis for economic development were accepted in the Western literature as constituting a particular "location theory" (Huzinec 1977). Their major drawback is the seemingly contradictory goals they have sought to achieve, which underscores both the importance and constraints of ideology in econ- omic policy making. Overall economic growth and regional and ethnic equality were, and still may be, concurrently sought, yet the guidelines established to attain the former goal invariably lessen the chances of meeting the latter goal, and vice versa. The so-called development principles that relate to overall economic growth have included (Wagener 1973: 63–102; Kistanov 1981: 14–27):

(a) Locate enterprises close to their required raw materials, fuels, and markets in order to reduce transport costs.
(b) Locate enterprises so that available natural resources are utilized most efficiently.
(c) Locate enterprises so that concentration, cooperation, and other progressive forms of socialist organization of production are furthered.
(d) Locate productive forces rationally so that the territorial complex forms the basis of economic regions.

These principles are not decidedly socialist, and in fact a few are Weberian (Rodgers 1974; Dyker 1985). The problem of consistency becomes more apparent when the guidelines for equalization are introduced:

(a) Locate facilities in such a way as to promote the rapid economic and cultural development of the relatively backward regions and nationalities to a level commensurate to those of the leading regions.
(b) Minimize the socioeconomic differences between the urban and rural areas.
(c) Seek an even distribution of production throughout the state to promote the effective participation in the national economy of the labor and resources from each region.
(d) Strengthen the entire state's defense potential.

Although these four principles are representative of the basic aims of a socialist state and are admirable goals for any state, they have rarely been compatible with the goal of maximizing overall economic growth. In fact, in most multinational states, maximum economic growth and regional equality are conflicting goals. Not surprisingly, Soviet regional development policies cannot meet both aims, and therefore often draw criticism from advocates of one strategy or the other. Pursuing a maximum growth strategy would result in a western USSR and west Siberian investment orientation that would, at least in the short term, work against regional and socioeconomic equalization. Alternatively, giving priority to policies designed to promote regional and national equality would favor the non-Slavic southern areas of the state, which, due to lower levels of capital and labor productivity, would work against the goal of maximizing economic growth. David Dyker (1983) argues that, even as "statements of aims," the dual goals seeking proportional development—which would require an "extensification" of the economy—and the steady expansion and improvement of production—which would require an "intensification" of productive forces—are so broad in their formulation that they are of dubious value.

Before the late 1960s, when growth rates for the national economy were impressive and consistent, one could have argued that continued high rates of economic growth could best be achieved by emphasizing the growth principles (intensification) of regional development. This strategy would mean giving less priority to the equalization principles (extensification), and would rely on indirect methods of financing the socioeconomic development of the relatively backward regions. Transfer payments, generated from surplus production attained by emphasizing the growth principles in the more developed and resource-rich regions, from the Ukraine, the Baltic republics, and some economic regions in the RSFSR, along with turnover tax revenues, could pass through the state's complex budgetary process and be redistributed to the relatively lesser developed regions.

Wagener (1973), among others, however, argues that transfer from wealthier to poorer regions in the USSR cannot be viewed as a convincing model of development, especially in the long term. Because the Soviet economy is strongly tied to the world economy, fluctuations in world prices for energy and natural resources would prevent consistent high rates of economic growth for an economy whose exports are dominated by primary resources.[7] And, even if resource extraction could support development in the backwaters of the Soviet Union, it is unlikely that citizens in the more developed regions would remain quiescent while the wealth they produced flowed to the lesser-developed republics. In fact, this very issue has become a rallying point for Russian nationalists, who claim that Russia's natural resource wealth is being extracted and "sold" to other republics at far less than world market prices.

Since declining economic growth rates have become more than a short-term phenomenon and the geographical imbalance has become more pronounced, it could be argued, albeit in a roundabout way, that the best long-term solution for reversing the economy's declining performance would be to stress the equalization principles—by accelerating the industrial development of the southern tier. With a shortage of skilled labor (and capital inputs) hindering economic growth in the more industrially productive parts of the state, the utilization of Central Asian surplus labor in labor-deficit regions of the state will soon be essential. However, while many demographers see an inevitable and sizable out-migration of Turkic-Muslim workers from Central Asia when economic conditions there decline beyond a certain point, little voluntary out-migration from the region has occurred, and to rely on the deterioration of economic conditions for Muslim out-migration might take too long.[8]

Despite the planned implementation of "khozraschet" (self-financing of enterprise operations) and Gorbachev's call for more rational investment decisions, it is doubtful that the leadership will abruptly change an investment strategy that has subsidized the non-Slavic south for decades. This conclusion is all the more probable given the volatile ethnic situation in the region between non-Russian Soviet nations; a reduction in the economic well-being of Central Asians could further destabilize the region. It is difficult, then, to believe the mobility of the indigenous population will increase significantly in the short term as a response to a reduction in economic supports to the region. Instead, an increase in Russian out-migration is likely to commence, which could significantly affect the Central Asian economy.

If economic factors eventually have an impact on Central Asian migration as they have had on other Soviet and Western nations, a method for increasing Central Asian labor mobility would be to accelerate the socioeconomic development of the population through the promotion of the equalization-oriented principles. The result of such a policy would provide skilled labor in the region and eventually a more mobile population that would conceivably move to labor-deficit regions. However, the "either–or" development options tied to the maximum growth and equalization strategies are virtually impossible to pursue due to the realities of the Soviet economy, the constraints created by Soviet ideology, the likelihood of harsh regional battles, and the pressures caused by the demands of the multinational population organized in long time national homelands. According to Bandera (1973: xx): "a predominant portion of current Soviet economic theory and almost all practice is not yet suitable for accurate economic analysis. . . . It is not only that the party and bureaucracy make purely 'voluntaristic' locational and allocational decisions. . . . The real process of spatial resource allocation is nothing but pure struggle."

Thus, the Soviet leadership often has been (and is) forced to make what appear to be irrational economic decisions—although, given ethnic, strategic, and other factors that must be taken into consideration, many decisions have been and are in fact politically rational. Political decisions, and good ones at that, however, are not necessarily ones that result in economic growth or can help alleviate the economy's spatial imbalance.

The political-administrative structure

The ethnically based political-administrative structure of the USSR contributes to the difficulty of solving the geographical imbalance, because the combination of "territory" and nationalism forces the leadership to grant some power to the secondary political units. The devolution of power reduces the control of the central government so it cannot simply dictate a policy that would best alleviate the imbalance. Beginning with Lenin's "federal compromise," which helped prevent the peripheral non-Russian areas of the former tsarist empire from seceding from the newly forming Bolshevik state, Soviet leaders recognized the rights of the non-Russian nations in their homelands and constructed the Soviet federation. In this self-proclaimed voluntary federation of semiautonomous political entities (union republics), the central government, in theory, forfeits some degree of power to the state's secondary political units. And while the federal arrangement was supposed to give way to a unitary state when national allegiances were superseded by class allegiances (Pipes 1968), the most recent Soviet constitution reaffirms the rights of the union republic (USSR 1977), and class associations are no closer to capturing the allegiances of the Soviet population than they were 60 years ago.

Originally, Lenin did not favor devolving power to nationality based secondary political units, but he reasoned that the granting of what had the appearance of political equality to national minorities would lessen the chances for political instability (Ziegler 1985). The leadership believed the need for the federal structure of the state would disappear once social and economic equality was attained, thereby eliminating the need for nationally and regionally based associations. However, after almost 70 years, the federal structure of the state remains, and while most power is concentrated in Moscow, the ethnically based union republic has become an object with which many ethnic groups strongly identify (Kaiser, this volume), and the numerous challenges to the center's control during the past two years are certainly changing the terms of union. At the least, official recognition of a system of national homelands provides a legitimate focus for national aspirations, but even more, it provides minority nationalities with a sense of power to use against the center in order to gain economic or political benefits.

There are several examples that reflect the added burden of the Soviet nationally based territorial structure. The siting of redundant productive forces as part of regional economic development schemes represents a general example of the burden, while the handling of the conflict between Armenia and Azerbaydzhan beginning in 1987 is a more specific example and one that reflects the potential volatility of the Soviet federation. If the constitutionally guaranteed Soviet (federal) territorial administrative structure were not depicted so often as a voluntary union with considerable power sharing, the central authorities would have more easily been able to impose a more effective settlement to the Nagorno-Karabakh crisis. Such a move, however, common in unitary states,[9] would have gone against the terms of the federation, and would have opened the proverbial Pandora's box to other national issues, some of which have the potential for even greater destabilizing effects than the Azeri–Armenian conflict.

The mere existence and repetitive institutionalization of the nationally based territorial administrative system do not ensure the kind of federation defined by the Soviet constitution, but they do prevent the center from exercising its power at will. Despite the totalitarian image of the Soviet state, the central authorities cannot always act decisively and unilaterally. More often than not they are forced to make concessions to the republics; the delayed removal of the party heads in Armenia and Azerbaydzhan reflects to some degree the circumspection with which Moscow must deal with the non-Russian republics. Moscow's reaction to the Baltic republics' declarations of independence reflects more than a desire not to strain relations with the West; it also reflects the limited options open to a central government in a federation.

In more tangible examples of how the administrative structure of the state limits the power of the central authorities, one can look at local politics and the lobbying power of the republics, autonomous republics, and oblasts. Regional biases within the Politburo are often linked to major economic issues, and even when the national composition of the Politburo has been found to have little impact on republican budgetary allocations, lobbying on the oblast (regional) level has been quite effective in changing directives from Moscow (Bahry 1987). Even if concessions from Moscow are less than one would expect from a federal arrangement, many have had to be made, and in the case of economic concessions, they often compromise the effectiveness of a "national" economic plan. In a state as large and diverse as the Soviet Union, the specific economic needs of individual regions cannot add up to a suitable all-union development policy that can alleviate economic imbalances like the ones that plague the Soviet economy.

The psychological importance and legitimacy of territorial homelands to Soviet Central Asian nations are apparent when one considers the limited migration from the region, which is a major contributor to the economy's imbalance. Had national divisions been eliminated and group allegiances transferred to a new class identity as Soviet nationality theory predicted, the labor component of the imbalance would be less of a problem, for the absence of national or cultural differences would diminish the importance of the homeland as a barrier to migration. Despite the claim by Brezhnev that the nationality question had been "resolved completely, resolved definitely and irrevocably," Soviet nationality policy has failed to produce the new class-based "Soviet man." The relatively backward nationalities have achieved impressive socioeconomic gains,[10] yet the anticipated merging of Soviet peoples into one nation has not occurred, and most non-Russian groups remain closely tied to their national homelands, viewing a move to other republics almost as unlikely as moving outside the USSR.

SOLUTIONS AND SOME PROBLEMS

Any strategy implemented to solve any one aspect of the economy's geographical imbalance is likely to promote a variety of other problems that could be even more problematic. Encouraging surplus labor from Central Asia to move to labor-deficit regions in the west and Siberia could raise ethnic tensions if the labor and interethnic relations followed a competitive course described by Connor (1972) or Nagel and Olzak (1982).[11]

Investing more industrial capital in Central Asia may also prove problematic. The costs of building and expanding transportation facilities and other infrastructure on the scale necessary to absorb the increasing labor supply would be exorbitant. Because the indigenous work force is relatively unskilled, large numbers of Slavs would probably have to be "imported" to fill management and some skilled positions, a development that would probably worsen national tensions between the indigenous population and the in-migrants. Additionally, some have argued that the Soviets will not increase strategic (industrial) investment into Central Asia because they still fear China (Newton 1976). This contention was more valid in the 1970s than it is today and is somewhat supported by capital investment data for the 1956–75 period. Investments in borderland tertiary units (on the Chinese and Mongolian borders) during this 20-year period were uniformly below the all-union per capita mean, while units one oblast to the north received allocations far in excess of the all-union mean (Liebowitz 1985). Of course, resource wealth and population distribution in the borderland oblasts can partially explain the significant differences in resource allocations from one unit to its neighbor, but the strategic factor has some validity. With regard to giving greater attention to Central Asia, however, the following is evident; the Soviet fear (military) of China may still be strong, but the costs of neglecting Central Asia, its growing work force, and an increasingly dissatisfied and rapidly growing population represent a much greater threat to the Soviet state. Mikhail Gorbachev's "Vladivostok address" and his speeches given shortly thereafter reflect a new pragmatism in dealing with China that suggests the chances of ignoring Central Asia because of security risks are unlikely.

Before the Gorbachev reform program, Leslie Dienes (1983) offered a "halfway-house" solution to the resource–labor–infrastructure imbalance in which the Soviet leadership would use the Transcaucasian republics as an "interim workshop" to alleviate the labor component of the imbalance. This strategy seemed logical because the region's population was more advanced (demographically and socioeconomically) than Central Asia's, its economy had become more diversified[12]—its industrial structure had expanded to include manufacturing and mineral production to go along with food production—its population was reasonably skilled, and the region was in relative close proximity to the European core of the state (markets and infrastructure). In addition to these positive features for further development, the region accounted for approximately 11 percent of the state's increase in the 20–59 working-age cohort in the 1980s,[13] and labor productivity in the region by some estimates is roughly equal to the all-union measure (Dienes 1983).

The four Central Asian republics lack many of the positive features for industrial development enjoyed by Transcaucasia, yet they do have a large and growing educated work force and substantial energy resources. Between 1979 and 1987, the region accounted for 25.7 percent of the state's *increments* to its working-age population (USSR 1988d), while accounting for less than 10 percent of the Soviet total population in 1979 and only 11.1 percent in 1987 (USSR 1988d). The shortage of skilled labor, however, remains a problem, reflected in the large percentage of the region's total workforce employed in nonindustrial branches of the economy (Table 5.1). The great majority of indigenous Central Asian labor still resides in rural areas and work largely in nonindustrial pursuits (Table 5.2).

The need to utilize Central Asian labor, however, goes beyond the problems of labor shortages and economic stagnation in the western USSR and Siberia. Living conditions in the southern tier will inevitably decline without large increases in capital investment needed to provide employment opportunities and social amenities for the increasing population. The strategy Dienes outlined may well help the national

Table 5.1   Work force by sector, 1987.

	% in industry	% in agriculture	% in services
USSR	38	19	6
RSFSR	42	14	6
Ukraine	40	20	5
Belorussia	40	22	4
Uzbekistan	24	38	4
Kirgiziya	27	34	4
Tadzhikistan	21	42	4
Turkmeniya	21	41	4
Estonia	42	13	6

Source: USSR 1988f: 16–17.

Table 5.2 Percentage of titular nationalities engaged in agriculture and industry, 1987.

	Agriculture	Industry
RSFSR	8	38
Ukraine	9	37
Belorussia	13	37
Uzbekistan	25	19
Kirgiziya	35	15
Tadzhikistan	25	18
Turkmeniya	14	15
Estonia	13	24

Source: USSR 1988f: 24–5.

economy obtain the labor it will require in the short term, but as a realistic solution to the geographical imbalance and the broader problems currently afflicting the Soviet economy it is limited. One must remember that when Dienes offered this halfway-house solution, the Soviet economy was in the earlier stages of what has become a sustained period of low growth. By the mid-1980s, the urgency of addressing the needs of Central Asia, its burgeoning population, and its environmental degradation[14] had become more readily apparent and widely known as a result of official disclosures about the overall health of the Soviet economy and society. It is unlikely that, given the near crisis situation in the national economy, special attention can be given to Soviet Central Asia any time in the near future. The growth rate for capital construction has declined over the past 15 years[15] (USSR 1988a), and one would expect regions with greater returns to the ruble (the Central Industrial Region and the Baltics) to receive the majority of what funds are invested. The political consequences of not establishing a more comprehensive development strategy than what has been followed the past four decades for Soviet Central Asia, however, could be as great as anything the Gorbachev leadership has thus far encountered.

Since the early 1950s, successive Soviet leaders have poured great amounts of capital into Central Asia. There are many angles from which one can assess investment flows in the USSR, and none is problem free. Some argue that if the kind of socioeconomic equalization described by Soviet ideology is to be attained, relatively backward regions need to receive significantly greater amounts of capital than the more developed regions in order to overcome great gaps in economic development. Others argue that since economic specialization precludes any realistic attempt to "equalize" the productive capacities or any other "objective" economic measure of disparate regions, comparing state investment allocations, in this case rubles, is meaningless. A third angle, perhaps best voiced by James Gillula (1978), is one that asks how much better or worse the relatively underdeveloped peripheral republics do in terms of receiving development funds under the current system than they would if they had to generate the funds solely from their local economies. This answer is virtually impossible to estimate due to the nature of the Soviet financial system and budgetary process; when (and if) "khozraschet"[16] is adopted and applied statewide, the answer may be obtainable. However, the social and political consequences of what "khozraschet" will probably do to enterprises in the poorer regions may delay its implementation or force a change in its a priori operating procedures.

Capital investment data for Central Asia since World War II reflect a Soviet development policy that, on the surface, appears to have favored the relatively backward regions of the state at the expense of the more developed areas, especially the Ukraine (Melnyk 1977; Gordijew and Koropeckyj 1981; Liebowitz 1985). But while investment allocations in the region as a whole have been favorable when measured on a *per worker basis*, a closer analysis reveals a narrowly directed investment history that has not been able to create the economic conditions needed to absorb surplus labor or develop the region's economy beyond its agricultural and resource-extraction orientation. A continuation of the investment strategy followed the past 20–30 years is likely to place added economic and social pressures on the region's indigenous population—a population that to date has resisted both significant rural to urban migration within the region's republics, and the kind of interrepublic migration needed to alleviate the labor component of the national economy's geographical imbalance.

TRENDS IN CAPITAL ALLOCATIONS

Between 1956 and 1985, per working-age capital investment[17] was more than 21 percent greater in the Uzbek SSR than in the Ukraine, 17 percent more in the Turkmen SSR than in the Russian Republic, and 67 percent more in the Turkmen SSR than in the Ukraine (Table 5.3). The Ukraine received virtually the same amount of rubles per worker as the Kirgiz SSR, which received the least amount of investment of the four Central Asian republics, despite the vast differences in the republic's economic infrastructures and levels of socioeconomic development. When these republic-level data are measured on a *per capita* basis (Table 5.4), disaggregated geographically, and compared over time, Central Asia's favorable position becomes qualified. The eroding position of the Central Asian republics and their constituent oblast-level units[18] becomes most obvious when using this second measure, primarily because of the substantially higher rates of natural increase in Central Asia than elsewhere, and the extremely young age structure of the region.

Table 5.3 Per worker investment allocations, 1956–85 (USSR = 100).[1]

	1956–60	1961–5	1966–70	1971–5	1976–80	1981–5
USSR	100	100	100	100	100	100
RSFSR	111	105	108	108	112	115
Ukraine	81	81	80	77	75	76
Belorussia	59	70	73	89	89	92
Uzbekistan	78	96	111	104	90	77
Kazakhstan	153	182	122	131	118	108
Georgia	67	69	75	64	70	78
Azerbaydzhan	103	97	90	80	74	76
Lithuania	64	87	103	104	96	99
Moldavia	50	64	71	84	77	78
Latvia	80	101	107	106	98	102
Kirgiziya	77	82	90	85	68	61
Tadzhikistan	73	83	95	89	71	55
Armenia	87	112	114	103	78	72
Turkmeniya	117	137	147	152	115	105
Estonia	103	119	118	109	104	104

[1] Per working age population figures were calculated using the 20–59 cohorts from the 1959 and 1970 Soviet censuses, plus 1979 census data and 1987 age data from USSR 1988d. For the 1956–60 and 1961–5 periods, 1959 population data were used; for the 1966–70 and 1971–5 periods, 1970 population data were used; for the 1976–80 period, 1979 population data were used; and for the 1981–5 period, the most recent 1987 population data were used.
Sources: for working age populations TsSU SSSR, *Itogi Vsesoyuznoy Perepisi Naseleniya 1959g, Itogi Vsesoyuznoy Perepisi Naseleniya 1970g* (Moskva: Statistika, 1962, 1972); for 1979 and 1987 age data, USSR 1988d; for capital investment figures, selected volumes of the all-union and republic-level *Narodnoye Khozyaystvo* statistical handbook series and USSR 1988a.

Table 5.4 Per capita investments, 1956–85 (USSR = 100).

	1956–60	1961–5	1966–70	1971–5	1976–80	1981–5
USSR	100	100	100	100	100	100
RSFSR	112	108	110	113	118	119
Ukraine	84	84	84	82	77	74
Belorussia	55	68	72	88	90	92
Uzbekistan	69	78	89	75	71	68
Kazakhstan	148	155	112	116	108	104
Georgia	64	67	73	62	68	77
Azerbaydzhan	93	82	74	61	63	74
Lithuania	61	86	104	104	96	101
Moldavia	49	59	70	81	76	74
Latvia	82	106	111	110	101	104
Kirgiziya	68	68	77	67	57	54
Tadzhikistan	65	69	76	64	54	47
Armenia	81	95	103	86	74	74
Turkmeniya	107	115	121	114	93	92
Estonia	104	125	127	117	108	106

Sources: investment figures, see sources, Table 5.3; population figures were from *Narodnoye Khozyaystvo SSSR* handbooks and USSR 1988d for the first year of each five-year period.

Investment allocations on the per capita basis have declined considerably throughout the region. On the republic level, the Turkmen SSR was the only Central Asian republic to maintain a level of investment that was near the all-union average through the 1981–5 period, reflecting the great investments made into the gas and energy industries since the 1960s.[19] By 1981–5, per capita investments in the other three Central Asian republics declined to levels between 68 (Uzbek SSR) and 47 percent (Tadzhik SSR) of the all-union mean.

Since 1970, all four Central Asian republics have lost ground to the rest of the USSR in per capita investment allocations for reasons other than their relative rapid population increases. A deemphasis on many large-scale projects throughout the country—especially in agriculture—has had profound effects on Central Asia's capital allocations. Agriculture accounted for about 20 percent of all investments in the national economy in both the 1971–5 and 1976–80 periods (and more than 30 percent of Central Asian investment), but its share declined to 16.7 percent of total investments by 1987.

The deemphasis on the region is also apparent when one considers the declining share of total capital investments allocated to each of the four republics and the region as a whole from the 1966–70 time period to the present (Table 5.5). Despite the great population growth in the Central Asian republics, the share of total state investment going to the region has declined and leveled since 1970, with Uzbekistan the only republic to have increased its share of all-union allocations. By 1981–5, the share of total state investment going to Central Asia leveled off at 6.6 percent, while over the same period the region's share of the Soviet population increased from 8.1 to 11.0 percent (USSR 1986, 1988d). With rates of natural population increase at least 2.5 times greater in each of the four Central Asian republics than in the country as a whole, and about four times greater than in the Russian Republic (USSR 1987), the lack of a significant increase in the share of state investment to compensate for the region's rapid population growth reflects the noticeable deemphasis on Central Asia within the state's framework for regional economic development.

Geographically disaggregated oblast-level capital investment data for 1956–85 (Table 5.6) reveal that only 3 of 12 tertiary-level units

Table 5.5   Share of all-union capital investments, 1956–85 (% of total investment).

	1956–60	1961–5	1966–70	1971–5	1976–80	1981–5	1956–85
Uzbekistan	2.57	3.13	3.98	3.76	3.93	4.13	3.80
Turkmenistan	0.73	0.86	1.00	1.04	0.93	1.00	0.96
Tadzhikistan	0.59	0.67	0.80	0.79	0.74	0.71	0.73
Kirgiziya	0.65	0.69	0.86	0.82	0.74	0.74	0.76
Central Asia	4.55	5.36	6.64	6.41	6.34	6.59	6.26

Sources: see Sources, Table 5.3.

received investment allocations greater than the all-union per capita mean in any of the six five-year periods for which comparable data were available. In every five-year period, at least one unit received less than one-twelfth of the all-union per capita norm. The great variability of investment allocations within the region reflects the backward state of the region's economy and the directed and largely specialized nature of "development" in the region, geared mainly, as was mentioned earlier, toward the natural gas industry, some light industry, and grandiose agricultural projects.

Table 5.6   Relative oblast-level per capita investments, 1956–85 (USSR = 100).

	1956–60	1961–5	1966–70	1971–5	1976–80	1981–5	1956–85
USSR	100	100	100	100	100	100	100
Kara-Kalpak ASSR	44	100	117	95	93	96	94
Andizhan[1]	45	44	49	50	45	46	47
Bukhara[2]	44	64	60	59	59	50	56
Dzhizak[3]	50	115	141	131	141	95	117
Surkhandar'ya[4]	37	57	85	90	89	77	78
Tashkent	126	121	142	91	84	88	101
Khorezm	26	44	68	67	66	62	60
Ashkhabad[5]	120	136	152	144	104	109	124
Chardzhou	79	89	89	99	94	84	90
Tashauz	50	63	75	72	54	50	60
Kirgiziya	63	66	76	66	56	54	62
Tadzhikistan	59	66	73	60	55	47	58
Central Asia	67	77	88	76	69	66	73

[1] Andizhan, Namangan, and Fergana oblasts.
[2] Bukhara, Navoi, and Samarkand oblasts.
[3] Dzhizak and Syrdar'ya oblasts.
[4] Surkhandar'ya and Kashkadar'ya oblasts.
[5] Ashkhabad, Krasnovodsk, and Mary oblasts.
Sources: oblast and ASSR investment and population figures are from oblast-level *Narodnoye Khozyaystvo* handbooks, various years, and *Uzbekistan za Gody Odinnadtsatoy Pyatiletki (1981–1985g.) Statisticheskiy Sbornik* (Tashkent: Izdatel'stvo TsK KP Uzbekistana, 1986); USSR and RSFSR data, see Sources, Tables 5.3 and 5.5; the data reflect 1961 territorial boundaries, so, for example, Navoi Oblast, created in 1982, has its territorial and capital investments allocated within the 1961 administrative boundaries.

Based on the more detailed oblast-level data, it appears as if Soviet investment practice has been geared more toward developing the economic capacity of Central Asia to meet the needs of the national economy than with developing and equalizing the productive capacities and socioeconomic conditions of the southern tier. Despite the "integrationist" ring to such a strategy, it is a strategy that, by creating a specialized economy, has prevented the indigenous work force from becoming diversified, skilled in industry, and ultimately more likely to migrate to urban areas within Central Asia and outside the southern tier. Again, this is not to suggest that great gains have not been made in the socioeconomic development of the Central Asian indigenous nations.

In fact, such gains have been documented by Soviet and Western researchers. Yet, despite the gains, labor and productive imbalances among regions persist and appear to be widening; the recent trends in the spatial allocation of capital do not suggest a narrowing in the development gap in the near future.

The oblasts that have fared well over the past 30 years (1956–85) have done so largely because of energy development (Krasnovodsk) and large-scale agriculture-related projects (Syrdar'ya and Ashkhabad), or, in the case of Tashkent, because of established industry, the great earthquake of 1966, and a relatively large service infrastructure. Typically, development in these oblasts have been capital- rather than labor-intensive, so even if investment levels were maintained, the problem of creating new jobs for the surplus labor would remain. Targeted investment into the few oblasts, some argue, may provide a pole or magnet for future development, but such poles will not be able to attract indigenous labor on the scale needed to compensate for the decline in investments made into the rest of the region. As researchers from the late seventies have noted, since World War II, investments in the region have been above the national mean (per worker) due in large measure to nonlabor-intensive industries—natural gas and nonferrous metal and chemical projects. Labor-intensive industries obviously need to be expanded in the region so that surplus labor will be absorbed and migration within the region can begin to balance the supply and demand for labor throughout the state.

The declining investment position of Central Asia extends to the

Table 5.7   Industrial investment, 1956–87 (Percentage of total investment).[1]

	1956–60	1961–5	1966–70	1971–5	1976–80	1981–5	1986–7
USSR	36	37	35	39	39	39	41
RSFSR	42	41	39	39	39	39	40
Ukraine	49	48	46	44	45	45	43
Belorussia	33	36	41	39	38	40	40
Uzbekistan	37	33	28	25	25	26	26
Turkmenistan	46[2]	42	37	33	33	44	46
Tadzhikistan	38	33	33	34	39	36	36
Kirgiziya	41	40	37	36	33	31	31

[1] Data do not include kolkhoz investments, which for the eleventh five-year plan accounted for 8% of total investments in the RSFSR, 16% in the Ukraine, 11% in Uzbek SSR, 14% in the Turkmen SSR, and 16% in Estonia. For the whole country, collective farm investments accounted for 10% of total investments. For the 1956–60 to the 1971–5 time periods, data for the share of investment by branch for some republics included collective farm investments. This inclusion minimizes the estimate for industrial investment, because collective farms typically contribute less than 2% of all industrial investments while they account for up to 16% of a republic's total investment. Therefore, for the Ukraine, Belorussia, and the Turkmen SSRs, the 1956–60 to the 1971–5 figures were derived by subtracting collective farm investments from total investments, and then dividing industrial investment by that total.
[2] Estimate. Sources USSR 1988a; republic-level *Narodnoye Khozyaystvo* handbooks, various years.

structure of capital allocations as well. The percentage of total investment going to industry has declined from one five-year period to the
next since 1956–60 in the four Central Asian republics with few exceptions (Table 5.7).[20] If enough jobs are going to be created to accommodate the rapidly growing population, industrial investments have to
increase substantially. When one considers the eroding position these
republics have experienced in their overall per capita investment allocations, the declining share given to industry becomes even more significant.
Thus far, there is nothing in current party or government pronounce-
ments to suggest there will be the kind of change in investment policy in
the near future needed to improve the situation. When one disaggre-
gates industrial investment into its productive and nonproductive
components,[21] it is clear the number of new industrial jobs will not
alleviate the labor surplus problem in Central Asia. The amount of
rubles allocated for productive investment declined in three of the
region's four republics (USSR 1988a), and the percentage of total
investment used for productive means has declined substantially since
the mid-1970s (Table 5.8).

Table 5.8 Productive investment (percentage of total investment).[1]

	1956–60	1961–5	1966–70	1971–5	1976–80	1981–5	1986–7
USSR	65	68	68	72	73	73	71
RSFSR	66	65	66	71	73	73	72
Ukraine	68	72	72	75	76	74	71
Belorussia	59	66	70	73	74	72	68
Uzbekistan	69	72	66	71	72	71	65
Turkmenistan	n.a.	77	78	80	79	78	75
Tadzhikistan	67	67	67	69	72	70	67
Kirgiziya	72	70	69	70	72	72	67

[1] Productive investments include capital allocated to industry, agriculture, forestry,
construction, transportation, communications, state food trade, material-technical
supplies and stocks, and computing services.
Sources: for 1971–5 to 1985–7, USSR 1988a: 17; for 1956–60 to 1966–70, republic-level
Narodnoye Khozyaystvo handbooks, various years.

In many ways, the 30-year investment strategy implemented by
central planners has left the region with many of the features associated
with what Michael Hechter (1975) described as an "internal colony."
Hechter (1975) used the term to describe British–Celtic economic
relations over the past four centuries, and parallels to the USSR–Central
Asian relationship are strong. Characteristics of the internal colony
include: (a) an economy dominated by primary economic activities; (b)
substantially lower levels of consumption than in other areas of
the state; and (c) a noticeable division of labor in which the state's
dominant nation retains the better-paying and higher-status jobs
and the indigenous nations hold the agricultural and service-related
positions.

The best example of the predominance of primary industry in Soviet Central Asia is the case of cotton and textiles. For years, local officials in Central Asia have complained about the exporting of cotton fiber out of the region—especially Uzbekistan—to be used for textile production, mostly in the Central Industrial Region of Russia. In 1987, the Uzbek and Turkmen SSRs produced more than 75 percent of Soviet cotton, yet the two republics produced less than 6 percent of the state's cotton fabrics. The Russian Republic, producer of no cotton, accounted for more than 70 percent of cotton fabric production (USSR 1988e: 30).

Per capita retail trade data for 1987 (USSR 1988c) reflect significant differences in consumption between Central Asia and the rest of the country. The four Central Asian republics recorded per capita retail trade measures between 54 (Tadzhikistan) and 68 (Uzbekistan) percent of the all-union mean and 35 and 45 percent of the Estonian figure (Table 5.9). Perhaps more significant is the widening gap in this measure; in 1960, the respective figures for Central Asian–USSR and Central Asian–Estonian comparisons were 67 (Tadzhikistan) and 86 percent (Uzbekistan) of the all-union mean, and 48 and 61 percent of the Estonian standard. Due to the settlement patterns of Soviet nations in Central Asia, urban–rural differentials in retail trade provide us with a rough surrogate measure of the differences in consumption by nation. Per capita retail trade in rural areas is on average at least 50 percent less than it is in urban areas; and within rural areas, Tadzhikistan's per capita retail trade is 38 percent of the RSFSR measure and 35 percent of Estonia's (Table 5.10). In urban areas, regional differentials are not as large; Uzbekistan has the lowest per capita retail trade in Central Asia, about 23 percent lower than the all-union measure, but all four Central Asian republics remain below the national mean. The greater inequality in retail trade in the rural areas reflects the lack of consumer industries in the country as a whole and of an inadequate distributional system

Table 5.9   Per capita retail trade (rubles).

	1960	1970	1980	1987
USSR	376	642	1,015	1,206
RSFSR	415	701	1,114	1,322
Ukraine	319	585	932	1,138
Belorussia	273	581	1,024	1,337
Uzbekistan	284	422	639	720
Kirgiziya	286	471	714	811
Tadzhikistan	253	402	577	648
Turkmeniya	324	472	697	825
Estonia	530	962	1,521	1,853

Sources: data on retail trade, *Narodnoye Khozyaystvo SSSR v 1987g*: 413; per capita measures derived by the author using population data from USSR 1988d and various volumes of *Narodnoye Khozyaystvo SSSR*.

Table 5.10   Urban and rural per capita retail trade (rubles).

| | Urban | | Rural | |
	1985	1987	1985	1987
USSR	1,429	1,482	675	666
RSFSR	1,436	1,490	882	851
Ukraine	1,366	1,413	557	577
Belorussia	1,525	1,640	712	780
Uzbekistan	1,137	1,139	447	417
Kirgiziya	1,145	1,176	575	571
Tadzhikistan	1,285	1,294	327	327
Turkmeniya	1,179	1,241	441	445
Estonia	2,079	2,213	862	937

Source: *Narodnoye Khozyaystvo SSSR v 1987g.*

outside urban areas. Since the indigenous Central Asians live pre-dominantly in rural areas, it is clear that living standards differ greatly by nationality both within and across republics.

Data on employment by branch of the economy for the 15 republics further support the internal colony analogy (see Table 5.1). Whereas in 1987, 38 percent of the all-union work force and 42 percent of the Russian Republic force were employed in industry, the corresponding figures for Uzbekistan, Kirgiziya, Tadzhikistan, and Turkmeniya were 24, 27, 21, and 21 percent. Similarly, the percentage of the Central Asian work force employed in agriculture in 1987 was about three times the Russian Republic share.

The growth of industry and associated industrial jobs in Central Asia, while exceeding what one would expect given the economic and social profile of the region, has not kept pace with the demand for jobs. Recent Soviet studies report that the Central Asian share of new industrial jobs throughout the USSR is less than 7 percent (Dienes 1987: 144–5) of the state total. Part of the problem is the kind of investment central planners allocated to the region—the fear of not making produc-tion targets has led planners to allocate industrial capital to more productive regions—but some of the problem is reportedly tied to the indigenous populations' resistance to move from rural to urban areas. Some Soviet and American studies have noted the importance of cultural factors in explaining the lack of enthusiasm for industrial jobs among Central Asians, yet others feel this is not a major factor in low rates of rural to urban migration among Central Asian indigenes (Sacks, Rowland, this volume). The resulting slow rate of industrial develop-ment in Central Asia, according to a growing number of analytical studies from the region, has increased the region's level of dependence on the Russian Republic for light industrial products and food (Dienes 1987), to the growing protests of local leaders.

Recently released data (USSR 1988e) reflect the lack of industrial production in Central Asia. Despite comprising about 11 percent of the state's population, the region's share of total Soviet industrial production in the following areas lags far behind: electric production (5.6 percent); steel (0.6 percent); metallurgical machinery (4.9 percent); chemical equipment and spare parts (4.2 percent); agricultural machinery (4.8 percent); paper (0.4 percent); cement (6.6 percent); bricks for construction (7.5 percent); and furniture (4 percent).

Perhaps the most indicative measure of Central Asia's status as an internal colony is the role its titular nations play in their own republics. The indigenous Central Asian nations remain largely employed in agriculture, while many of the better-paying industrial and managerial positions are held by nonindigenous (predominantly Russian) inmigrants. This phenomenon can be seen from the data in Table 5.11. The percentage of all industrial jobs in the Central Asian republics filled by their titular nations is far lower than the corresponding figures for the Slavic republics. While Russians hold 83 percent of all industrial positions in the Russian Republic, Kirgiz hold only 25 percent of their republic's industrial jobs, the Tadzhiks 48 percent, and the Uzbeks and Turkmens 53 percent. Despite meaningful gains in this measure over the past ten years (see Table 5.2) the employment structure of Soviet Central Asia remains consistent with the internal colonial model.

Table 5.11 Percentage of republican jobs in industry and agriculture held by titular nationalities, 1977 and 1987.

	Industry		Agriculture	
	1977	1987	1977	1987
RSFSR	86	83	72	75
Ukraine	65	68	76	79
Belorussia	75	77	89	89
Uzbekistan	38	53	67	76
Kirgiziya	15	25	63	69
Tadzhikistan	35	48	62	63
Turkmeniya	34	53	58	81
Estonia	48	43	86	84

Source: USSR 1988f: 22–3.

CONCLUSION

The geographical imbalance afflicting the Soviet economy has become more acute during the past decade, the effects of which are reflected in declining growth rates for the national economy. The role of the four Central Asian republics in alleviating the imbalance is an important one, largely because of the region's growing work force. The cumulative effects of Soviet development strategy since World War II, however, leave the Soviet leadership in a quandary over how to utilize the labor resources most effectively while at the same time minimizing potential destabilizing political problems.

Past development "strategy" in Central Asia, condemned by many for being exploitative, has resulted in some great socioeconomic gains for the indigenous population and impressive growth of selected sectors of the region's economy. However, most of the industrial development that has taken place in Central Asia has been resource-related, and the economy remains dominated by the agricultural sector. With a rapidly growing population that remains largely immobile, the need for more labor-intensive industries is apparent and becoming more important. Surplus labor from the region, virtually all of which is now educated, is needed in many of the industrially developed western economic regions of the state and in resource-rich, but labor-short, Siberia. The strain on the Central Asian economy and state transfer payments and subsidies to support a rapidly growing population with rising expectations is likely to exacerbate any already surfaced political problems if expectations are not met and if living standards continue to fall.

For the national economy, the geographical imbalance highlights labor shortages in the west and Siberia, and the lack of a sizable industrial infrastructure in the non-Slavic south. For all the planning mechanisms and "scientific" methodologies supposedly available to economic planners and policy makers, the Soviets have yet to establish a clear-cut policy for alleviating the geographical imbalance in the national economy. With regard to the Central Asian role in a potential solution, two broad options appear open to the Soviets: (a) to promote out-migration of Central Asians to labor-deficit regions; or (b) to invest greater amounts of capital in labor-intensive industries in rural and urban areas of Central Asia.

Both of these solutions are wrought with problems if they prove successful, for they are both likely to result in the migration of labor. If successful, promoting out-migration from Central Asia could cause problems due to an increased presence of non-Slavs in the predominantly Slavic labor-deficit regions. Competition for goods, services, and housing is already very fierce in the Slavic areas; an influx of southerners would increase competition, which could easily take on nasty ethnic overtones. The second option, increasing the number of industrial enterprises and jobs in Central Asia, aside from the great costs and finding a way to stem typically high rates of labor turnover among indigenous workers (Gleason 1986), could cause a problem if it once again results in a substantial influx of Slavs into the region. Finally, parochial jealousies among the Slavic and Baltic populations in reaction to what would appear to be the unfair investment of rubles into non-Slavic areas at the expense of their homelands' more productive economies could present another destabilizing situation for Moscow.

Since World War II, many factors have prevented significant

industrial development in Central Asia. Today, the state of the national economy will hinder the region's development, and some see Central Asia remaining on the "backburner" for some time because of increasing capital shortages in the Soviet economy (Dienes 1983). The need to continue developing and expanding the state's energy resources for domestic use and for hard currency earnings will require further Siberian development. However, the lack of capital available for both a western USSR and Siberian and a Central Asian development emphasis will rule out the significant industrialization of Central Asia in favor of a continuation of the status quo. Such a strategy assumes or rests upon the hope that the increasing number of Central Asia's underemployed will not become restless and serve as yet another catalyst for discontent, and labor shortages in western regions and in targeted areas of development (especially in parts of Siberia) will not intensify, so as to limit resource extraction and production below externally determined break-even levels.[22] The status quo option envisioned by Dienes may be the only viable one for Soviet planners from an economic point of view, but the political consequences of this option may be too great. While recent national tensions in Kazakhstan, Uzbekistan, Kirgiziya, and the Transcaucasus cannot be attributed solely to poor economic conditions, there is no doubt that economic conditions added to the frustrations and tensions of the demonstrators. A report from *Trud* (24 June 1989) explains the national conflict in Mangyshlak Oblast, Kazakhstan, between Kazakhs and people from the Caucasus by pointing to social and economic problems.

> Having visited the site, leaders of the party, soviet, and law enforcement organs of the republic ... talked about the social causes of the events straightforwardly and honestly at a press conference. While developing the oil peninsula through the use of ultramodern equipment, the union ministries and offices have ignored the needs of the local population and "maintained" extra-hard living conditions. ... Everything in the city is an insoluble problem dating a long way back: housing, water, supplies. ... The city authorities cannot give the exact number of unemployed young people—either 1,200 or 2,000 persons. (Foreign Broadcast Information Service, 1989: 71–2)

Similarly, the much worse national violence between Uzbeks and Meskhetians in Fergana Oblast was no doubt inflamed by poor and declining socioeconomic conditions in the region. The Moscow television service on 1 July 1989 made the following report. "The situation here is getting back to normal. ... Even the most superficial analysis shows that the greatest excesses occurred in places where there is high unemployment, a difficult ecological situation, and a serious socioeconomic situation among the population. ... Today, about 200,000 people here are unable to find work."[23]

It is clear that the relatively backward nature of the Central Asian economy needs to be addressed not only for economic reasons but for political reasons as well. The twelfth five-year plan and the basic tenets of Gorbachev's policy of perestroyka called for a move away from past economic policies in fundamental ways. The new (domestic) economic strategy was supposed to stress the more effective use of the economy's productive capacity by emphasizing retooling and modernizing the state's existing infrastructure. Approximately 50 percent of investments were targeted for production facilities so as to upgrade the existing infrastructure (Leggett and Kellogg 1988), a significant increase over the previous two five-year plans. The emphasis on retooling and modernizing the industrial infrastructure was and is supposed to increase worker productivity, reduce the demand for labor—an important development given the heretofore limited in-migration from labor-surplus to labor-deficit regions—and save on capital expenditures. In fact, between 1980 and 1988, the share of state investment going to retrofitting existing enterprises increased from 33 to more than 45 percent, while the share allocated for new construction declined from 38 to 34 percent (USSR 1989b: 264). Any positive developments in the Soviet economy due to this change in strategy have thus far gone unnoticed. The commonly heard complaint of pro-perestroyka forces— that the "old economic mechanism" has been dismantled but the "new one" is not yet working—is perhaps an optimistic view of why a changing investment strategy has done little to reverse so many of the downward trends in the national economy.

Because a disproportionate share of the state's industrial infrastructure is concentrated in the western parts of the state, the kind of change in investment strategy outlined by Gorbachev as early as 1985–6 will mean that the non-western regions (including Central Asia) will receive smaller shares of industrial investment than in the past. As Ed Hewett (1985: 295) correctly projected at the start of Gorbachev's economic reform program, more investment will likely flow into the machine-building industry at the expense of the agricultural and selected resource sectors. Such a shift would be another loss for Central Asia, since agriculture has accounted for a significantly larger portion of total investment in Central Asia than in the western and Siberian economic regions.[24]

Without even considering the implications of state-wide "khozraschet," which is likely to hurt Central Asian industries and enterprises more than others due to the region's low productivity and profitability, it is likely that Central Asia will receive relatively smaller shares of total investment in the 1990s. Therefore, judging from the reforms under consideration and the state's changing investment strategy, the Soviet leadership has apparently chosen not to tackle or actively address the Central Asian labor situation,[25] and has only recently pledged to address the poor socioeconomic conditions as a result of the ethnic unrest in Uzbekistan.[26] While the modernization of the existing industrial infrastructure in the west might result in short-term improvements in capital–output ratios and overall economic growth, it is doubtful that the technological innovations envisioned by soviet planners in the coming decade—innovations that are hardly guaranteed and perhaps very unlikely—will make up for current and future labor deficits, or address the needs of the growing work force in the southern tier.

It is difficult to determine how the Gorbachev leadership will attempt to solve the geographical imbalance, especially when in many ways the problem now appears dwarfed by ethnic unrest throughout the union and the very integrity of the union is in question. One thing is certain; conditions in Soviet Central Asia must be addressed and the Soviet leadership can no longer view Central Asia as a residual when it develops its national economic plan. The region needs a well-defined policy for its own economic development as well as for the utilization of its much needed labor resources in the national economy. The political stability as well as the economic prospects of the state depend on such a policy.

NOTES

1. It should be noted that this decline was relative and not absolute.

2. These rates, along with those for industrial labor productivity, were estimated by dividing the total indexed rates of growth by the number of years from one time period to the next.

3. The "non-Slavic south" refers here to the four Central Asian republics plus Kazakhstan, Armenia, Azerbaydzhan, and Georgia. Central Asia will account for over half the non-Slavic total, or about 40 percent of all new entrants.

4. Working age, according to the Soviet definition, differs for men and women. For men, 16–59 years is considered of working age, while for women it is 16–54. I have computed the 20–59 cohort to compensate for rising educational attainments and for delayed retirement, which has become more common in the last two decades.

5. Data were available for comparisons to be made by republic only for 1960–8 and 1968–75.

6. The "khozraschet" system of self-financing and accountability may have been the first sign of the loosening of ideological constraints described here. Some may argue that the Gorbachev reform package is a *de jure* if not a *de facto* annulment of the equalization precepts of Soviet development strategy.

7. More than 70 percent of Soviet hard currency is earned through the sale of energy and natural resources.

8. Here the Soviet "system" might work against itself in not allowing the economic conditions in rural Central Asia to decline enough to induce rural out-migration. Subsidies and transfer payments to Soviet Central Asia have apparently helped delay the out-migration of the indigenous population by artificially bolstering the local economy. Also, the kind of agricultural development in the region serves to inhibit rural out-migration. See Gleason (1986).

9. China is a unitary state whose government's actions are not circumscribed by a regionally and ethnically defined federal administrative structure. The Chinese crackdown on the student protests would have been more difficult to defend, justify, and sell to the Chinese people if the territorial administrative structure of the state was not unitary.

10. There is a considerable literature on the "equalization" of Soviet nationalities. See Silver (1974); Bahry and Nechemias (1982); Jones and Grupp (1984); Liebowitz (1987).

11. In Nagel and Olzak's (1982) competition model, ethnic identification is created or maintained as a basis for collective action when there are clear advantages attached to an ethnic identity. A reduction in equality among ethnic groups in multinational states leads to competition among ethnic groups, which, in turn, brings on ethnonationalist movements.

12. By the late 1970s, 19, 22 and 33 percent of the indigenous populations of Azerbaydzhan, Georgia, and Armenia were employed in industry, while corresponding figures for Uzbekistan, Turkmenistan, Tadzhikistan, and Kirgiziya were 16, 14, 16, and 12 percent (USSR 1988b: 24–5).

13. Estimated from 1979 and 1987 age data from USSR 1988d.

14. See DeBardeleben (1985); Darst (1988); Micklin (1988).

15. Absolute investments have increased at a decreasing rate since the early 1970s. Investments increased 27.5 percent from 1971–5 to 1976–80, and 17.6 percent from 1976–80 to 1981–5.

16. This cost-accounting system will force enterprises to be profitable or face closure. It leaves enterprises responsible for generating investments from profits, and will reduce state subsidies to unprofitable firms.

17. Capital investments ("kapital'nyye vlozheniya") include budgetary and nonbudgetary outlays from the state, cooperatives, and population for new construction, reconstruction, for the expansion and reequipment of existing industrial, agricultural, transportation, trade, and other enterprises, as well as outlays for the construction of housing, municipal service facilities, and facilities for rendering cultural and everyday services to the public.

18. The term "oblast level" refers to the tertiary-level administrative units (kray, ASSR, oblast) in the USSR, but the two small republics of Kirgiziya and Tadzhikistan were treated as tertiary units.

19. From 1971 to 1985, the oil and gas industry accounted for 67 percent of all industrial investments in Turkmeniya compared to 18 percent for the whole country and 20 percent for the Russian Republic.

20. Turkmen SSR, for example, increased its share of industrial to total investment from 33 to 44 percent between 1976–80 and 1981–5, and Tadzhikistan increased its share between 1966–70 and 1981–5, yet their improved shares are

still below the 1956–60 levels, and these percentages do not take into account these republics' great population growth.

21. Productive investments refer to capital allocated for industry, agriculture, forestry, construction, transportation, communications, state food trade, material-technical supplies and inventories, and computing services. Nonproductive investments refer to expenditures for housing, municipal services, and culture.

22. The profitability of Siberian gas and oil and other resources is determined by world market prices over which the Soviets have limited control.

23. The transcripts of the report could be read in FBIS (1989: 74).

24. For example, in Uzbekistan during the seventh, eighth, ninth, tenth, and eleventh five-year plans, agriculture received between 24 and 42 percent of total state investments. In comparison, the corresponding figures for the entire USSR did not exceed 20.3 percent. These figures do not include collective farm investments, which, if included, would increase the percentage of agricultural investment to well over 40 percent in most five-year plans in Uzbekistan since the mid-1950s.

25. The government has sponsored some limited pilot out-migration programs, including sending rural construction teams to the non-Chernozem zone of the RSFSR and establishing low-technology industrial and manufacturing activities in rural Central Asia (Gleason 1986). On the whole, the programs have been on a small scale and have not been significant in terms of inducing out-migration.

26. During the Moscow television report cited above, it was reported that the Politburo and the government adopted a program for the rapid socioeconomic development of Uzbekistan during the thirteenth five-year plan. The program is to create more than 150 industrial enterprises in the Fergana Valley alone, the majority of which are to be labor-intensive industries including electronics, and light industries, such as textiles, shoemaking, and food production.

REFERENCES

Bahry, Donna 1987. *Outside Moscow*. New York: Columbia University Press.

Bahry, Donna and Nechemias, Carol 1981. Half full or half empty? The debate over Soviet regional equality. *Slavic Review* 40 (September) 366–83.

Baldwin, Godfrey 1980. *Population Projections by Age and Sex for the Republics and Major Economic Regions of the USSR: 1970 to 2000*. US Bureau of the Census, International Population Reports, Series P-91, No. 26. Washington, DC: US Government Printing Office.

Bandera, V. N. 1973. Introduction. In V. N. Bandera and Z. L. Melnyk (eds), *The Soviet Economy in Regional Perspective*, pp. XVII–XXI. New York: Praeger.

Connor, Walker 1972. Nation-building or nation-destroying? *World Politics* XX: 319–55.

Darst, Robert G. Jr 1988. Environmentalism in the USSR: the opposition to the river diversion projects. *Soviet Economy* 4, 30: 253–74.

DeBardeleben, Joan 1985. *The Environment and Marxism-Leninsm: The Soviet and East German Experience.* Boulder, Colo.: Westview Press.

Dienes, L. 1982. The development of Soviet regions—economic profiles, income flows, and strategies for growth. *Soviet Geography: Review and Translation* 23 (May): 205–44.

Dienes, L. 1983. Regional economic development. In A. Bergson and H. Levine (eds), *The Soviet Economy: Toward the Year 2000*, pp. 218–58. London: Allen & Unwin.

Dienes, L. 1987. *Soviet Asia: Economic Development and National Policy Choices.* Boulder, Colo.: Westview Press.

Dyker, David 1983. *The Process of Investment in the Soviet Union.* Cambridge: Cambridge University Press.

Dyker, David 1985. *The Future of Soviet Planning.* New York: M. E. Sharpe.

Foreign Broadcasting Information Service 1989. *FBIS Daily Report, Soviet Union*, 3 July: 71–2, 74.

Gillula, James 1978. Regional inerdependence and economic development in the USSR. PhD dissertation, Department of Economics, Duke University.

Gleason, Gregory 1986. *Migration and Agricultural Development in Soviet Central Asia.* Kennan Institute for Advanced Russian Studies Occasional Paper No. 218. Washington DC: Wilson Center.

Gordijew, I. and Koropeckyj, I. S. 1981. Ukraine. In I. S. Koropeckyj and Gertrude E. Schroeder (eds), *Economics of Soviet Regions*, pp. 267–99. New York: Praeger.

Hechter, Michael 1975. *Internal Colonialism: The Celtic Fringe in British National Development, 1536–1966.* Berkeley, California: University of California Press.

Hewett, ed 1985. Gorbachev's economic strategy: a preliminary assessment. *Soviet Economy* 1: 285–305.

Huzinec, George 1977. A reexamination of Soviet industrial location theory. *Professional Geographer* 29 (November): 259–65.

Jones, Ellen and Grupp, Fred 1984. Modernization and equalization in the USSR. *Soviet Studies* 36 (April): 159–84.

Kistanov, V. V. 1981. *Territorial'naya Organizatsiya Proizvodstva.* Moscow: Ekonomika.

Koropeckyj, I. S. 1972. Equalization of regional development in socialist countries: an empirical study. *Economic Development and Cultural Change* 21 (October): 68–84.

Leggett, Robert and Kellogg, Robert 1988. The Soviet Union: an economy in transition and its prospects for economic growth. In Ronald D. Liebowitz (ed.), *Gorbachev's New Thinking: Prospects for Joint Ventures.* Cambridge, Mass.: Ballinger.

Lewis, Robert A. 1983. Regional manpower resources and resource development in the USSR: 1970–1990. In Robert Jensen, Theodore Shabad, and Arthur K. Wright (eds), *Soviet Resources in the World Economy*, pp. 72–97. Chicago: University of Chicago Press. Washington, DC: AAG.

Liebowitz, R. D. 1985. The spatial and ethnic dimensions of Soviet regional investment: 1956–75. PhD dissertation, Department of Geography, Columbia University.

Liebowitz, R. D. 1987. Soviet investment strategy: a further test of the "equalization hypothesis." *Annals of the AAG* 77, 3: 396–407.

Lubin, Nancy 1984. *Labour and Nationality in Soviet Central Asia: An Uneasy Compromise.* Princeton, NJ: Princeton University Press.

Melnyk, Z. L. 1977. Capital formation and financial relations. In I. S. Koropeckyj (ed.), *The Ukraine within the USSR: An Economic Balance Sheet*, pp. 268–98. New York: Praeger.

Micklin, Philip 1988. Desiccation of the Aral Sea: a water management disaster in the Soviet Union. *Science* 241 (September): 1170–5.

Nagel, Joane and Olzak, Susan 1982. Ethnic mobilization in new and old states: an extension of the competitive model. *Social Problems* 30: 127–43.

Nechemias, Carol 1980. Regional differentiation of living standards in the RSFSR: the issue of inequality. *Soviet Studies* 32 (July): 366–78.

Newton, Francis 1976. Soviet Central Asia: economic progress and problems. *Middle Eastern Studies* 12, 3: 87–104.

Pipes, Richard 1968. *The Formation of the Soviet Union.* New York: Atheneum.

Rodgers, Allan 1974. Location dynamics of Soviet industry. *Annals of the AAG* 64 (June): 226–40.

Schiffer, Jonathan R. 1989. *Soviet Regional Economic Policy: The East–West Debate over Pacific Siberian Development.* New York: St Martin's Press.

Silver, Brian 1974. Levels of sociocultural development among Soviet nationalities: a partial test of the equalization hypothesis. *American Political Science Review* 68 (December): 1618–37.

Taaffe, Robert 1984. The conceptual, analytical and planning framework of Siberian development. In George Demko and Roland Fuchs (eds), *Geographical Studies on the Soviet Union: Essays in the Honor of Chauncy Harris*, pp. 157–87. Chicago: University of Chicago Press.

United States, Central Intelligence Agency 1982. *Soviet Statistics on Capital Formation: A Reference Aid.* Washington, DC: CIA.

USSR, various years. *Narodnoye Khozyaystvo SSR.* Moscow: Statistika; and *Narodnoye Khozyaystvo* yearly series for each republic, 1956–88. Published in republic capitals.

USSR, various years. *Narodnoye Khozyaystvo* volumes for several oblasts between 1956 and 1975. Published in oblast or republic capitals.

USSR 1977. *Konstitutsiya (osnovnoy zakon) Soyuza Sovetskikh Sotsialisticheskikh Respublik.* Moscow: TsK VLKCM.

USSR 1983. *Narodnoye Khozyaystvo Uzbekskoy SSR v 1982.* Tashkent: Uzbekistan.

USSR 1986. *Narodnoye Khozyaystvo SSSR v 1985g.* Moscow: Finansy i Statistika.

USSR 1987. *Narodnoye Khozyaystvo SSSR za 70 Let.* Moscow: Finansy i Statistika.

USSR 1988a. *Kapital'noye Stroitel'stvo SSSR.* Moscow: Finansy i Statistika.

USSR 1988b. *Narodnoye Khozyaystvo Uzbekskoy SSR v 1987.* Tashkent: Uzbekistan.

USSR 1988c. *Narodnoye Khozyaystvo SSSR v 1987g.* Moscow: Finansy i Statistika.

USSR 1988d. *Naseleniye SSSR 1987.* Moscow: Finansy i Statistika.

USSR 1988e. *Promyshlennost'.* Moscow: Finansy i Statistika.

USSR 1988f. *Trud v SSSR.* Moscow: Finansy i Statistika.

USSR 1989a. *Ekonomicheskaya Geografiya SSSR.* Moscow: Izdatel'stvo Moskovskogo Universiteta.

USSR 1989b. *Narodnoye Khozyaystvo SSSR v 1988g.* Moscow: Finansy i Statistika.

Wagener, Hans-Jurgen 1973. Rules of location and the concept of rationality: the case of the USSR. In V. N. Bandera and Z. L. Melnyk (eds), *The Soviet Economy in Regional Perspective*, pp. 62–102. New York: Praeger.

Wheeler, Geoffrey 1975. The Russian presence in Central Asia. *Canadian Association of Slavicists* 17, 2–3: 189–201.

Ziegler, Charles, E. 1985. Nationalism, religion, and equality among ethnic minorities: some observations on the Soviet Case. *Journal of Ethnic Studies* 13, 2: 19–32.

Chapter 6 ———————————————————————————

Agricultural Change, Labor Supply, and Rural Out-Migration in Soviet Central Asia

Peter R. Craumer

Agriculture in Central Asia and southern Kazakhstan plays a special role within both the region and the agricultural system of the Soviet Union. No other major part of the country is so dependent upon agriculture for its livelihood. Not only does 59 percent of the population still live in rural areas, working mainly in agriculture, but within the industrial sector industries devoted to supplying the agricultural sector with its tools of production and to processing its output predominate.

As a result of its long growing season, Central Asia has traditionally been assigned a role of supplier of subtropical crops to the markets in the rest of the country. Central Asia and Kazakhstan provide 91 percent of all of the Soviet Union's cotton, 47 percent of the rice, and 35 percent of the fruits, grapes, vegetables, and melons (Voropayev 1984; USSR, Goskomstat 1988d: 149, 151).

Central Asia's vital agricultural sector, however, is increasingly beset by a number of serious problems, including rapid rural population growth, increasing limits on irrigation water, decline in soil quality, and pollution of water supplies. These problems, which threaten the productivity of both agricultural land and labor, may force substantial changes in the structure of Central Asian agriculture.

The purpose of this chapter is to examine the changes taking place in Central Asian agriculture in the postwar period, especially in the period beginning in the 1960s. The chapter will first briefly assess problems in irrigation and water resources, changes in agricultural land use and the cotton economy, and problems of food supply, before focusing on what is probably the leading problem of Central Asian agriculture, that of surplus labor. The labor problem will be analyzed in relation to rural population growth, levels of mechanization, labor productivity and wages, and prospects for relieving the labor surplus by rural out-migration. Finally, the changing role of the private agricultural sector in rural incomes will also be assessed to determine its possible effect on migration.

Study of Central Asian agriculture on a regional basis should ideally include not only the four republics of Central Asia, but also the southern oblasts of Kazakhstan. Although Kazakhstan is not classified by the Soviets as part of Central Asia, the irrigated agriculture in the oases of southern Kazakhstan and the extensive livestock grazing on desert and mountain pastures make this region similar in many ways to the other four republics. Unfortunately, most of the agricultural data used here, especially on labor resources and private agriculture, are not readily available below the republic level within Kazakhstan. The extensive dryland grain farming of the northern part of Kazakhstan is so different from the southern oblasts that using data for all of Kazakhstan is generally meaningless in describing the conditions of that irrigated zone. Only in cases where certain types of production are found only in the south (most importantly, cotton) can all-republic data be used. Therefore, Kazakhstan will be included here only when those crops are discussed or when data for the south can be assembled by aggregating oblast statistics.

Study of spatial change in Central Asian agriculture is also hampered by the scarcity of oblast-level time-series data for republics other than Uzbekistan. Most available data are limited to basic land-use or crop and livestock inventories, and the frequent administrative reorganizations have made comparison over long time periods for comparable oblasts nearly impossible. Nevertheless, analysis of oblast-level change in rural population, sown area, grain and cotton production, livestock production, and irrigated area is possible for most republics since about 1965 or 1970. Other agricultural subjects, including labor, will have to be examined here at the republic level only. Finally, data by oblast are also limited to the period up through 1987 or 1988; more recent data, including those of the 1989 census, represent a quite different set of oblasts, because of several major administrative reorganizations beginning in 1988 or 1989 in all of these republics.

OVERVIEW OF CENTRAL ASIAN AGRICULTURE

The agricultural heart of Central Asia and southern Kazakhstan is a discontinuous chain of irrigated oases, extending from south of Lake Balkhash in Taldy-Kurgan Oblast of Kazakhstan southwestward along the base of the Tien Shan, Pamirs, and Kopet Dag mountains almost to the Caspian Sea. With precipitation in the desert lowlands of the Aral and Balkhash basins ranging from less than 100 millimeters up to about

200 millimeters, little crop production is possible without irrigation. Most of the soils in these oases are sierozems or meadow soils, although patches of salinized solonetz and solonchak soils are widespread, especially in the areas most recently put into irrigated agriculture.

Along the base of the mountains especially on the northern side of the ranges, extensive deposits of loess, formed from windblown silt carried from the deserts of the Aral–Balkhash lowlands, provide a fertile base for the only large area of nonirrigated cropping in the region. Here the rainfall is high enough (more than 200 millimeters) and the evapotranspiration at these elevations of generally 1,000–1,600 meters low enough for dryland farming, most importantly of wheat. Orchards, vineyards, and vegetable production are also important in this zone, which includes about 80 percent of all dryland farming in Central Asia. Samarkand and Kashkadar'ya oblasts in Uzbekistan, Osh Oblast in Kirgiziya, and the rayons of republic subordination of Tadzhikistan contain the bulk of this dryland agriculture (*Sredneaziatskiy . . .* 1972: 164).

In the vast deserts of the lowlands, agriculture is restricted mainly to extensive grazing, with sheep, karakul, and goats. Grazing is also the main agricultural activity in the mountains of Central Asia above about 1,600 meters in elevation. Many of these mountain pastures are used only for summer grazing, while cattle and sheep are wintered on lowland desert pastures.

Map 6.1 Irrigated area, 1988.
Source: modified from Trochenov et al. 1983: 152.

IRRIGATION AND WATER RESOURCES

Agriculture in Central Asia is especially distinguished by its dependence upon irrigation (Map 6.1). In 1985, the region and southern Kazakhstan together contained about 41 percent of all irrigated land in the Soviet Union, with only 5.3 percent of the country's sown area (USSR, TsSU 1985: 208, 228; Kazakh SSR, TsSU 1985: 82, 99).

Ninety-two percent of the surface water resources of the region are in the Aral Sea drainage basin: Amudar'ya basin, 62 percent; Syrdar'ya basin, 30 percent; and smaller rivers of Kirgiziya and southern Kazakhstan such as the Chu and Talass (which are not in the Aral basin), 8 percent (Kes' et al. 1982). As recently as the late 1950s, most surface water for irrigation came not from the large rivers, the Amudar'ya and Syrdar'ya, but from smaller streams. About 15 to 17 percent or less of these two rivers' flow was used for irrigation (Lewis 1962).

Since the mid-1960s, however, expansion of irrigated lands has caused water use not only to equal, but to exceed, stream flow on most rivers, so that much of the water used for irrigation is drainage water from irrigation upstream. This near total use of water enabled the overall area of irrigated lands to grow by more than 3.1 million hectares, or about 60 percent, between 1950 and 1985 (Table 6.1), and another 493,000 hectares by 1988. Most of the new lands were in Uzbekistan and in Turkmeniya. Building of the Kara Kum Canal westward from the Amudar'ya beginning in 1954 transformed irrigated agriculture in Turkmeniya; the Kara Kum Canal now provides water for about 60 percent of all of the sown area in the republic (Mamedov 1984: 9). Most of the rest of the expansion of irrigation in the republic was along the lower Amudar'ya, where irrigated land area doubled between 1970 and 1986. In Uzbekistan, the long-irrigated areas of the Fergana basin (the single most important cotton-growing region of Central Asia) and the Hungry Steppe were joined by numerous projects extending westward from the Tashkent Steppe, and in Tadzhikistan and Kirgiziya the Vakhsh and Chu valleys received intensive irrigation development, adding to the already important irrigation projects those republics had in the Fergana basin. Not all of the oblasts of the region are dominated by irrigation, however. While nearly all croplands in Turkmeniya and the plains areas of Uzbekistan, Tadzhikistan, and Kirgiziya are irrigated, irrigated agriculture is only about a third of cropland in southern Kazakhstan.

Increases in newly irrigated lands in the region (including southern Kazakhstan) have slowly diminished since the early 1970s (Table 6.2), with the decline becoming steeper by the mid-1980s. The great capital intensiveness of irrigation developments and the increasing shortage of

Table 6.1   Area of irrigated land (thousand hectares).

	1950	1960	1965	1970	1975	1980	1985	1987	1988
Uzbek	2,276	2,571	2,639	2,696	3,006	3,476	3,930	4,109	4,149
Kirgiz	937	929	861	883	910	955	1,009	1,028	1,037
Tadzhik	361	427	468	518	567	617	653	675	686
Turkmen	454	496	514	643	819	927	1,107	1,224	1,258
S. Kazakh	1,130	1,191	1,129	1,187	1,310	1,478	1,568	1,654	1,630
Total	5,158	5,614	5,611	5,927	6,612	7,453	8,267	8,690	8,760

Sources: Kazakh SSR, TsSU, various years; USSR, TsSU, *Narodnoye Khozyaystvo SSSR*, various years; USSR, Goskomstat 1988d: 206, 1989c: 102.

suitable irrigable soils are important reasons for the decline. In the four Central Asian republics, capital investment in water reclamation projects peaked in 1981–5 at an average of 1.527 billion rubles per year, or 25.9 percent of all USSR water reclamation investment (USSR, Goskomstat 1988a: 66). This level of investment was less effective than smaller levels in the earlier years of irrigation development, because expansion of irrigation was on more marginal lands. By 1987, however, Central Asian water reclamation investment had declined by 12 percent, while investment in the rest of the country grew by 18 percent. Most of the decline in Central Asia was in Uzbekistan, which had accounted for 66 percent of Central Asian water reclamation funds in 1981–5.

Table 6.2   Area of new irrigated land (thousand hectares).

	1966–70	1971–5	1976–80	1981–5	1986	1987	1988
Uzbek	270	514	482	430	62	36	22
Kirgiz	36	45	54	70	10	12	6
Tadzhik	60	70	57	59	7	11	6
Turkmen	82	111	107	132	22	11	4
S. Kazakh	n.a.	169	159	117	31	27	14
Total	na.	909	859	808	132	97	52

Sources: Kazakh SSR, TsSU, various years; USSR, TsSU, *Sel'skoye Khozyaystvo SSR* 1971: 365; USSR, Goskomstat 1988d: 383, 1990: 148.

An even more important factor in the slower development of new irrigated lands was increasing shortages of water—exacerbated by the very low efficiency of irrigation water use—as the irrigated area grew. Although average withdrawals of water for irrigation in the Amudar'ya and Syrdar'ya basins declined from 15,500 cubic meters per hectare in 1970 to 11,700 in 1985 (Strategiya . . . 1988), the average efficiency of use of the withdrawn water remained by the early 1980s at about 60 percent, the lowest efficiency in the USSR (Micklin 1988). The low efficiency of Central Asian irrigation causes overuse, and diminution in both quality and quantity to other users downstream. Excessive water withdrawals from the Syrdar'ya for irrigation have caused river flow in the lower course of the river to cease in most recent years, beginning in the late 1970s (Akramov and Rafikov 1990: 59), thus depriving the irrigated lands of the delta of water. The salinity of the river throughout its course has been steadily increasing, especially in the lower reaches, where average salinity now approaches levels at which it can cause severe crop damage. The main cause of this increase is the return of saline drainage waters from irrigation to the river (Starodubtsev and Bogachev 1983: 90–3). In the last 15 years the seasonal salinity of the lower Amudar'ya has increased by more than two to four times, reaching 2.1 to 2.2 milligrams per liter periodically. The main cause of the increase in the Amudar'ya salinity is also saline drainage waters (Redzhepov and Ibragimova 1986: 27). At a level of about 3 milligrams per liter, use of the lower Amudar'ya will be unsafe for irrigation.

In 1987, a resolution of the CPSU Central Committee ("On the unsatisfactory use of the natural-economic potential of the agro-industrial complex in the Uzbek SSR, Tadzhik SSR, and Turkmen SSR"), referring to the severe problems of the current irrigation network in the region, directed the Central Asian republics to devote a minimum of 70 percent of all land reclamation funds to improvement of the existing systems (KPSS 1987b). Previously, the amount spent on improvement of irrigation systems, at least in Uzbekistan, was at most 10 percent of water reclamation investment (Akramov 1988). Rebuilding of the irrigation network will increase efficiency of water use by reducing filtration, enable better—including automated—control of water distribution, substitute more efficient methods of irrigation (such as drip irrigation), and provide adequate drainage systems to prevent waterlogging and secondary salinization.

Probably the most important impetus for the decision to emphasize increased efficiency of water use was the cancellation of the long-planned diversion of water from Siberian rivers to Central Asia and Kazakhstan (Micklin 1986, 1987). Without the prospect of new water supplies from the north, the only possible source of water for expansion of irrigation is conservation of already available water resources. In fact, arguments by some water resources experts that more efficient use of existing supplies in Central Asia could save as much water as would be diverted, and at lower cost, were instrumental in the decision to put the project in abeyance (Shabad 1986).

Measures to use local water supplied more efficiently may be cheaper than bringing water from Siberian rivers, but they will still be enormously expensive. Reconstruction of irrigation systems in the Aral Sea basin will cost at least 25 billion rubles, or an amount slightly greater than all Central Asian water reclamation investment from 1971 to 1987, and will require the temporary loss of about 0.5 million hectares of land per year, if it is done over 20 years (Sapayev 1987a and 1987b; USSR, Goskomstat 1988a: 66). Assuming that 1987 levels of investment in Central Asia are maintained, nearly all water reclamation investment in the four republics will be required in order to accomplish this, leaving very little left for developing new irrigated lands. Furthermore, these new irrigation systems tend to be in areas of poorer soils, which are often more subject to salinization, and require even more capital to build and maintain, than existing systems (Effektivnoye . . . 1987).

Prospects for expansion of the irrigated area are further limited, not only by the necessity to increase efficiency of current irrigation before water can be released for new irrigation, but also by the catastrophic condition of the Aral Sea. Rapidly declining water levels, collapse of the fisheries, salinization of the delta, and decline in water quality in the lower reaches of the Amudar'ya and Syrdar'ya are among the problems cited in the September 1988 Communist Party resolution on the Aral Sea (USSR, CPSU 1988). This resolution directs the appropriate governmental ministries and agencies to work out a plan to ensure that the water flow to the delta of the Amudar'ya and Syrdar'ya and the Aral Sea reaches at least 8.7 cubic kilometers a year in 1990, 11 in 1995, 15 to 17 in 2000, and 20 to 21 by 2005. Although part of this water will be drainage water, providing so much water through increase in the efficiency of irrigation systems will require at least a 15 percent reduction in water use per hectare by 1990 and a 25 percent reduction by 2000. Furthermore, the planned expansion of irrigated land in the Aral basin for the 1986–90 period will be cut back by 160,000 to 170,000 hectares, out of the original 1986–90 plan of 570,000 hectares in Central Asia and 410,000 hectares in all of Kazakhstan (Micklin 1987). Most importantly, the resolution states: "The construction of large-scale irrigation systems and the commissioning of new areas of irrigated land in the Aral Sea basin on the basis of the water resources of the Amudar'ya and Syrdar'ya river basins are to be halted from 1991" (USSR, CPSU 1988: 57). Further on it specifies that future agricultural development in the basin should be based "mainly" on more efficient use of water and land resources.

Based on this resolution, it appears that prospects for future expansion of the irrigated area are poor, at least in terms of major new irrigation schemes. Some reports on Central Asian irrigation issued after the resolution, however, have continued to stress the need for new irrigated lands, to keep up with the rapid population growth (Akramov 1988). If some expansion is allowed, it still will be constrained by the shortage of capital and the physical limits on water resources, regardless of how strictly the guidelines on the amount of water that is supposed to be sent to the Aral Sea are followed.

Salinization

Aside from the low efficiency of water use in most of the Central Asian irrigation systems, the biggest irrigation problem is the lack of adequate drainage systems and the associated high levels of soil salinity.

Salt problems occur because of both the high salt content of many irrigated soils, especially newly reclaimed ones, and the high level of salinity in irrigation waters. Newly irrigated soils often require high volumes of water to flush out accumulated salts, and the flushing process usually requires many years. As irrigated lands in Central Asia have expanded so rapidly in the past 30 years the new lands are increasingly more and more marginal ones with salt accumulations. Therefore, the larger irrigated area has, on the average, a poorer quality for irrigated agriculture. In addition, adequate drainage systems are required to keep the water table far enough from the surface (typically, below a depth of 2.5 to 3 meters) to prevent secondary salinization and waterlogging. Increasing shortages of water in Central Asia have caused a steady increase in the proportion of these drainage waters which are re-used for irrigation. High salt content of these drainage waters further adds salts to the soil and requires higher than normal use of irrigation water to keep salts flushed out of the soil. Furthermore, the declining flow of the Amudar'ya and Syrdar'ya—which barely reach the Aral Sea in most years now—means that nearly all of the salts which they once deposited in the sea are now deposited in the basins of the rivers, including on irrigated lands. In 1985–7, for example, the net accumulation of salts in the basins of these two rivers was about 20 to 25 million tons per year (Voropayev et al. 1988).

Of the Central Asian republics, Turkmeniya has suffered especially seriously from salinization and waterlogging of irrigated lands. Only 53 percent of the irrigated lands have any drainage systems, and 83 percent of that is low-efficiency surface drainage canals instead of covered horizontal or vertical drainage. Lands watered by the Kara Kum Canal are the most poorly drained—only 32.3 percent of those in Ashkhabad Oblast and 40 percent in Mary Oblast, while in Chardzhou and Tashauz oblasts the figures are 90 percent and 70 percent. As a result of the poor development of drainage systems, Turkmeniya had moderate to severe salinization on 37 percent of the irrigated arable lands in 1980, with about 190,000 hectares of cotton planted on salinized soils. Between 1976 and 1980 about 8,000 hectares in the republic were taken out of production, mainly in the Kara Kum Canal zone of Ashkhabad and Mary oblasts (Saparlyyev 1983: 21). Until the mid-1960s there was practically no drainage network in Turkmeniya, and other republics had low levels of drainage use also. Surveys of Central Asia on the extent of salinization estimated that, in the two republics where it is the greatest problem, Uzbekistan had 1.34 million hectares of salinized irrigated lands (57 percent of it moderately to strongly salinized) and Turkmeniya 0.36 million hectares (60 percent of it moderately to strongly salinized). In Turkmeniya the biggest problems were in Chardzhou and Tashauz oblasts, where irrigated lands were 75–85 percent salinized (Gerardi 1967: 17).

Development of vast new irrigation schemes in Central Asia since then has changed somewhat both the severity and the location of salinization. As discussed above, improvements in drainage in the Turkmen oblasts watered by the lower and middle course of the Amudar'ya and the development of new lands watered by the Kara Kum Canal without adequate drainage have shifted the problem somewhat to those Kara Kum Canal oblasts (Tedzhen and Murgab oases in particular). In Uzbekistan, soils in new irrigation developments in the Karshi Steppe, Hungry Steppe, Dzhizak Steppe, and Surkhandar'ya Oblast have been especially difficult to reclaim. They have retained large concentrations of salts and as a result yields have been below expectations. Amudar'ya delta lands in Khorezm Oblast are experiencing severe problems with secondary salinization, as are lands of the Chu Valley in Kirgiziya. Lands in oases where adequate drainage systems have been built, however, including the Tashkent and Samarkand regions and the Fergana basin, are in good condition. In Tadzhikistan it is mainly some newly irrigated soils which have been subject to salinization (Pankova et al. 1986: 144–5).

Therefore, the effects of salinization in Central Asia are very widely divergent from one oasis to another, causing great regional differentiation in its effect on crop yields. Slight conditions of soil salinity are estimated to cause about a 20 percent decline in cotton yields, while severe salinity may depress yields by up to 50 percent. In Uzbekistan the yearly losses to salinization are about half a million tons of cotton production (Imamaliyev 1979: 5), or the equivalent of about 9 percent of production overall, and in Tadzhikistan loss of yield is about 45,000 tons of cotton per year (Akramov et al. 1984: 34). Losses in Turkmeniya are proportionally greater than in either of these republics, however, based on the area of land affected by salinization (Table 6.3). From these figures, we can estimate that cotton losses from Turkmeniya's smaller crop may be of the order of three times the percentage lost in Uzbekistan, or more than 300,000 tons per year.

Comparison of the levels of salinization presented in Table 6.3 with

Table 6.3   Area of irrigated lands by level of salinization (percentage of irrigated lands).

	Severe	Moderate	Slight
Uzbek	14.9	45.8	39.3
Kirgiz	16.6	22.4	61.0
Tadzhik	3.6	31.6	64.8
Turkmen	46.0	43.4	10.6

Severe: more than 50 percent of land severely salinized.
Moderate: 20–50 percent of land severely salinized.
Slight: less than 20 percent of land either severely or moderately salinized.

Source: Pankova et al. 1986: 143.

those reported above for the mid-1960s are difficult because the definitions of salinization are slightly different between these two reports. Clearly, however, salinization as a major problem in Central Asian irrigated agriculture has not been significantly reduced.

Agricultural pollution

The problem of salinization of agricultural lands and of drainage waters is closely related to the pollution of water supplies by salts and by runoff from croplands. Most rural inhabitants of Central Asian irrigated areas receive their water from irrigation canals, which are increasingly polluted with salts, fertilizers, biocides, and other agricultural runoff or drainage waters (Maksumov 1988a; Ovezgel'dyyev 1988; Salakhitdinov 1988). High morbidity rates for typhoid, hepatitis, and gastrointestinal illness, along with high infant mortality, have been linked to this pollution, especially in locations downstream from other irrigated areas, such as parts of the Kara Kum Canal zone and along the lower Amudar'ya and Syrdar'ya (Bohr 1988; Shadimetov 1988; Elpiner 1990). Infant mortality rates in Karakalpakya increased 20 percent from 1980 to 1989 (Elpiner 1990), while in adjacent Tashauz Oblast infant mortality rose an astounding 43 percent from 1980 to 1986, reaching 76.3 before declining to 63.3 per thousand by 1988 (Turkmen SSR, TsSU 1988: 28). Improvements in reporting probably account for a part of this rise.

LAND USE AND THE COTTON ECONOMY

Most of Central Asia's agricultural lands are desert or mountain pastures for livestock, typically with very low productivity. Arable lands in the four republics in 1987 ranged from only 3.6 percent of all agricultural lands in Turkmeniya to 20.2 percent in Tadzhikistan (USSR, Goskomstat 1988d: 49–50), but they produced all but a very small part of the value of agricultural output. Within the arable land base, emphasis has been shifting steadily toward use of irrigation, both to expand the area of cultivation and to increase productivity and reduce the effects of droughts.

Changes in the sown area since the 1960s have been mainly the result of this irrigation development (Tables 6.1, 6.4). Units which had the greatest increases in sowings during 1970 to 1986 are those where major new irrigation projects were implemented: Turkmeniya (along both the Kara Kum Canal and the Amudar'ya), Karakalpak ASSR, Khorezm Oblast, Surkhandar'ya Oblast, Syrdar'ya Oblast, and Naryn

Oblast (Map 6.2). Meanwhile, some older areas of irrigation such as the Fergana Valley, Tashkent Oblast, Chimkent Oblast, and Dzhambul Oblast had relatively little new land development.

Table 6.4 All sown area (thousand hectares).

	1940	1950	1960	1965	1970	1980	1985	1988
Uzbek	3,036.5	2,804.0	3,038.3	3,336.2	3,476.0	3,994.6	4,080.2	4,349.0
Kirgiz	1,055.5	1,061.0	1,195.8	1,170.1	1,264.2	1,271.8	1,292.0	1,314.0
Tadzhik	807.1	837.0	727.4	764.7	764.9	763.6	803.0	848.0
Turkmen	411.0	368.0	446.0	517.0	636.0	895.0	1,028.0	1,243.0
S. Kazakh	n.a.	n.a.	n.a.	3,318.5	3,579.1	4,062.8	3,875.8	3,865.0
Total	n.a.	n.a.	n.a.	9,106.5	9,720.2	10,987.8	11,079.0	11,619.0

Sources: Kazakh SSR, TsSU, various years; USSR, TsSU, *Narodnoye Khozyaystvo SSSR*, various years; USSR, Goskomstat 1988d: 60–1.

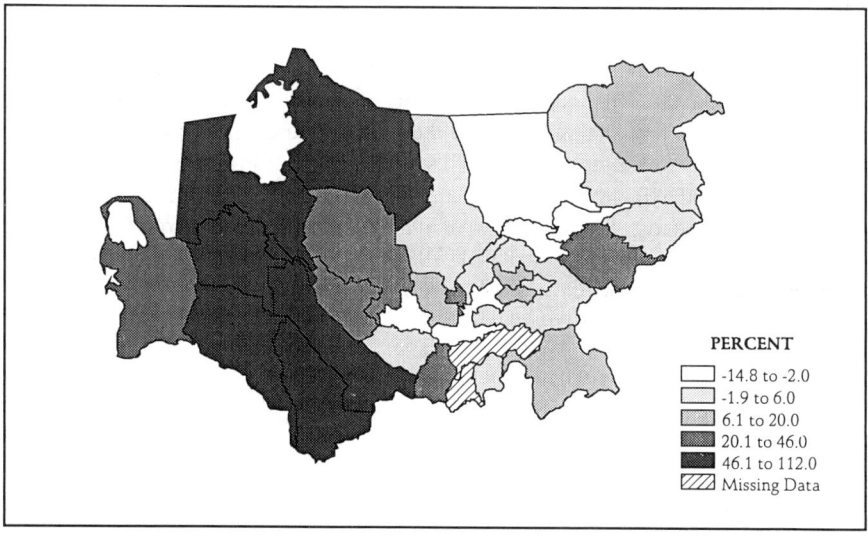

Map 6.2 Change in sown area, 1970–86.
Sources: USSR, Goskomstat 1988d: 60–1; TsSU and Goskomstat of Central Asian republics and Kazakhstan, *Narodnoye Khozyaystvo*, various years.

The process of irrigation expansion greatly reduced the importance of nonirrigated (dryland) sowings in total agricultural production. Dryland crops had occupied 31.1 percent of sown area in 1970, but they declined to 17.4 percent in 1986 (56 percent to 34 percent including southern Kazakhstan). In 1986, irrigated sowings as a percentage of all sown area were: Turkmeniya, 97.8; Uzbekistan, 85.8; Tadzhikistan, 71.7; Kirgiziya, 65.4; and southern Kazakhstan, 35.1. Furthermore, dryland cropping declined not only relatively but also absolutely in every republic, from 1.91 million hectares in 1970 to 1.25 million in 1986 (4.47 million to 3.74 million hectares including southern Kazakhstan).

Grain sowings have also declined dramatically across the region. The grain area of the four republics in 1986 was about half of the 1940 level, and only 72 percent of the area that had been planted in 1970 (Table 6.5). Therefore, as the sown area increased the importance of grain dropped, from 57 percent of sown area in 1940 to 35 percent in 1970 and 22 percent in 1986.

Table 6.5   Grain sowings (thousand hectares).

	1940	1950	1960	1965	1970	1980	1985	1986
Uzbek	1,479.7	1,102.0	894.8	1,252.6	1,159.8	1,173.8	969.3	700.4
Kirgiz	777.9	704.0	593.0	606.7	582.8	552.5	533.7	535.6
Tadzhik	567.4	552.0	360.6	397.0	320.5	195.0	210.0	151.0
Turkmen	183.0	128.0	71.0	133.0	84.0	132.0	143.0	164.0
S. Kazakh	n.a.	n.a.	n.a.	2,237.1	2,258.8	2,474.0	2,260.8	2,150.9

Sources: Kazakh SSR, TsSU, various years; USSR, TsSU, *Narodnoye Khozyaystvo SSSR*, various years; USSR, Goskomstat 1988d: 66–7.

In some regions, such as the Fergana Valley and most of Turkmeniya, by 1986 grain occupied only 10 percent or less of the sown area. By 1988, however, although grain area had increased dramatically, it remained at only two-thirds of the 1940 level in the four republics. Changes in the type of grain grown—away from spring wheat and toward winter wheat, barley, and corn—helped to compensate for the decline in grain area by replacing lower-yielding with higher-yielding grains, and the replacement of dryland with irrigated cropping also helped boost grain output. The 1986–7 yearly grain production in the four republics was 58 percent higher than it had been in 1971–5. Unfortunately, most of the increase was in the 1970s; yields in Tadzhikistan and Kirgiziya have continued to increase in the 1980s, but Uzbekistan and Turkmeniya made most of their advance in the second half of the 1970s and have leveled off or declined since then.

The other leading feed crop of Central Asia, fodder, has shown large increases in recent years, with a 76 percent larger area in 1986 than in 1970. Fodder production, however, which made significant gains in the 1970s, also began to level off in the 1980s.

In spite of some shift of grain and fodder production to irrigated lands, the big change during the period of irrigation expansion has been in cotton production. Cotton was only 26.7 percent of the sown area in Central Asia (without southern Kazakhstan) in 1950, but had increased to 34.3 percent by 1960 and to 39.7 percent by 1970 (Tables 6.4, 6.6; Map 6.2). After 1970, however, the proportion of cotton stabilized (while the area grew rapidly), increasing only slowly to 40.4 percent in 1988, although it was an unusually high 42.5 percent in 1986. Since virtually all cotton in Central Asia is irrigated, cotton occupied about 47 percent of

Central Asian irrigated sowings in 1987, but 53 percent in Turkmeniya, 56 percent in Tadzhikistan, and 60 percent in Uzbekistan. In some oblasts the percentage is even higher: 64–72 percent in the Fergana Valley of Uzbekistan, 70 percent in Bukhara Oblast, 68 percent in Mary Oblast, and 65 percent in Tashauz Oblast. Individual rayons in the Fergana Valley have up to 85–90 percent of their sown area in cotton (Inoyatov 1988).

Table 6.6 Cotton sowings (thousand hectares).

	1940	1950	1960	1965	1970	1980	1985	1988
Uzbek	923.5	1,098.0	1,386.6	1,549.9	1,709.2	1,877.7	1,989.8	2,017.0
Kirgiz	64.0	65.0	71.2	73.0	74.9	76.3	28.4	32.0
Tadzhik	106.1	126.0	172.4	228.0	254.0	308.5	311.0	320.0
Turkmen	150.0	153.0	222.0	257.0	397.0	508.0	561.0	636.0
S. Kazakh	101.8	96.9	105.6	112.1	117.5	126.5	131.0	128.0
Total	1,345.4	1,538.9	1,957.8	2,220.0	2,552.6	2,897.0	3,021.2	3,133.0

Sources: Kazakh SSR, TsSU, various years; USSR, TsSU, *Narodnoye Khozyaystvo SSSR*, various years; USSR, Goskomstat, 1988d: 74.

Although cotton does dominate agricultural production in the region as a whole, many parts of Central Asia and southern Kazakhstan are not cotton-growing regions. Of the 33 oblast-level units in 1986, 10 had no cotton production (Map 6.3). In southern Kazakhstan, only Chimkent Oblast produces cotton, and in Kirgiziya cotton growing is restricted to the margins of the Fergana basin, in Osh Oblast. Without

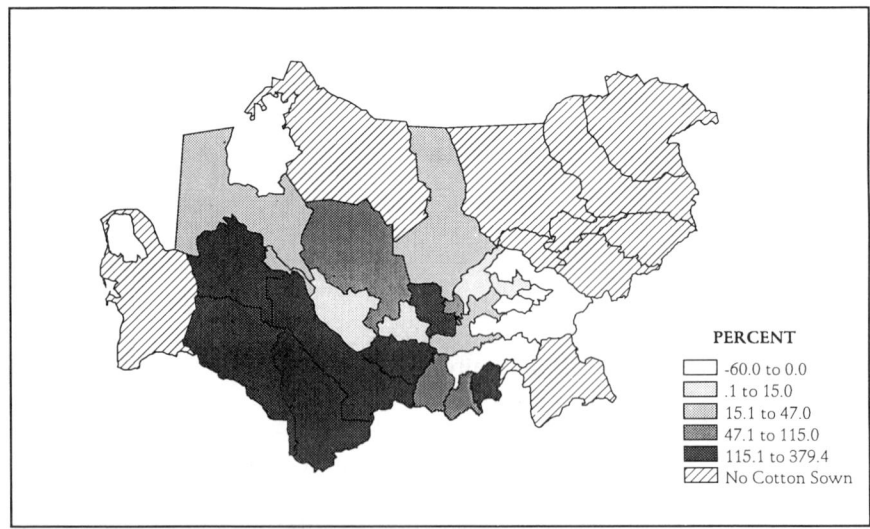

PERCENT

-60.0 to 0.0
.1 to 15.0
15.1 to 47.0
47.1 to 115.0
115.1 to 379.4
No Cotton Sown

Map 6.3 Change in cotton sowings, 1965–86.
Sources: TsSU and Goskomstat of Central Asian republics and Kazakhstan, *Narodnoye Khozyaystvo*, various years.

cotton production, these regions often have very different agricultural problems from the cotton-dominated oblasts that are the focus of much of this chapter.

The great growth in cotton production in Central Asia has occurred increasingly through expansion of irrigated area devoted to cotton, rather than through gains in yields, especially since the 1970s (Table 6.7). From the early 1960s to the early 1970s, large advances in cotton yields occurred in all republics, although a slight downturn in yield was already noticeable in Turkmeniya. By the late 1970s, however, yields had nearly leveled off in most republics, while the decline in Turkmeniya continued. Overall, cotton yields in the 1980s in the region were generally as low as or lower than they had been 10 or 15 years earlier.

Table 6.7   Cotton yields in Central Asia (centners per hectare).

	1961–5	1966–70	1971–5	1976–80	1981–5	1986–8
Uzbek	21.9	25.1	28.5	29.4	26.7	24.7
Kirgiz	20.6	23.5	27.6	28.3	19.1	23.9
Tadzhik	24.2	27.1	30.7	30.7	29.8	28.8
Turkmen	17.8	23.9	23.1	22.4	21.4	19.6
S. Kazakh	n.a.	n.a.	26.6	27.0	23.3	25.2

Sources: *Narodnoye Khozyaystvo* of Central Asian republics, various years; USSR, TsSU, *Narodnoye Khozyaystvo SSSR*, various years; USSR, Goskomstat 1988d: 151.

Problems with land quality as a result of massive influxes of poorly reclaimed salinized land into the land base may be a major cause of this stagnation, especially in Turkmeniya and Uzbekistan, where the greatest increases in new irrigated lands have taken place (Kurbangel'diyev 1987; Nikonov 1988). In addition to the decline in crop yields as a result of salinization of irrigated lands and irrigation waters, however, yields have also fallen due to decline of soil quality from the cotton mono-culture. In Uzbekistan, in the past ten years the share of long-irrigated lands which have high soil humus levels declined from 45 percent to 36 percent (Rashidov 1988); while in Turkmeniya, where the decline appears to be especially serious, the area of irrigated land with low humus levels increased from 58.1 percent in 1965–70 to 74.4 percent in 1986 (Dyuzhev 1988).

The declining humus levels are usually blamed on insufficient application of organic fertilizer and lack of crop rotations. Rotations in cotton-growing areas are most commonly cotton–alfalfa rotations, which adds nitrogen and organic matter to the soil, in addition to helping to control weeds, reduce plant diseases, and raise yields. The small percentage of fodder in the rotations also limits the supply of livestock feed, the size of livestock herds and therefore the supply of manure.

Lack of rotations has led to the development of cotton wilt, which has become a major factor in the declining yields. Almost 40 percent of cotton sowings in Uzbekistan are now affected by wilt, causing a yield loss of 20 percent in recent years (Inoyatov 1988). Inoyatov also claims that cotton losses due to insufficient rotations in the USSR from 1981 to 1986 were more than 6 million tons, along with a "loss" of 40 million tons of fodder which could have been produced.

Statistics on the use of cotton rotations, especially of the area of cotton which is in completed rotations, fluctuate so much from year to year and from source to source that it is difficult to determine exactly how widespread these rotations are. According to Khusanov (1989), the percentage of completed cotton–alfalfa rotations in Uzbekistan has not increased in the 1980s. A gauge of how far from completing cotton rotations Uzbekistan must be is provided by Rashidov (1988); in 1987 alfalfa was 14.5 percent of sown area, although 25–30 percent is needed to complete the cotton–alfalfa rotations.

In addition to the declining or stagnating yields of raw cotton, the output of cotton fiber per ton of raw cotton has also suffered. In the USSR (including Azerbaydzhan), cotton fiber was 32.3 percent by weight of the raw cotton in 1971–5, 30.6 percent in 1976–80, and 29.5 percent in 1981–5 (calculated from USSR, Goskomstat 1988d: 151–2). Every Central Asian cotton-producing republic experienced this decline, although the decrease in raw cotton quality was greatest in Kirgiziya, Uzbekistan, and Kazakhstan. In all republics, there was some improvement in the mid-1980s, but the cotton fiber percentage remained lower than the average level of the early 1970s. An important factor in stopping the steady decline in fiber yield from raw cotton may be the recent shift in method of cotton procurement; the price for purchased cotton is being changed to include not only the total tonnage, but also the fiber content of the cotton (Mogilevets 1987; Pestryakov 1988; Usmankhodzhayev 1988). Some specialists, however, argue that this may have a negative effect on the use of cotton combines, because machine-picked cotton is often of lower quality than that harvested by hand.

The great emphasis on cotton output led not only to the cotton monoculture, but also to the now famous cotton scandal, when it was revealed in 1984 that officials at all levels in Uzbekistan had participated in massive padding of figures on production and processing of cotton (Sheehy 1988). According to Ovcharenko (1988), the falsification of statistics caused Uzbekistan to be paid for nearly one million tons of nonexistent cotton a year, for a total loss to the state of more than 4 billion rubles. By comparing cotton production data published before the padding was uncovered with newly corrected data, Severin (1987) estimated that USSR raw cotton production figures were exaggerated by a total of nearly 6 million tons between 1976 and 1985, or 7 percent of the total reported. Not all of this padding was in Uzbekistan; in nearby Chimkent Oblast of Kazakhstan, for example, in 1979–84, 138,000 tons of raw cotton were falsely reported (Kolbin report . . . 1987), and during the last five-year plan in Tadzhikistan raw cotton production was padded by more than 700,000 tons (Ponomarev 1990).

In the wake of the scandal, it became obvious that if part of the ambitious cotton production plan was being fulfilled only on paper, the plan would have to be reduced if it were to be met. Furthermore, expansion of alfalfa in the rotations would have to come at the expense of some cotton production. Therefore, in 1985 the Uzbekistan raw cotton procurement plan was reduced by 300,000 tons to 5.7 million tons (Usmankhodzhayev 1985), followed by further reductions to 5.25 million tons by 1988 (Nazarov 1988). Unfortunately, the decline in planned procurement does not represent an actual easing of the pressure for the cotton monoculture, but only sets the level closer to the real production level that was being achieved while the figures were padded (Nishanov 1989). Further reductions will be needed to allow more alfalfa area in the rotations.

Food supplies and consumption

The effect that excessive emphasis on cotton production continues to have on local food production and consumption has become one of the leading social and agricultural issues in Central Asia (Fierman 1989; Shpolyanskaya 1989). Rapid population growth has created rising demand, at the same time that the emphasis of local agriculture has shifted away from food production and toward cotton production. In Uzbekistan, for example, food crops (including fodder for livestock) occupied 61 percent of sown area in 1950, but since 1970 they have declined to 47–50 percent of sowings (Khamrayev and Bedrinstev 1989: 5). According to Karchikyan (1977: 68), in 1975 all four republics of Central Asia had only between 81 percent and 86 percent of the USSR average per capita caloric consumption. Furthermore, from 1958 to 1975 per capita caloric consumption in three of the four Central Asian republics (except Tadzhikistan) declined as a percentage of USSR average consumption. The Central Asian diet also remained primarily starch-based; while in the Soviet Union the share of grain products and potatoes declined from 63.2 percent of calories in 1958 to 47.7 percent in 1975, in Central Asia by 1975 these starches still constituted 57 to 62 percent of the republics' food consumption, higher than anywhere else in the country (Karchikyan 1977: 78). Analysis by Stebelsky (1988) of food consumption data from 11 republics, including the Turkmen and Kirgiz SSRs, showed that in the 1960s and 1970s these Central Asian republics ranked at or near the bottom of republics in their per capita consumption of livestock products (meat, milk, and eggs). In the 1980s, all Central Asian republics remained at the bottom of the rankings in consumption of animal products (USSR, Goskomstat 1990: 102–3). Lack of fodder, due especially to the low percentage of alfalfa in cotton rotations and to low fodder and pasture yields, has been the primary limitation on increasing animal production.

Calculation of the relationship between food consumption, food production, and population in Uzbekistan from 1970 to 1987 shows that while there were very large increases in absolute production of most food products, in most cases production barely kept even with population growth, and for meat, potatoes, and vegetables the per capita consumption declined after 1980. Furthermore, except for meat, there was little or no increase in the percentage of consumption that was supplied by local production (Uzbek SSR, Goskomstat 1988). By 1987, Central Asia was a large net importer of most food products, except vegetables and fruits (USSR, Goskomstat 1989e: 163–81). Although cultural factors, the young age structure, and the warmer climate in Central Asia may certainly cause different demand for the type and amounts of food consumption here compared with other parts of the country, the great gap in per capita consumption between Central Asia and the entire USSR must primarily reflect lower food availability rather than lower demand. Therefore, the declines in a republic such as Uzbekistan in meat consumption may be a result of failure of the authorities to provide sufficient imports to keep up with population growth. According to the former general secretary of the Uzbek Communist Party (Nishanov 1989), plans are to double food production in the republic by the year 2000, which will make possible self-sufficiency in milk production but not in meat production. Rutgayzer and Artykov (1988) assert that decreasing the share of cotton in Turkmeniya to 35 to 40 percent of sowings, which would mean a decline of up to 0.9 to 1.0 million tons of cotton, would make possible raising meat and milk consumption per person by 15 to 30 percent, but would still leave consumption at less than 75 percent of the norms for those products. They conclude that satisfying demand by local production would require total cessation of cotton production.

Thus, with rapid population growth, increasing constraints on irrigation water supplies, problems maintaining land quality and improving yields, and, at the same time, pressure on Central Asia to continue its role as supplier of cotton not only to the country as a whole but also for export, it is unlikely that Central Asian demand for food products will be met in the near future.

POPULATION GROWTH AND CENTRAL ASIAN AGRICULTURE

The rapid rates of population growth in Central Asia and southern Kazakhstan are a primary reason why the region is increasingly unable to produce enough food to feed itself adequately. Furthermore, greater assertiveness by the growing number of both urban and rural Central Asians in demanding improved food supplies will place stress on the ability of the region to serve as a cotton plantation for the rest of the USSR and for export. In addition, population growth in rural areas in particular has important effects on the productivity of agricultural labor, levels of agricultural mechanization, and incomes of agricultural workers. All of these are key factors in enhancing or retarding migration from the rural areas of Central Asia. Although population growth and its social and economic effects are discussed elsewhere in this volume, here it is necessary to look in more detail at the rural population itself and its relationship to agricultural labor.

Rural population change

The phenomenal growth of the Central Asian population, especially the rural population, is probably the most important and widely known of all of the social and economic problems of the region. During the period of 1951 to 1989, when the rural population of the USSR declined by 10.7 million (9.8 percent), the rural population of Central Asia and southern Kazakhstan increased by 13.7 million, or 149 percent (USSR, Goskomstat 1988c: 16–33, 1989b). Growth rates varied widely across the region, however, with percentage increases in Kirgiziya and southern Kazakhstan about half or less of that in Tadzhikistan (Table 6.8). Variation among the oblast-level units was even greater; the rural population of Kzyl-Orda Oblast of Kazakhstan grew by only 23 percent, while Kashkadar'ya and Surkhandar'ya oblasts of Uzbekistan and Kurgan-Tyube Oblast of Tadzhikistan experienced rural population increases of more than 200 percent (Map. 6.4). Thus, the southern oblasts of Uzbekistan and the western parts of Tadzhikistan contrast sharply with the more moderate growth trends in eastern Kirgiziya and parts of southern Kazakhstan. Even the more slow-growing oblasts, however, gained rural population more rapidly

Table 6.8 Rural population (thousands).

	1951	1961	1971	1981	1987	Percent Change 1951–87	1981–7
Uzbek	4,459	5,686	7,696	9,452	11,052	+148%	+17%
Kirgiz	1,261	1,454	1,1867	2,235	2,497	+ 98	+12
Tadzhik	1,130	1,400	1,871	2,631	3,204	+184	+22
Turkmen	756	860	1,159	1,512	1,760	+132	+16
S. Kazakh	1,558	1,890	2,471	2,780	2,858	+ 83	+ 3
Total	9,164	11,290	15,064	18,610	21,371	+133	+15

Source: USSR, Goskomstat 1988c: 28–32.

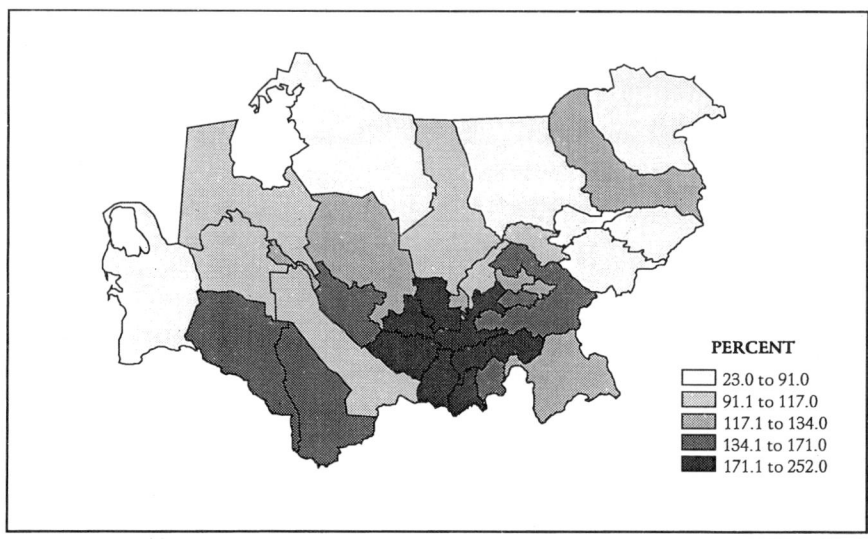

PERCENT

☐ 23.0 to 91.0
▨ 91.1 to 117.0
▧ 117.1 to 134.0
▩ 134.1 to 171.0
■ 171.1 to 252.0

Map 6.4   Rural population increase, 1951–87.
Source: USSR, Goskomstat 1988c: 28–32.

than all but a few regions outside of Central Asia (most notably, Azerbaydzhan).

When growth rates are calculated for subperiods since 1951 (Table 6.9), most oblasts show their most rapid increases during 1959–70, with an average oblast growth rate of 2.8 percent per year. This average oblast rate dropped to 2.2 percent during 1970–9 and then nearly leveled off at 2.1 percent in 1979–87. Among these oblasts, until recently only Naryn Oblast ever showed an actual period decline in rural population, from 1951 to 1959. Then, in 1970–9 and 1979–87, units in southern Kazakhstan such as Kzyl-Orda and later Dzhambul and Taldy-Kurgan began to experience rural population decline. Some units, however, had even higher rural growth rates in the 1980s than in the previous period: Dzhizak, Kashkadar'ya, Navoi, Samarkand, Surkhandar'ya, Karakalpak ASSR, Leninabad, and the Tadzhikistan rayons of republic subordination among them. The high growth rates in some of these most populous regions actually raised the overall (weighted average) Central Asia and south Kazakhstan rural growth rate from 2.3 percent in 1970–9 to 2.5 percent in 1979–87.

These rapid rural growth rates in most of the oblasts are the result of high rates of natural increase, as is discussed elsewhere in this volume. In 1988, the rural rate of natural increase in Kashkadar'ya and Surkhandar'ya oblasts of Uzbekistan and Khatlon Oblast and the rayons

Table 6.9   Rural population growth rates (percent per year).

	1951–9	1959–70	1970–9	1979–87
Uzbek				
Andizhan	2.4	2.7	2.0	1.6
Bukhara	3.4	3.4	2.3	1.8
Dzhizak	1.7	4.4	2.3	3.2
Fergana	2.0	3.0	2.6	2.3
Karakalpak	2.5	1.8	1.6	1.7
Khorezm	1.1	3.2	3.0	1.4
Kashkadar'ya	2.3	3.8	2.5	3.7
Namangan	2.1	2.9	2.0	2.0
Navoi	1.6	3.3	1.7	2.6
Samarkand	2.4	3.5	1.0	4.5
Surkhandar'ya	2.4	4.0	2.9	3.3
Syrdar'ya	2.1	2.4	2.7	2.1
Tashkent	4.1	1.5	1.5	1.5
Kirgiz				
Issyl-Kul	0.9	2.3	1.1	0.7
Naryn	−2.3	3.4	1.8	1.2
Osh	0.8	3.3	2.6	2.5
Talass	1.1	2.7	1.8	1.8
Kirgiz rayons	3.1	1.3	0.9	0.7
Tadzhik				
Gorno-Badakhshan	1.2	2.5	2.6	2.2
Kulyab	1.2	1.9	3.5	3.3
Kurgan-Tyube	3.5	3.0	4.2	3.4
Leninabad	1.9	3.0	2.9	3.1
Tadzhik rayons	1.8	3.2	3.1	3.5
Turkmen				
Ashkhabad	0.2	4.1	2.9	2.7
Chardzhou	1.3	2.5	2.4	2.2
Krasnovodsk	1.8	2.3	1.5	1.5
Mary	1.9	3.1	3.1	2.8
Tashauz	0.1	2.4	2.6	2.7
S. Kazakh				
Alam-Ata	3.5	2.7	1.9	0.5
Chimkent	2.0	2.9	2.0	1.3
Dzhambul	2.2	2.6	0.9	−0.2
Kzyl-Orda	0.6	2.2	−0.5	−0.4
Taldy-Kurgan	1.9	1.4	0.80	−0.5

Sources: *Narodnoye Khozyaystvo* of Central Asian republic and Kazakhstan, various years; USSR, Goskomstat 1988c: 28–32; 1959, 1970, and 1979 data are from the post-census estimates made by TsSU and Goskomstat to correct for numerous oblast boundary changes since the censuses.

of republic subordination in Tadzhikistan even exceeded 3.8 percent (USSR, Goskomstat 1989a: 107), a level higher than in all but a few developing countries.

Throughout the region, declines in natural increase can be the

primary cause of the slower rural growth rates between 1979 and 1987 in
a minority of the Central Asian oblasts, because only eight units
reported decreases in the rate of natural increase. The changing pattern
of rural population growth, therefore, must also be explained by the
trends in rural out-migration or in-migration.

The rapid population growth of rural Central Asia has maintained
such large rural populations that the relationship between rural and
urban population, conventionally expressed as the percent urban, can
best be expressed as the percent rural (Map 6.5). By 1987, only Kzyl-
Orda and Dzhambul oblasts in southern Kazakhstan, and Krasnovodsk
Oblast in Turkmeniya (an oblast with little agriculture) had larger urban
than rural populations. Furthermore, the "ruralness" of the region is
decreasing slowly in most oblasts, with Mary, Osh, Samarkand, and all
oblasts of Tadzhikistan even showing increases in the percent rural from
1979 to 1987. The effect of this ruralness is often cited in studies of the
low rates of rural out-migration in Central Asia, which note the presence
of a "rural outlook" in the population, fostered by a lack of contact
with urban influences (Mukomel' 1988). Equally important, the
paucity of nearby urban centers where produce from the private plots
may be sold, and the limited size of most of those markets, may
impede the generation of cash income from private agriculture in many
oblasts.

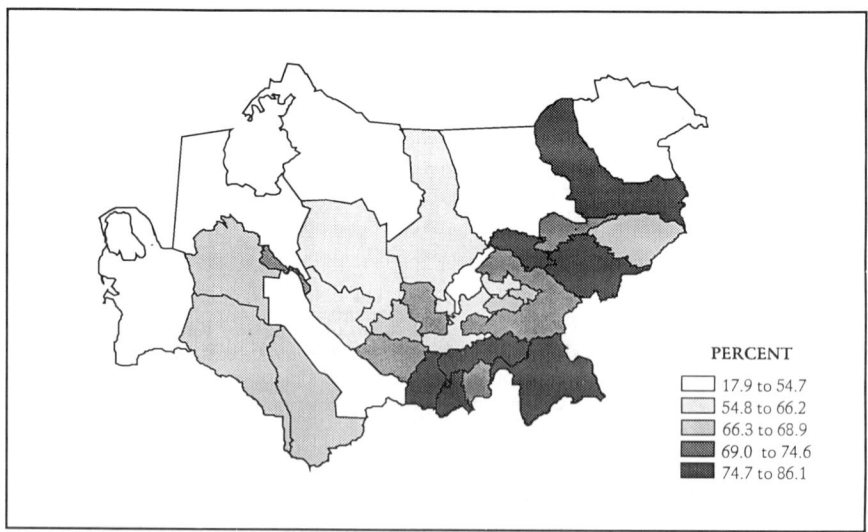

Map 6.5   Rural population as a percentage of total population, 1987.
Source: USSR, Goskomstat 1988c: 28–32.

Rural population growth and agricultural labor

Growth in work force. The overall rural population growth rates in Central Asia of 2.0 percent per year in 1951–9 and 2.9 percent in 1959–70 began to have a large effect on the work force only in the 1970s, when the youngest cohorts began to reach working ages. Although in 1959–70 the rural population growth rates in the four republics ranged from 2.7 to 3.0 percent per year, the growth in the rural population of working ages was only 0.8 to 1.3 percent per year. By 1970–9, however, when rural population growth rates had declined to 1.8 to 2.7 percent per year (except in Tadzhikistan, where they increased to 3.4 percent), the rural working-age population was growing at 3.3 to 4.9 percent per year (Table 6.10). Thus, the increase in the Central Asian potential rural labor pool of 0.5 million people from 1959 to 1970 was dwarfed by the 1.98 million added in 1970 to 1979, when the rural working-age population grew by 43 percent. Furthermore, in spite of declining growth rates of the working-age population in all of these republics between 1979 and 1987, as a result of the larger population base the potential rural labor pool grew by another 1.81 million (USSR, Goskomstat 1988c: 60–93). In a period of only 17 years, then, the Central Asian rural labor supply increased by 82 percent (virtually the same as the increase in urban areas of 81 percent), while the rural labor pool in the other republics together decreased by 12 percent. The beginning of this period was also the point at which labor surpluses started to be felt in Central Asian agriculture (Orazmuradov 1988).

Table 6.10   Growth in rural working-age
population (percent per year).

	1959–70		1970–9		1979–87	
	(a)	(b)	(a)	(b)	(a)	(b)
Uzbek	1.1	3.0	3.9	2.1	3.0	2.5
Kirgiz	1.3	2.7	3.3	1.8	2.2	1.8
Tadzhik	0.8	2.8	4.9	3.4	3.8	3.2
Turkmen	1.2	2.9	4.4	2.7	3.2	2.5
Central Asia	1.1	2.9	4.0	2.3	3.0	2.5

(a) growth in working ages.
(b) growth in all rural population.

Sources: calculated from USSR, Goskomstat 1988c: 60–93;
and sources, Table 6.9.

As the leading employer, agriculture was forced to absorb much of this huge increase in the rural work force (Table 6.11). Although average yearly agricultural employment decreased slightly in the 1950s (except in Turkmeniya), and rose only about 11 percent in the 1960s, it increased by 12.5 percent in 1970–5, 11.3 percent in 1975–80, and another 10.7

percent by 1987. In the period since 1970, the most rapid growth of agricultural labor was in Turkmeniya (57 percent), followed by 38 to 39 percent in Uzbekistan and Tadzhikistan, and a relatively slow (only by comparison) 24 percent in Kirgiziya. From 1985 to 1987, however, the size of the work force appeared to stabilize or even decline slightly.

Table 6.11   Workers in agriculture (average yearly number, thousands).[1]

		1950	1960	1970	1975	1980	1985	1987
Uzbek	all	1,416	1,261	1,406	1,588	1,782	1,957	1,946
	kolkhoz %	94	78	73	66	56	54	53
Kirgiz	all	303	282	331	347	370	426	410
	kolkhoz %	90	75	65	59	49	46	46
Tadzhik	all	338	318	323	361	400	454	450
	kolkhoz %	96	92	81	73	60	54	56
Turkmen	all	235	239	267	323	363	423	419
	kolkhoz %	95	92	91	91	81	81	80
Central Asia	all	2,292	2,099	2,327	2,619	2,914	3,260	3,226
	kolkhoz %	94	82	75	69	59	57	56

[1] The sum of yearly average number of workers per month (who work at least one labor day in a month) on kolkhozes and blue- and white-collar workers in agriculture, includes all workers engaged in agriculture, but, for non-kolkhoz workers, excludes employees of agricultural enterprises who are engaged in nonagricultural activities. Totals may not add because of rounding.
Source: USSR, Goskomstat 1988f: 71–4, 76.

Overall, except for a modest decline in Kirgiziya (14 percent) and a large increase in Turkmeniya (38 percent), the number of kolkhozniks in Central Asia by 1987 was about the same as had worked in agriculture in 1970. The biggest change, then, in Central Asian agricultural employment occurred in the state sector. Conversion of kolkhozes into sovkhozes and formation of sovkhozes in areas of new lands (mainly new irrigation projects) caused state agricultural employment to grow rapidly, while collective farm employment stabilized. From 1970 to 1987 the number of state agricultural workers increased by 829,000, or 144 percent. Thus, by 1987 kolkhoz employment was little more than half of Central Asian agricultural labor, whereas it had been three-quarters of employment as recently as 1970.

In spite of the great expansion of the agricultural work force, it could not grow fast enough to keep up with the growth in rural labor supply, especially by the 1970s. Based on the growth in the average yearly number employed in agriculture (Table 6.11) and the growth in the rural working population (from the censuses), approximately 62 percent of the increase of the Central Asian rural work force must have been in agriculture in the 1960s, but this figure dropped to about 35 percent in the 1970s. After 1970, the percentage of rural labor that was employed in agriculture began to decline (Table 6.12). By 1987, only in Turkmeniya was more than half of the rural working-age population engaged in agriculture in the collective or the state sector.

Table 6.12 Agricultural workers (percentage of rural working-age population).[1]

	1960	1970	1980	1987
Uzbek	51	51	46	39
Kirgiz	45	46	38	35
Tadzhik	51	48	38	32
Turkmen	64	63	57	51
Central Asia	51	51	44	39

[1] Calculated by dividing the average yearly number of agricultural workers by the rural working-age population.

Sources: USSR, Goskomstat 1988c: 60–92 (rural working-age population), 1988f: 76 (number of kolkhoz workers), 71–4 (number of other agricultural—mainly Sovkhoz—workers estimated using number of blue- and white-collar workers in agriculture); 1960 and 1980 figures are estimates, based on population data for 1959 and 1979 (census years) but workers for 1960 and 1980.

Not only did the percentage of entrants into the rural work force who went into the state and collective agricultural sectors begin to decline, but overall participation rates in the public sector declined also. According to most reports, the share of rural labor that is employed only in domestic and private agricultural activities is growing (Kononenko 1988; Orazmuradov 1988). Sychev et al. (1989) claim that in 1971–5 in Central Asia about one-fourth of the growth in available labor did not go into the public sector, and about one-third in 1976–80. In Turkmeniya, the share of working population in domestic or private plot activities increased from 16.1 percent in 1970 to 25.3 percent in 1986 (Dvoryadkina 1988). In Uzbekistan, about 600,000 of the rural population of working ages do not work in the public sector (Valiyev 1988: 105); based on the 1987 rural working-age population, this must be about 12 percent (USSR, Goskomstat 1988c: 63). In some regions of Uzbekistan, however, about 20 to 25 percent of the available labor does not work in the public sector (Valiyev 1988: 80). The implications of this trend—either that the sovkhozes and kolkhozes are unable to employ all of this excess labor or that private agriculture is more lucrative—will be discussed further below.

Agricultural labor density. The ability of Central Asian agriculture to support about 900,000 more workers in 1987 than it had in 1970, and 1.1 million more than the level of 1960, occurred as the area of agricultural lands, especially irrigated cropland, was also expanding. Nevertheless, the labor density of the region's agriculture increased substantially (Table 6.13). The only exception to this pattern was the increases in both sovkhozes and kolkhozes in Turkmeniya, where the growth of irrigated land from the Kara Kum Canal and other new projects more than

doubled the area of irrigation. All other republics on the whole were
unable to increase the arable lands as fast as they were allowing
additional labor into the agricultural work force. In the 1980s, however,
continued increases in irrigated land and slower growth in agricultural
employment caused the steady decline in area per worker to ease
somewhat in most republics, or to even reverse slightly.

Table 6.13   Sown area per kolkhoz and sovkhoz worker (hectares).

		1965	1970	1975	1980	1985	1987
Uzbek	k	1.94	1.88	1.62	1.45	1.38	1.49
	s	3.92	3.63	3.28	2.70	2.39	2.59
Kirgiz	k	3.27	3.49	3.40	3.13	2.74	2.90
	s	3.21	3.21	2.79	2.86	2.68	2.79
Tadzhik	k	2.22	2.10	1.65	1.55	1.47	1.51
	s	3.31	3.03	2.42	2.18	2.06	2.07
Turkmen	k	2.11	2.27	2.44	2.35	2.29	2.77
	s	2.29	2.64	2.83	2.63	2.60	2.93

Sources: calculated from average yearly number of kolkhoz workers in public labor and
total number of sovkhoz workers, plus the area of public sowings in these farms; data from
Narkhoz of Central Asian republics, various years; 1987 data from USSR, Goskomstat
1988d: 445, 456.

Although labor data are not available by oblast for all of Central
Asia, the relationship between rural population growth and irrigated
area at the oblast level shows how widely labor density must vary within
these republics (Map 6.6). Especially in the older irrigated areas of
Central Asia, since 1970 rural population growth has outpaced irrigation
development. Regions such as the Fergana Valley, and Tashkent,
Surkhandar'ya, and Mary oblasts, plus Kirgiziya and Tadzhikistan, have
suffered up to 25 percent declines in the per capita availability of
irrigated lands. Meanwhile, in zones of new irrigation development
such as along the Kara Kum Canal, the lower Amudar'ya and Syrdar'ya
and the Karshi and Hungry steppes, expansion of irrigated land has
actually been faster than population growth. In fact, many of those
regions have long had shortages of agricultural workers. By 1986, less
than one-third of a hectare of irrigated land per person was available to
support the dense rural populations in the older irrigated zones, while
more than a hectare was available in the regions where irrigation
expansion had been the greatest in the 1970s and early 1980s.

The limited prospects for large increases in irrigated area in the near
future, because of the water resources constraints already discussed,
and the continuing high growth rates of the rural population, will cause
the density of rural labor to continue to increase, barring massive out-
migration. If the sown area stabilizes at a 10 percent greater area than in
1987, and the rural population growth rate remains at the 1979–87 level,
by the year 2000 the Central Asian (without southern Kazakhstan) rural
population in relation to sown area will increase about another 25

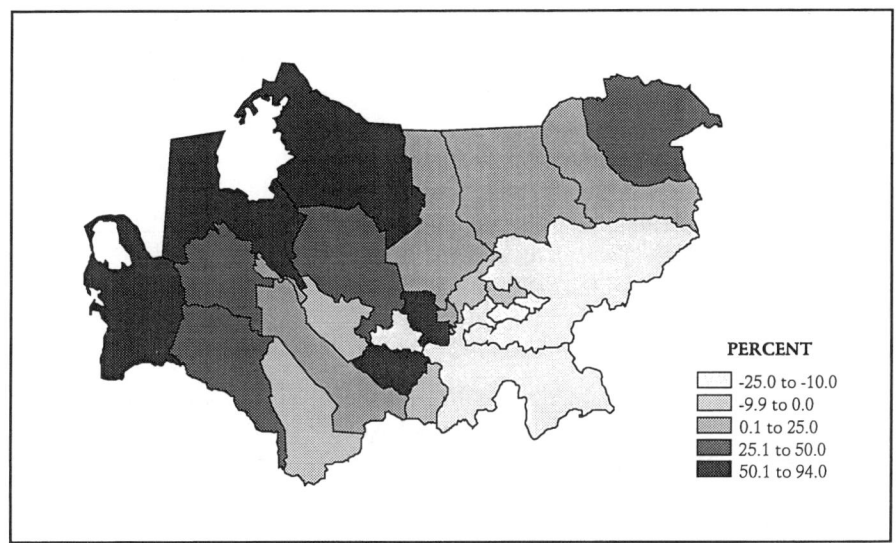

Map 6.6 Change in irrigated area per rural population, 1970–86.
Sources: USSR, Goskomstat 1988c: 28–32 (1986 population); TsSU and Goskomstat of Central Asian republics and Kazakhstan, *Narodnoye Khozyaystvo*, various years (irrigated land area and 1970 population, adjusted by TsSU and Goskomstat to 1986 borders).

percent. Even if fertility rates decline substantially, however, the rural labor force will continue to increase, because entrants to the work force in the year 2000 have already been born.

Labor productivity

Based on the value of agricultural output per worker, the productivity of the expanding work force in Central Asian agriculture has improved at a considerably slower rate than for the whole USSR (Table 6.14). By 1988, each worker in these four republics was producing only 23 to 42 percent more than in 1965, whereas the countrywide increase in agricultural labor productivity was 123 percent. In fact, during the 1981–5 period productivity per worker actually declined in all of these republics compared with the previous five-year period, and, in spite of improvement by 1988, productivity in that year remained below the level of 1980 in Uzbekistan and Tadzhikistan. Although these data are based on the average yearly number of workers participating in agriculture each month, and thus slightly exaggerate the amount of full-time labor, data on labor productivity by man-hour show virtually the same patterns (USSR, Goskomstat 1988d: 438–9).

Low labor productivity appears to be an especially serious problem

Table 6.14   Rate of growth in agricultural labor productivity (per worker; 1965 = 100).[1]

	1970	1975	1980	1985	1986	1987	1988
USSR	137	146	168	193	209	216	223
Uzbek	111	112	125	114	116	116	123
Kirgiz	120	125	128	118	128	131	138
Tadzhik	118	127	139	125	133	123	133
Turkmen	131	135	139	129	129	141	142

[1] Number of workers is based on the average number working each month.

Source: USSR, Goskomstat 1989c: 150.

in crop production; all four republics maintained the same or a higher level of labor efficiency in the livestock sector in 1985 compared with 1970, while only Uzbekistan managed any increase in crop labor productivity between these two years (4 percent) (USSR, TsSU 1985: 318). In 1987, labor costs as a percentage of all costs of crop production in Uzbekistan were 57.1 percent in kolkhozes and 39.2 percent in sovkhozes (Uzbek SSR, Goskomstat 1988: 158, 164), but they were only 25.3 percent in kolkhozes of the whole Soviet Union in 1986 (USSR, Goskomstat 1988b: 110). In Uzbekistan livestock production, labor costs were 46.3 percent of the total in kolkhozes and 20.2 percent in sovkhozes, versus 23.1 percent in all Soviet kolkhozes. Thus, at least for the Uzbek SSR, the labor burden is the most acute in crop production, particularly in that of kolkhozes.

The main reason for the low labor productivity in Central Asian crop production overall is the extremely high labor use in cotton production. Between the early 1970s and the mid-1980s most republics achieved some improvement in the labor efficiency of meat, milk, and grain production, but the amount of labor used in cotton production actually increased (Table 6.15). Compared with 1970, labor expenditure per centner of raw cotton rose during 1981–5 in both state and collective farms of all republics except Kazakhstan. Although there was some improvement in 1986–7 in most republics, labor use remained above levels of the 1970s throughout the region. Based on the figures in Table 6.11, and the output of raw cotton in Central Asia, in 1986–7 cotton required approximately 2.6 billion man-hours per year to produce. This represents, at the recommended 230 labor days per worker, 1.4 million full-time workers, or at the very least 43 percent of the average yearly labor supply in Central Asian agriculture. Although some of this labor comes from off-farm workers, and not all of the average yearly labor supply is full-time, clearly the labor productivity problem of the region cannot be solved without reducing the tremendous labor burden of cotton production. The only way to solve this problem, short of reducing cotton output, is to increase the use of mechanization.

Table 6.15   Labor expenditure per centner of raw cotton (person-hours).

		1960	1965	1970	1971–5	1976–80	1981–5	1986–7
Uzbek	kolkhoz	48	38	37	36	35	38	39
	sovkhoz	46	30	30	29	29	38	32
Kirgiz	kolkhoz	34	31	35	34	39	56	54
	sovkhoz	62	38	44	36	33	45	30
Tadzhik	kolkhoz	62	42	38	40	39	44	44
	sovkhoz	37	32	29	31	33	40	36
Turkmen	kolkhoz	65	45	32	32	34	38	37
	sovkhoz	59	44	29	22	27	36	34
Kazakh	kolkhoz	42	37	37	24	28	34	30
	sovkhoz	32	33	33	18	22	25	21

Sources: calculated from *Narodnoye Khozyaystvo* of Central Asian republics and
Kazakhstan, various years; and USSR, TsSU and Goskomstat, *Narodnoye Khozyaystvo SSR*,
various years; Yakubov 1976: 45.

Labor supply and mechanization

Soviet agricultural policy toward Central Asian labor has long
operated on two contradictory levels; on the one hand, progress in
socialist agriculture is seen to require the freeing of labor through
mechanization in order to provide labor for the shortages in nonagricultural sectors; on the other, in Central Asia the rapid rural population
growth has far outstripped the ability of the state to expand agriculture
fast enough to employ adequately the surplus labor resources. Mechanization only serves to free agricultural labor into conditions of
unemployment or underemployment as long as the rural labor supply
continues to grow faster than nonagricultural rural employment.

In spite of this contradiction, improvement of labor productivity in
agriculture, which is necessary to reduce the cost of production, requires
the state to press on with the mechanization campaign. During the past
two decades major investments in mechanization have raised the level of
agricultural horsepower per hectare of sown area to the highest in the USSR.
For example, in 1986 the whole USSR level was 358 horsepower per 100
hectares, while in Central Asia it ranged from a low of 582 in Turkmeniya
to a high of 828 in Tadzhikistan (USSR, Goskomstat 1989e: 147).

When levels of horsepower are measured per worker, however,
Central Asia is not only far short of the all-union level, but it has fallen
even further behind since 1960. The Soviet Union increased its mechanization (horsepower) per worker by six times while the increase in
Central Asia was about three to four times. By 1986, per worker levels
were highest in Kirgiziya, at about 54 percent of the USSR level, while
they were only 43–46 percent as high as the all-union level in Tadzhikistan,
Uzbekistan, and Turkmeniya (USSR, Goskomstat 1989e: 147).

The cause of the widening gap between all-USSR and Central Asian
levels of mechanization per worker is not only the rapid rise in the

Central Asian work force, but also a declining share of the all-union machinery investment. While agricultural machinery investment capital rose substantially in the 1970s and 1980s in the region, growing 44.1 percent in 1976–80 compared with 1971–5 (greater than the USSR increase), in 1981–5 it increased by only 24.2 percent, although in 1986–7 the rate of growth rose slightly to 12.3 percent for the period. Meanwhile, growth in the country as a whole in 1986–7 was much greater than in Central Asia; it rose 17.7 percent compared with the 1981–5 levels (USSR, Goskomstat 1988a: 66).

Mechanization in cotton production. As already discussed, cotton production requires more labor than any other sector of Central Asian agriculture, not only because it is the largest crop by sown area, but also because of the large number of operations required to grow and harvest it. Valiyev (1988: 50) estimates that a hectare of cotton produced by hand labor requires 3,400 man-hours, but a fully mechanized hectare requires only 500 to 600 man-hours. As of the late 1970s, the average labor use for cotton production in the kolkhozes of Uzbekistan was (hours per hectare): pre-sowing preparations and sowing, 149; cultivation, 395; harvest, 436. Thus, harvesting used about 44 percent of all labor, but at a level of harvest mechanization of 62 percent. Based on the example of all kolkhozes in the most labor-productive rayon of Uzbekistan, Galabinskiy, where 96 percent of cotton was machine harvested, more than 400,000 Uzbekistan agricultural workers (about 20 percent) could be released from agriculture with the level of mechanization already achieved by leading farms (Yunusov 1983: 8). Since about 90 percent of the hand labor in Uzbekistan agriculture is by women, this would free up mainly young women (Mirzayev 1981: 84).

This average kolkhoz level of 62 percent machine harvesting by the late 1970s reflects the great advances in harvest mechanization achieved from 1965 to 1980. The successes, however, have been considerably weakened in recent years (Table 6.16). The percentage harvested by combine in all republics except Kazakhstan since 1980 has generally declined. Thus, from a peak of 68 percent in 1981, Uzbekistan levels declined to only 31 percent in 1984. By 1985 or 1986, use of cotton harvesters was rising again in all republics, but even by 1988 levels of machine harvesting in all except Turkmeniya and Kazakhstan were still lower than the levels that had been reached by about 1980.

The decline in the percentage of cotton that was machine harvested in recent years does not fit with the apparent steady increase overall in the number of cotton harvesters (Table 6.17). In Central Asia and southern Kazakhstan there were 5.6 times as many cotton harvest combines in 1986 as there were in 1960, and 6.4 percent more combines than were in agriculture in 1980, but their effectiveness had obviously

Table 6.16 Percentage of cotton harvested by machine.

	1965	1970	1975	1980	1981	1982	1983	1984	1985	1986	1987	1988
Uzbek	23	33	46	63	68	54	34	31	40	42	45	47
Kirgiz	33	39	62	51	49	41	43	26	23	47	n.a.	n.a.
Tadzhik	14	22	28	36	29	22	12	11	13	22	20	24
Turkmen	n.a.	32	47	52	51	45	43	39	52	n.a.	57	65
Kazakh	21	41	69	63	61	63	49	60	76	69	69	78

Sources: *Narodnoye Khozyaystvo* of Central Asian republics and Kazakhstan, various years; Kasyyev and Ylyasov 1986; Agadzhanov 1988; Molofeyev 1989.

greatly decreased. Possibly the most important factor, other than weather conditions in a particular year, is the great number of non-working harvesters. For example, in 1982 in Turkmeniya 19 percent of cotton harvesters were not available for the harvest because of lack of parts or repairs (Fateyev 1983: 3), while later in the 1980s it was reported that almost half of harvesters in Turkmeniya do not actually work in the harvest (Totskiy 1988). In Uzbekistan, estimates of the share of non-working cotton harvesters range from 46 percent in 1984 (Usmankhodzhayev 1985: 3) to about a third by the mid- to late 1980s (KPSS 1987a; Usmankhodzhayev 1988). The considerable mechanical complexity of cotton combines requires a sophisticated maintenance system, yet availability of parts and skilled repair personnel has fallen behind the increases in the number of machines.

Table 6.17 Number of cotton harvest machines (thousands).

	1960	1965	1970	1975	1980	1986
Uzbek	8.30	21.80	26.10	28.70	36.60	37.90
Kirgiz	0.14	1.64	1.74	1.27	1.05	0.74
Tadzhik	0.17	2.23	2.85	2.91	3.42	4.11
Turkmen	0.68	3.30	3.50	6.50	9.30	10.90
Kazakh	0.70	n.a.	2.80	n.a.	2.50	2.60
Total	9.99	n.a.	36.99	n.a.	52.87	56.25

Sources: *Narodnoye Khozyaystvo* of Central Asian republics and Kazakhstan, various years.

The low use rate of combines, however, is not only the result of poor-quality machines or lack of parts and skilled repair personnel. Deficiencies in design of cotton harvest machines have long been blamed for low use of machine harvesting (Hodnett 1974; Dzhurabekov 1986: 28). Until recently, machines were not available for the harvest of fine-fiber cotton (USDA 1977: 11), but by the early 1980s at least 33 percent of this variety in Turkmeniya was machine harvested (Yazberdiyev 1984: 2). Combine harvesting also has been blamed for causing declines in cotton quality, a factor which could be of greater importance as the region converts to a procurement system based on differential prices by quality. Many experts assert, however, that the quality issue is only an excuse by farms to avoid mechanization for other reasons.

Other problems which have been cited in many regions as key impediments to greater use of combines include improper or untimely defoliation of cotton plants in preparation for the harvest, and later-maturing varieties which must be harvested in a short period, requiring supplemental hand labor (Mogilevets 1987; Maksumov 1988b). In addition, shortages of skilled machine operators are reported on many farms, in spite of the large numbers who are being trained each year. Finally, perhaps the most powerful incentive of all not to use combines fully is the labor surpluses on most farms (Totskiy 1988).

Partial use of combines, for whatever the reasons, means that not only do farms have to pay harvest wages, but they also must pay for the depreciation of the already acquired machinery. According to Ergasheva (1987: 84), at least 70 tons must be harvested per cotton combine in order for machine harvesting to be effective, and skilled operators in 1985 achieved from 120 to 350 tons per combine. Although in 1975 and 1980 all republics in Central Asia met this minimum requirement (except Turkmeniya in 1980), in 1985 through 1987 only Kazakhstan reached this level (USSR, Goskomstat 1988d: 412). Based, however, on Totskiy's (1988) claim that a combine must harvest from 100 to 120 tons of cotton to be profitable, no republic has yet consistently achieved use of cotton combines. By the mid-1980s, then, throughout Central Asia farms had huge investments in cotton harvesters which were not being used.

Nonfarm labor in harvest

One of the main consequences of the poor use of available machinery, in addition to the low capital productivity of machinery investment, is the need for large numbers of temporary nonfarm laborers, especially in the cotton harvest. In 1982 in Uzbekistan 9 percent of the entire agricultural labor supply consisted of temporary workers from non-agricultural enterprises or urban areas (Usmanov 1984: 4). It is likely that the numbers of extra laborers for the cotton harvest increased in the 1980s, as a result of the declining percentage of cotton that was picked by machine and the growth in the harvest itself. For example, in Uzbekistan the data on tons of cotton produced and the percentage that was machine harvested indicate that, while about 2 million tons were hand-picked in 1980 and 1981, the figure rose to nearly 4 million tons in 1983 and 3.7 million in 1984, then declined to about 3.2, 2.9, and 2.7 million in 1985–7 (Uzbek SSR, TsSU and Goskomstat various years). In 1986 and 1987, in addition to urban workers, 650,000 to 700,000 schoolchildren and as many as 140,000 students of colleges and vocational schools worked in the cotton harvest in Uzbekistan (Artemenko 1987; Sokolov 1988). The majority of these outside workers in Uzbekistan probably are used in kolkhozes, where levels of cotton harvest mechanization are far below those in sovkhozes. The great economic losses in nonagricultural enterprises and the effects of interruption of education prompted Tadzhikistan and Uzbekistan in 1986 to declare that participation of college students and schoolchildren in the cotton harvest would be ended (KPSS 1987a; Artemenko 1988), but it apparently has continued. Gavrichkin (1988), however, declared that in 1988 in Tadzhikistan, Uzbekistan, and Turkmeniya the harvest was accomplished without urban workers or students.

According to Usmankhodzhayev (1985) and Totskiy (1988), existing labor and machinery are sufficient for the cotton harvest, without using urban workers or schoolchildren. The main reason for the use of outside labor in the harvest may be the low wages received by farm workers for picking cotton, which makes work on the private plot far more lucrative than cotton harvesting (Gladkov and Somov 1985; Sokolov 1988). While kolkhoz and sovkhoz workers labor on their own plots, however, their chidren are often let out of school to pick cotton. Even those modest harvest wages are important supplements to many rural family budgets (Abakumov and Karpov 1987). In an effort to increase participation in the harvest, wages for cotton picking were increased sharply in 1988 (Critchlow 1989).

Agricultural wages

As the numbers of agricultural workers in Central Asia have increased, the surplus labor has retarded the development of mechanization and caused a decline in labor productivity, with greater cost of labor per unit of production. In Uzbekistan, for example, labor payments per ruble's worth of agricultural production rose 40.7 percent between 1966–70 and 1981–5 in kolkhozes and 19.4 percent in sovkhozes (Ergasheva 1987: 81–2). As a result, according to *Razvitiye* (1988: 146) and Khan et al. (1988), these rising labor costs from excess labor have directly depressed agricultural wages.

Although it is impossible to determine what wages would be if labor productivity had increased at all-union rates, clearly Central Asian agricultural wages have lagged dramatically compared to the country as a whole, especially in the 1980s (Table 6.18). In kolkhozes, where republic average monthly wages were 24 to 55 percent higher than the USSR average in 1965 and up to 29 percent higher as recently as 1980 (in Turkmeniya), in 1988 wages fell to 6 to 23 percent less than the countrywide average. Sovkhozes, except in Turkmeniya, have usually had wages below the all-union average, but in the 1980s the gap widened in all of these republics.

Furthermore, the greater surplus of labor in kolkhozes compared with sovkhozes in Central Asia is apparently now having a larger effect on kolkhoz wages; while the traditionally lower kolkhoz wage

Table 6.18   Agricultural wages in Central Asia (percentage of USSR average).

	1965	1970	1975	1980	1985	1988
Kolkhoz wages						
Uzbek	137.2	122.7	104.3	110.2	85.0	77.0
Kirgiz	135.3	110.7	108.7	100.0	104.6	94.4
Tadzhik	123.5	114.7	103.3	110.2	92.2	80.9
Turkmen	154.9	168.0	135.9	128.8	105.9	85.4
Sovkhoz wages						
Uzbek	90.7	98.0	95.3	102.0	84.8	80.5
Kirgiz	86.7	83.2	88.2	82.7	82.1	78.6
Tadzhik	100.0	88.1	85.8	80.7	72.3	69.0
Turkmen	120.0	134.7	146.5	111.3	98.9	93.8

Source: calculated from USSR, Goskomstat 1989c: 188.

has been steadily catching up to the sovkhoz wage in the USSR as a whole, increasing from 68.0 to 84.8 percent of the sovkhoz wage from 1965 to 1988, here all republics except Kirgiziya showed a growing kolkhoz–sovkhoz wage gap since 1980. Kolkhoz wages in Kirgiziya and Tadzhikistan were still at or above sovkhoz wages by 1988, but wages in Uzbekistan and Turkmeniya were only 81 percent and 77 percent, respectively, of sovkhoz levels in that year. The primary reason for the decline in the wage position of the kolkhoz in the 1980s was the virtual stagnation of kolkhoz wages compared with the large increases of previous years (USSR, Goskomstat 1989e: 188).

While kolkhoz wages in the USSR have been catching up with sovkhoz wages, sovkhoz wages have now reached parity with the average Soviet wage of all blue- and white-collar workers (including sovkhoz workers) (USSR, Goskomstat 1989e: 187–8). In Central Asia, however, the advances in sovkhoz wages relative to these wages, which occurred in most republics of the region from 1965 to 1975 or 1980, slowed or slipped back in the 1980s. In 1987, sovkhoz workers earned only 94.1 percent as much as all blue- and white-collar workers in Uzbekistan, 89.5 percent in Kirgiziya, and 84.3 percent in Tadzhikistan, although they had the same average wages in Turkmeniya. With agricultural labor productivity barely increasing, or even declining, while productivity of industrial labor was growing at 8 to 24 percent from 1980 to 1987 (USSR, Goskomstat 1987: 98), it is easy to see why sovkhoz wage gains fell behind the increases of all blue- and white-collar wages in Central Asia.

Thus, in spite of wage increases of 79 to 143 percent in kolkhozes and 93 to 154 percent in sovkhozes from 1965 to 1988, the republics of Central Asia failed to keep up with the even greater increases in the USSR as a whole. The declining agricultural labor productivity, especially in the 1980s, may be sufficient to explain this shift. Khan et al. (1988) claim that wages in the period 1970 to 1985 in Uzbekistan did not show an absolute decline, in spite of the falling labor productivity, only as a result of higher state wage scales and an increase in procurement prices for agricultural products.

In addition, it is important to stress that the wage statistics just presented represent republic or USSR averages, while actual wages vary considerably by type of production and by farm. In Uzbekistan, for example, in 1985 the average wage per man-hour in kolkhozes varied from 56 kopeks in horticulture, 69 kopeks in grape production, 85 kopeks in livestock production, and 91 kopeks in cotton, to 2.29 rubles in tobacco production, and average kolkhoz monthly wages fluctuated from less than 70 rubles to more than 200 rubles (*Razvitiye* . . . 1988: 148). A sample of five kolkhozes in Uzbekistan and Tadzhikistan analyzed by Khan and Ghai (1979: 98), based on 1973 and 1976 data, showed that average wages per man-day from kolkhoz labor were 74 percent higher at the highest- as compared with the lowest-paying, but total household incomes and per capita incomes were only 45 to 46 percent higher at the wealthiest farm. Larger incomes from private plots and off-farm labor in the kolkhozes where wages were lower narrowed the gap. Thus, if the experience of these farms is typical, the lag in agricultural wages in general may be an important factor in keeping potential workers at home or encouraging them to engage in more lucrative private agriculture, just as low cotton harvest wages reduce the participation of regular farm workers in the harvest.

Labor supply and migration

The large and growing surplus of rural labor in many Central Asian oblasts, and the associated stagnation or decline of agricultural productivity and agricultural wages, may be viewed as "push" factors which could stimulate rural out-migration. Lewis et al. (1976) determined that by the early 1970s a growing labor surplus was already evident which would inevitably lead to out-migration to urban areas. As is discussed elsewhere in this volume, that assumption aroused considerable controversy, on the grounds that large increases in rural out-migration were not occurring and that cultural factors among the indigenous nationalities in rural areas would keep those populations from migrating.

Although data on rural net migration by oblast are not available, sufficient data exist to estimate the impact of rural out-migration at the republic level (Table 6.19), for intercensal periods and five-year periods since 1959. These figures, which represent the estimated net rural out-migration as a percentage of the rural natural increase, show that in no period did any republic lose even half of its rural population growth to out-migration. Rural out-migration in the most important republic,

Uzbekistan, steadily increased from 1959 to the late 1970s and then declined in the 1980s. The other republics had a more mixed pattern, but all three had increasing rural outflow in the early 1980s as compared with the previous five-year period. Nevertheless, rural out-migration remains low in all republics, especially Tadzhikistan, where in the late 1970s there was even net rural in-migration, with almost all cities except Dushanbe and Leninabad losing population to rural areas (Makarova et al. 1986: 85).

Table 6.19 Net rural out-migration as a percentage of natural increase.

	1961–5	1966–70	1971–5	1976–80	1981–5	1959–70	1970–9	1959–86
Uzbek	5.2	14.3	28.7	45.9	24.5	15.0	36.2	25.3
Kirgiz	15.7	8.0	35.6	26.4	34.6	10.4	31.1	24.4
Tadzhik	0.7	18.0	4.5	–8.5	9.7	11.3	–4.3	5.7
Turkmen	20.6	6.6	13.8	15.3	18.3	20.7	13.7	16.6

Source: calculated from USSR, Goskomstat 1988c: 10–15, 115–25.

These calculations generally agree with those of Soviet demographers. According to Kulakov (1988), between 1951 and 1980, 18.5 percent of the rural population in Uzbekistan migrated, but only 4.6 percent of the rural Tadzhikistan population, while in Central Asia as a whole, out-migration was not more than 15 percent of the rural natural increase (Makarova et al. 1986: 84). On kolkhozes in Uzbekistan, less than one percent of the work force left for off-farm employment in 1985 (Chebotareva 1987). Rybakovskiy and Tarasova (1989), however, report that Central Asian rural migration rates in 1981–5 increased by 30–40 percent and were the greatest in 15 years, with the exception of rates in Tadzhikistan.

Even if mobility of the rural population is increasing, it still accounts for only a fraction of the rural population growth in these republics. The causes of this low mobility have been the subject of many surveys of the rural population and of recent migrants to urban areas (Zyuzin 1983, 1985; Savurov 1986; Kubasova 1987). Insufficient employment opportunities in the cities, lack of needed skills to fill industrial jobs, poor ability to speak Russian, dissatisfaction with problems of housing and other living conditions in urban areas, inability to have a private agricultural plot, and especially preference for the rural way of life and associated family and community relationships are all common reasons why rural out-migration may be considered undesirable. Lack of skills among women, who perform most of the manual agricultural labor, has an especially important effect on their mobility; they are less than 30 percent of the rural out-migrants (*Razvitiye . . .* 1988: 50).

Zyuzin (1985) reports that levels of satisfaction with the family standard of living are actually highest in urban-type settlements where the family has an agricultural plot, followed by rural areas, capital cities, and then other cities. The role of the private plot in family incomes is considered by many researchers to give rural families a considerable economic advantage over urbanites in Central Asia, more than compensating for the larger rural family sizes (Zyuzin 1983, 1985; Lubin 1984; Kononenko 1988). Zyuzin (1985) concludes, therefore, that higher incomes among large rural families are sufficient to explain why migration is low. If it is true that the plot does play the pivotal role in keeping rural incomes high in spite of rapid population growth and stagnating agricultural wages, then the state of the private agricultural sector will be a key factor in rural to urban migration in Central Asia.

In addition to rural-to-urban migration as a means of lessening agricultural labor surplus, migration of labor between rural areas in Central Asia—especially to areas of new irrigation projects—has also been a major way by which labor resources have been redistributed among regions. Rural population growth rates (Table 6.9) of more than 4.0 percent per year have been observed in many of these new irrigated regions. Unfortunately, living conditions are often harsh there and return migration is very common (Murakayev 1980: 69; Valiyev 1988: 19, 115–16). According to Yakubov (1976: 126–31), in Uzbekistan only Syrdar'ya Oblast and Karakalpak ASSR had overall labor shortages, but all regions had some enterprises or rayons with shortages.

Oversupply of agricultural labor in the irrigated plains of Tadzhikistan has also caused rural-to-rural migration, but there it is often to the mountain regions, where increasing overuse of mountain forests and pastures for fuel and grazing is reported. Partial surveys estimate that about 300 mountain villages, abandoned in the 1930s to the 1950s as people went to the plains for agricultural work, have now been resettled (Umarov 1987).

Finally, transfer of part of the agricultural work force to nonagricultural employment by locating industries in rural areas is another widely advocated means of avoiding the problem of rural to urban migration. Although industry is not yet well developed in the countryside, commuting of the rural population to urban employment while still living on farms may be an important means of reducing agricultural labor surpluses in some areas; in the 1970s only about 1.9 percent of the rural population in Uzbekistan commuted to the city, but 4.4 percent in Kirgiziya and 3.0 percent in Tadzhikistan did (Mukomel' 1988). The figure for Uzbekistan rose to about 4–5 percent in the 1980s. Poor rural public transportation, however, limited most migrants to trips of less than ten kilometers (*Razvitiye* . . . 1988: 151).

PERSONAL AGRICULTURAL PLOTS AND RURAL INCOMES

Although the modest rates of rural out-migration, rural–rural migration, development of rural industry, and rural–urban commuting may be easing labor surpluses in some regions, in Central Asia as a whole the surplus continues to grow. If the excess labor cannot be accommodated in the state and collective agricultural sectors, then the private agricultural sector must fill the gap or rural unemployment will increase. Furthermore, with agricultural wages in Central Asia declining in relation to all-union agricultural wages and to Central Asian industrial wages, private agriculture could be the key factor in maintaining rural incomes and retarding rural out-migration.

Reports based on personal experience or interviews with rural inhabitants in Central Asia often emphasize those persons who are making large profits from private agriculture, even many thousands of rubles per year (Zyuzin 1983: 2). These impressive figures have fostered a common perception among many Soviets and foreign researchers alike that rural Central Asians are actually relatively well off as a result of their private production. Available data, however, do not support the claim that, in general, rural incomes in the region are substantially higher than urban incomes as a result of lucrative private agriculture, or that Central Asian farmers are better off than their counterparts in the country as a whole.

Results of the USSR family budget surveys, which as of 1988 included 90,000 families, provide income data by republic by social group (blue- and white-collar workers and kolkhozniks) (USSR, Goskomstat, Byuro . . . 1990a). Per capita incomes in kolkhoz families in Central Asia as compared with per capita incomes in kolkhoz families of the whole USSR declined from 1975 to 1988, to only 47 to 74 percent of the all-union level in these four republics. Only Kirgiziya managed to improve its position slightly during that period.

Central Asian kolkhoz family incomes per person in comparison with incomes in families of blue- and white-collar workers (which include sovkhoz workers) also showed no improvement from 1975 to 1988, except, again, in Kirgiziya. By 1988, while an average person in a kolkhoz family in the Soviet Union had an income 79 percent as high as in a family of blue- or white-collar workers, in Uzbekistan and Tadzhikistan it reached 70 percent, in Turkmeniya 76 percent, and in Kirgiziya 84 percent.

The role of private agriculture in these Central Asian family incomes was no more significant than in the country as a whole, contrary to popular perceptions. In 1988, between 16.9 and 26.5 percent of kolkhoz family income in the Central Asian republics came from private produc- tion, versus the USSR level of 24.5 percent, and this relationship was essentially the same as it had been in 1975. Furthermore, the absolute level of income from the plots per kolkhoz family member, which was only 51 to 68 percent as high as the USSR average in 1975, increased faster than the all-union level only in Kirgiziya. Uzbekistan matched the increase in USSR kolkhoz plot output per capita, but Tadzhikistan and

Turkmeniya fell even further behind, to only 42–45 percent of all-union levels.

Therefore, based on available income data for kolkhozniks, not only do Central Asian farm families have far lower per capita incomes than Soviet kolkhozniks as a whole, but private agriculture has generally not been able to fully compensate for the lower incomes from the collective sector.

This is not to say, however, that the output from Central Asian private agriculture is not increasing. In fact, the yearly value of private agriculture increased from about 1.5 billion rubles in the four republics in 1961–5 to about 4.2 billion rubles in 1986–8 (USSR, Goskomstat 1989f: 6). For each family having a plot, value of production rose dramatically as compared with the whole USSR, although by 1986–8 the value

Table 6.20 Indicators of private plot production (per each family having a plot).

(a) Average yearly value of plot production (rubles).

	1961–5	1986–8	Change (1986–8/1961–5)×100
USSR	1,415	1,437	102
Uzbek	702	1,018	145
Kirgiz	651	1,260	194
Tadzhik	716	1,050	147
Turkmen	1,182	1,595	135

Source: calculated from USSR, Goskomstat 1989f: 6, 19; 1961–5 number of plots estimated using average of 1960 and 1965; 1986–8 number of plots estimated using 1987.

(b) Average plot size (hectares).

	1965	1987	Change (1987/1965)×100
Uzbek	0.114	0.104	91
Kirgiz	0.146	0.148	101
Tadzhik	0.122	0.113	93
Turkmen	0.154	0.134	87

Source: calculated from USSR, Goskomstat 1989f: 19.

(c) Livestock per plot.

		1966	1989	Change (1989/1966)×100
Uzbek	cattle	0.97	1.00	103
	sheep, goats	1.11	1.31	118
Kirgiz	cattle	0.62	0.76	123
	sheep, goats	1.31	3.22	246
Tadzhik	cattle	1.10	1.38	125
	sheep, goats	1.99	2.83	142
Turkmen	cattle	1.29	1.46	113
	sheep, goats	5.78	5.23	90

Source: calculated from USSR, Goskomstat 1989f: 19, 25, 26.

of output per plot was still less than the USSR average, except in Turkmeniya (Table 6.20a). Unfortunately, while the output of each plot was increasing, so were the already much larger Central Asian family sizes, especially in rural areas. From 1959 to 1979, an average rural family in Uzbekistan and Tadzhikistan grew by 29 percent, in Turkmeniya by 30 percent, and in Kirgiziya by 20 percent (Vasil'yeva et al. 1985: 72). By 1988 the average kolkhoz family that the plot had to support contained from 5.7 (in Kirgiziya) to 7.7 members (in Tadzhikistan), compared with 3.3 to 3.5 members in the Ukraine and the RSFSR (USSR, Goskomstat, Byuro . . . 1990a: 195–210).

These data on the value of private agricultural production have been criticized by Western researchers on numerous grounds, including the weakness that the statistical system which accounts for sales of food in the kolkhoz markets underestimates the amounts substantially (Lubin 1984; Shenfield 1986). Therefore, the proportion of rural incomes coming from marketing of agricultural products may be greater than the official figures suggest. Furthermore, the amount of food consumed in the family from its private production may also be undervalued, since it is not priced at the kolkhoz market price, which is much higher than official state prices.

An additional way of estimating possible changes in the role of the plots, therefore, is to use data on the inventories of plot land and livestock over time. These data should be subject to less error than are the estimates of market sales from private agriculture, and they will not be affected by changes in the proportion of plot output which is sold.

Because of limitations on the size of individual private plots, it is to be expected that the main cause of expansion in the area of plot lands will be an increase in the number of workers authorized to have them. Extreme land pressure in the irrigated areas of Central Asia as the population has increased has not only restricted plot sizes to the smallest in the Soviet Union, but has even caused some decline in average plot area per family (Table 6.20b). From 1961–5 to 1986–8 the average plot size declined in all Central Asian republics except Kirgiziya. Although in Uzbekistan, for example, the decline was only by 0.01 hectares per plot, this represents a 9 percent decline in plot area.

Livestock holdings per plot, however, have actually increased during the same period (Table 6.20c), in all Central Asian republics, except for sheep and goats in Turkmeniya. Larger livestock herds have been partly supported by increases in the proportion of plot area devoted to fodder production, but mostly they have depended upon transfer of fodder or grazing rights from the state and collective sectors. Severe fodder shortages have developed as private herds have grown, to the extent that many peasants no longer keep livestock because insufficient fodder is available (USSR, Goskomstat, Byuro . . . 1990b).

Higher yields and, most importantly, larger private livestock herds have made possible increasing production from private agriculture in Central Asia, in spite of constraints on expansion of plot area. Incomes per person from private agriculture have also risen, but not fast enough, apparently, to narrow the gap significantly between per capita farm incomes in Central Asia and in the country as a whole. While private agriculture has been unable to expand fast enough to provide, on a per capita basis, high rural incomes in a time of rapid population growth, it has been able to provide reasonably high family incomes. Value of plot output per kolkhoz family in 1988 exceeded the all-union level in Uzbekistan and Kirgiziya, and was approximately equal to the lower RSFSR level in Tadzhikistan and Turkmeniya (USSR, Goskomstat, Byuro . . . 1990a: 244–59). Since large Central Asian families benefit more from economies of scale in use of consumer goods, and have many children, for whom average consumption is less, family income may be as important a measure of economic welfare as per capita income.

Nevertheless, the claim by many analysts that large incomes from private agriculture raise rural incomes so much that the rural population is economically better off than the urban population, regardless of the often low wages in kolkhozes and sovkhozes, does not seem to be supported. This rural advantage is undoubtedly real in some cases, especially near large urban markets, but many farms are so remote that very little of the private output is sold off the farm. Furthermore, the small share of the urban population in many oblasts means that the ability of the urban population to subsidize the far more numerous rural families is limited. If much of the output is sold in rural areas, the income from those sales does not represent a transfer of urban to rural income, although it may represent a transfer of income from the rural nonfarm to the farm population.

CONCLUSION

The agricultural revolution in Soviet Central Asia during recent decades brought a major expansion of the area of irrigated land, increases in yields through use of fertilizers, biocides, and improved cultivation techniques, and large improvement in levels of mechanization for most types of crop and livestock production. Growth in cotton production in Central Asia enabled the Soviet Union to become not only one of the world's largest cotton producers, but one of the largest exporters as well.

In the 1970s and 1980s, however, a series of accumulating problems began to erode the earlier successes. Irrigation expansion has now become increasingly limited, both by inefficient use of water and by the need to preserve the Aral Sea. Salinization of irrigated lands, partly as a result of the incorporation of large areas of salt-plagued soils, not only reduces crop yields but pollutes irrigation water for other users downstream. Agricultural pollution from these salts and from fertilizers and pesticides in drainage waters has become a major health threat to rural populations, especially in the lower Amudar'ya and Syrdar'ya Valleys.

State pressure to increase cotton production has caused a virtual cotton monoculture in many regions, with a lack of the proper cotton–alfalfa rotations. The result was stagnating or declining cotton yields in the 1980s and a severe shortage of the fodder needed to increase animal production. Overemphasis on cotton production also limits the ability of the region to produce food, keeping per capita consumption of many food products—especially animal products—at the lowest levels in the USSR. Local outrage over the cotton monoculture and its effects on the food situation has become an important issue of nationalism among the indigenous population.

In rural areas, rapid population growth has outstripped the ability of agriculture properly to employ the growing work force. Labor that should ideally have been displaced from agriculture as mechanization increased remains on the farm, often in only a semiemployed state. As a result, labor productivity declined in the 1980s. Cotton mechanization, the most important means of raising labor productivity, also declined in the 1980s, after major improvements in the late 1970s. With the large inventories of machinery on farms, especially cotton harvesters, while cotton harvest mechanization lagged, capital productivity has been cut.

Rising labor costs are directly related to the fall in Central Asian agricultural wages as compared with USSR levels. In spite of all of these stresses on the rural economy and rural incomes, however, very little rural out-migration has taken place. The share of the rural population in the total population of many of these oblasts continues to increase.

It is difficult to determine whether the critical situation in Central Asia itself has awakened the Soviet leadership to the need for change there, or whether recent developments are largely a reflection of the changes occurring in the country as a whole, but a series of increasingly more important policy shifts under perestroyka have begun to address the agricultural problems outlined above. An additional incentive for Moscow to deal with this crisis is the integral relationship between the state of the rural economy and issues of nationalism. Massive ethnic riots in the summers of 1989 and 1990 in the Fergana Valley were caused partly by the difficult rural economy, with heavy unemployment and shortages of agricultural land among the root causes.

Official recognition of the seriousness of the situation in Central Asia was finally given by Prime Minister Ryzhkov in June 1989, when he agreed that all of the key problems of the agricultural sector relating to the overdependence upon the cotton economy were legitimate issues that must be resolved (Ryzhkov 1989).

Although this recognition of the crisis in Central Asian agriculture by the Soviet leadership has been belated, many steps have already been taken to try to alleviate the problems. Emphasis on rebuilding the irrigation systems to conserve water, introduction of water use fees, change of cotton procurement prices to reflect cotton quality, and decline in the cotton procurement quotas are all important developments. In addition, cotton procurement prices—the key element in the profitability of Central Asia's cotton economy—will be raised substantially, beginning with the 1989–90 crop (USDA 1989; Bohr 1990). Of major importance for rural incomes, the severe shortages of plot land will be alleviated by a doubling of plot area in Uzbekistan (Mirkasymov 1989). During my travels in rural areas of Central Asia in June 1990, it became evident that in many regions food and fodder crops have replaced cotton fields, and numerous rural inhabitants claimed that increases in food supplies were already noticeable. Furthermore, by the spring of 1990 Kirgiziya, Uzbekistan, and Tadzhikistan had banned the export from their republics of most or all food products, supposedly on a temporary basis.

A new group of less docile Central Asian leaders seem increasingly unwilling to accept the cancellation of the water diversion projects which would have brought water from Siberian rivers into Central Asia. The report of the meeting of the leaders of the Central Asian republics and Kazakhstan, held in June 1990, specifically requested the reopening of the diversion projects, which they considered vital to the future of the region (Statement . . . 1990).

By August 1990, when Tadzhikistan and Turkmeniya declared their sovereignty, the only republics remaining in the USSR which had not yet done so were Kirgiziya and Kazakhstan (*New York Times*, 25 August 1990). In a climate of rapid decentralization of the economy and shift to local control, the Central Asian republics will most likely soon be free to determine their own correct balance of cotton versus food production, to expand further the area and type of private agriculture, and to make their own decisions on the proper use of water resources, including the acceptable amount to allocate to saving the Aral Sea. This shift to republic self-financing or even economic sovereignty, however, will make generation of the capital needed to accomplish many of the agricultural reforms more difficult. In addition, interest in the Russian Republic in allocating capital investment to the diversion of some of its water resources southward into Central Asia will certainly be even less in a situation of sovereign republics. In spite of the new freedom to control their agricultural economies for Central Asian needs, then, these republics will find that the structural problems outlined in this chapter—including rural population growth and labor oversupply—are not so easily solved.

REFERENCES

Abakumov, I. and Karpov, A. 1987. Hands and machines. *Current Digest of the Soviet Press* 39, 14: 18–19. Translated from *Izvestiya*, 31 March: 2.

Agadzhanov, Ya. 1988. Progressivnyye tekhnologii ne nakhodyat shirokogo rasprostraneniya v khlopkovodstve respubliki. *Agropromyshlennyy Kompleks Turkmenistana* 5: 20.

Akramov, E. 1988. Povysheniye effektivnosti ispol'zovaniya oroshayemykh zemel'. *Kommunist Uzbekistana* 12: 3–9.

Akramov, Z. M. and Rafikov, A. A. 1990. *Proshloye, nastoyashcheye i budushcheye Aral'skogo Morya*. Tashkent: Mekhnat.

Akramov, Yu., Sultanov, M. S., and Kireyev, V. K. 1984. Kapital'noy planirovke i promyvke oroshayemykh zemel'-osoboye vnimaniye. *Sel'skoye Khozyaystvo Tadzhikistana* 2: 33–8.

Artemenko, V. 1987. Khop, khop, and . . . forget it: on the value of certain promises and assurances. *FBIS Daily Report, Soviet Union*, 14 May: s3–s8. Translated from *Pravda*, 25 April: 3.

Artemenko, V. 1988. Own correspondent's column: deception. *FBIS Daily Report, Soviet Union*, 15 January: 41. Translated from *Pravda*, 12 January: 2.

Bohr, Annette 1988. Infant mortality in Central Asia. *Radio Liberty Research Bulletin* 32, 33: 1–8.

Bohr, Annette 1990. Turkmenistan under *perestroika*: an overview. *Report on the USSR* 2, 12: 20–30.

Chebotareva, V. G. 1987. Aktual'nyye problemy povysheniya sotsial'noy aktivnosti zhenshchin v sovremennykh usloviyakh. *Obshchestvennyye Nauki v Uzbekistane* 5: 9–18.

Critchlow, James 1989. How solid is Uzbekistan's support for Moscow? *Report on the USSR* 1, 6: 7–10.

Dvoryadkina, A. N. 1988. O zanyatosti sel'skogo naseleniya. *Agropromyshlennyy Kompleks Turkmenistana* 10: 11–12.

Dyuzhev, G. 1988. Uluchsheniye oroshayemykh zemel'. *Agropromyshlennyy Kompleks Turkmenistana* 5: 31–2.

Dzhurabekov, I. 1986. Agropromyshlennyy kompleks Uzbekskoy SSR v dvenadtsatoy pyatiletke. *Ekonomika Sel'skogo Khozyaystvo* 9: 27–32.

Effektivnoye ispol'zovaniye proizvodstvennogo potentsiala v sel'skom khozyaystve 1987. *Sel'skoye Khozyaystvo Uzbekistana* 1: 37–41.

Elpiner, L. I. 1990. Medical-ecological problems in the eastern Aral Region. Paper given at conference on the Aral Sea Crisis: Environmental Issues in Central Asia, Indiana University, Bloomington, 14–18 July.

Ergasheva, S. R. 1987. *Sovershenstvovaniye Organizatsii Upravleniya Regional'nym Agropromyshlennym Proizvodstvom (na materialakh UzSSR)*. Tashkent: Fan.

Fateyev, V. N. 1983. Obraztsovo provesti mashinnuyu uborku khlopka-syrtsa. *Sel'skoye Khozyaystvo Turkmenistana* 8: 2–4.

Fierman, William 1989. Glasnost' in practice: the Uzbek experience. *Central Asian Survey* 8, 2: 1–45.

Gavrichkin, V. 1988. Agricultural review: the unalloyed glitter of "White gold." *FBIS Daily Report, Soviet Union*, 15 November: 69. Translated from *Izvestiya*, 10 November: 1.

Gerardi, I. A. 1967. Meliorativnoye sostoyaniye oroshayemykh zemel' v khlopkovoy zone SSSR i meropriyatiya po yego uluchsheniyu. In A. N. Askochenskiy, *Bor'ba s Zasoleniyem Oroshayemykh Zemel'*, pp. 15–26. Moscow: Kolos.

Gladkov, N. and Somov, V. 1985. The cotton season is not for schoolchildren. *Current Digest of the Soviet Press* 37, 10: 4–5, 24. Translated from *Pravda*, 6 March.

Hodnett, Grey 1974. Technology and social change in Soviet Central Asia: the politics of cotton growing. In H. W. Morton and R. L. Tokes (eds), *Soviet Politics and Society in the 1970s*, pp. 60–117. New York: Free Press.

Imamaliyev, A. 1979. Land, water, and the harvest. *Current Digest of the Soviet Press* 31, 14: 5. Translated from *Pravda*, 7 April: 2.

Inoyatov, I. 1988. Nekotoryye problemy razvitiya khopkovodstva v SSSR i SShA. *Obshchestvennyye Nauki v Uzbekistane* 9: 16–24.

Karchikyan, O. Kh. 1977. *Proizvodstvo i Potrebleniye Sel'skokhozyaystvennykh Produktov*. Yerevan: Izdatel'stvo AN Armyanskoy SSR.

Kasyyev, B. and Ylyasov Sh. 1986. Reservy povysheniya urozhaynosti. *Sel'skoye Khozyaystvo Turkmenistana* 3: 11–12.

Kazakh SSR, Tsentral'noye Statisticheskoye Upravleniye and Goskomstat 1960–88 (annual volumes). *Narodnoye Khozyaystvo Kazakhstana. Statisticheskiy yezhegodnik*. Alma-Ata: Kazakhstan.

Kes' A. S., Krenke, A. N., Minayeva, Ye. N., Reshetkina N. M. and Tsigel'naya, I. D. 1982. Vodnyye resursy Sredney Azii i yuzhnogo Kazakhstana. *Gidrologiya i Melioratsiya* 2: 67–70.

Khamrayev, N. R. and Bedrintsev, K. N. 1989. Voprosy strategii vodoobespecheniya v razvitii narodnogo khozyaystva Uzbekistana. *Obshchestvennyye Nauki v Uzbekistane* 1: 3–12.

Khan, A. and Ghai, D. 1979. *Collective Agriculture and Rural Development in Soviet Central Asia*. New York: St Martin's Press.

Khan, N., Mazalovskiy, N., Isamukhamedov, M., Muradov, Ch. and Isachenko, G. 1988. Dlya uskoreniya progressa na sele. *Sel'skoye Khozyaystvo Uzbekistana* 5: 51–2.

Khusanov, R. Kh. 1989. Osnovnyye problemy povysheniya plodorodiya pochvy v zone khlopkoseyaniya. *Vestnik Sel'skokhozyaystvennoy Nauki* 4: 47–52.

Kirgiz SSR, Tsentral'noye Statisticheskoye Upravleniye and Goskomstat 1960–88 (annual volumes). *Narodnoye Khozyaystvo Kirgizskoy SSR. Statisticheskiy Yezhegodnik*. Frunze: Kirgizstan.

Kolbin report to Kazakh central committee plenum. *FBIS Daily Report, Soviet Union*, 1 April. Translated from *Kazakhstanskaya Pravda*, 15 March: 1–3.

Kononenko, B. 1988. Zanyatost' naseleniya: sotsial'no-demograficheskiye aspekty. *Kommunist Uzbekistana* 12: 15–21.

KPSS Tsentral'nyy Komitet 1987a. V Tsentral'nom Komitete KPSS. *Pravda*, 24 February: 1.

KPSS Tsentral'nyy Komitet 1987b. V Tsentral'nom Komitete KPSS. *Pravda*, 20 June: 1–2.

Kubasova, Ye I. 1987. Problemy migratsionnoy podvizhnosti sel'skogo naseleniya v Turkmenskoy SSR. *Izvestiya AN Turkmenskov SSR. seriya obshchestvennykh nauk* 2: 14–21.

Kulakov, V. 1988. Planirovaniye ispol'zovaniya trudovykh resursov. *Planovoye Khozyaystvo* 11: 110–16.

Kurbangel'diyev, S. 1987. Po-khozyayski otnosit'sya k zemle. *Sel'skoye Khozyaystvo Turkmenistana* 9: 17–18.

Lewis, Robert A. 1962. The irrigation potential of Soviet Central Asia. *Annals of the Association of American Geographers* 52, 1: 99–114.

Lewis, Robert A., Rowland, Richard H. and Clem, Ralph S. 1976. *Nationality and Population Change in Russia and the USSR*. New York: Praeger.

Lubin, Nancy 1984. *Labor and Nationality in Soviet Central Asia*. Princeton, NJ: Princeton University Press.

Makarova, L. V., Morozova, G. F. and Tarasova, N. V. 1986. *Regional'nyye osobennosti Migratsionnykh Protsessov v SSSR*. Moscow: Nauka.

Maksumov, A. N. 1988a. Ob ekologicheskoy obstanovke v Tadzhikistane. *Vestnik Akademii Nauk SSSR* 11: 139–43.

Maksumov, A. N. 1988b. O korennoy perestroyke khlopkovodstva, obespechivayushchey vnedreniye industrial'nykh tekhnologiy. *Agropromyshlennyy Kompleks Tadzhikistana* 1: 2–8.

Mamedov, Kh. 1984. Rabotayet reka zhizni. *Sel'skoye Khozyaystvo Turkmenistana* 7: 9.

Micklin, Philip 1986. The status of the Soviet Union's north–south water transfer projects before their abandonment in 1985–86. *Soviet Geography* 27, 5: 287–329.

Micklin, Philip P. 1987. The fate of "Sibaral": Soviet water politics in the Gorbachev era. *Central Asian Survey* 6, 2: 67–88.

Micklin, Philip P. 1988. Desiccation of the Aral Sea: a water management disaster in the Soviet Union. *Science* 241: 1170–6.

Mirkasymov, M. M. 1989. (Speech at USSR Congress of Peoples Deputies.) *FBIS Daily Report, Soviet Union*, 19 December. Translated from *Pravda*, 15 December: 4–5

Mirzayev, T. 1981. Problemy vysvobozhdeniya rabochey sily iz sel'skogo khozyaystva Uzbekistana. *Ekonomika sel'skogo Khozyaystva* 5: 83–7.

Mogilevets, Yo. 1987. Kursom intensifikatsii. *Khlopkovodstvo* 12: 17–23.

Molofeyev, Yu. N. 1989. Chtoby ne prostaivali kombayny. *Agropromyshlennyy Kompleks Turkmenistana* 8: 22–3.

Mukomel', V. I. 1988. Urbanizatsiya Sredney Azii i problemy sotsial'no-ekonomicheskogo razvitiya regiona. *Problemy Osvoyeniya Pustyn'* 5: 45–9.

Murakayev, M. I. 1980. *Trudovyye Resursy Uzbekistana*. Tashkent: Uzbekistana.

Nazarov, R. 1988. Uzbek cotton plan reduced, made "realistic." *FBIS Daily Report, Soviet Union*, 29 January: 70. Interview with R. Nazarov, entitled "No more window dressing," from *Moscow News in English*, 24 January: 8.

Nikonov, V. P. 1988. APK: na rel'sy novykh ekonomicheskikh otnosheniy. (Speech at plenum of the Central Committee of the Uzbek Communist Party, 24 August.) *Kommunist Uzbekistana* 10: 26–40.

Nishanov, R. N. 1989. I khlopok, i prodovol'stviye. Interview of R. N. Nishanov by V. Artemenko and A. Kaipbergenov. *Pravda*, 30 March: 2.

Orazmuradov, B. 1988. Regional'nyye aspekty zanyatosti naseleniya v Sredney Azii. *Problemy Osvoyeniya Pustyn'* 5: 36–41.

Ovcharenko, G. 1988. Cobras standing watch over gold. *Current Digest of the Soviet Press* 40, 3: 1–4. Translated from *Pravda*, 23 January: 3.

Ovezgel'dyyev, O. G. 1988. Voprosy ekologizatsii sotsial'no-ekonomicheskogo razvitiya Turkmenskoy SSR. *Vestnik Akademii Nauk SSSR* 11: 147–51.

Pankova, Ye. I., Golovina, N. N. and Ventskevich, S. D. 1986. Opyt otsenki zasoleniya pochv oroshayemykh territoriy Sredney Azii po materialam kosmicheskoy s"yemki. *Pochvovedeniye* 3: 138–46.

Pestryakov, G. 1988. Sovershenstvovat' zagotovku, pererabotku i realizatsiyu khlopka. *Agropromyshlennyy Kompleks Tadzhikistana* 9: 33–5.

Ponomarev, V. 1990. The bells of hope. *FBIS Daily Report, Soviet Union*, 14 May: 130–3.

Rashidov, U. 1988. Intensifikatsiya proizvodstva v agropromyshlennom komplekse. *Kommunist Uzbekistana* 7: 12–20.

Razvitiye narodonaseleniya i Problemy Trudovykh Resursov Respublik Sredney Azii (sotsial'no-demograficheskiy analiz). 1988. Tashkent: Fan.

Redzhepov, O. and Ibragimova, L. G. 1986. Berezhno ispol'zovat' vodu. *Sel'skoye Khozyaystvo Turkmenistana* 9: 27–8.

Rutgayzer, Ye. M. and Artykov, S. A. 1988. Problemy prodovol'stvennogo obespecheniya. *Agropromyshlennyy Kompleks Turkmenistana* 11: 4–5.

Rybakovskiy, L. L. and Tarasova, N. V. 1989. Sovremennyye problemy migratsii naseleniya SSSR. *Istoriya SSSR* 2: 68–81.

Ryzhkov, N. I. 1989. A serious lesson for everyone. *Current Digest of the Soviet Press* 41, 24: 5–7. Translated from *Pravda*, 16 June: 3, 5.

Salakhitdinov, M. S. 1988. Problemy ekologii i okhrany prirody v Uzbekskoy SSR. *Vestnik Akademii Nauk SSSR* 11: 117–20.

Saparlyyev, D. 1983. Luchshe ispol'zovat' oroshayemyye zemli. *Sel'skoye Khozyaystvo Turkmenistana* 2: 20–3.

Sapayev, K. 1987a. Sovremennyy podkhod k resheniyu problem melioratsii. *Khlopkovodstvo* 7: 2–7.

Sapayev, K. 1987. Meliorirovannyye zemli—rezerv intensifikatsii. *Sel'skoye Khozyaystvo Uzbekistana* 6: 59–61.

Savurov, M. D. 1986. Nekotoryye voprosy uchastiya Uzbekskoy sel'skoy sem'i v obshchestvennom proizvodstve. *Obshchestvennyye Nauki v Uzbekistane* 7: 45–8.

Severin, Barbara 1987b. Special report on Soviet cotton production data. *Newsletter for Research on Soviet and East European Agriculture* 9, 4: 1–2.

Shabad, T. 1986. Soviet decree officially cancels north–south water transfer projects. *Soviet Geography* 27 (October): 601–3.

Shadimetov, Yu. 1988. Sotsial'no-ekologicheskiye problemy zdorov'ya naseleniya. *Kommunist Uzbekistana* 12: 34–40.

Sheehy, Ann 1988. Cultivation of cotton to remain "internationalist duty" of Uzbekistan. *Radio Liberty Research Bulletin* 32, 21: 1–6.

Shenfield, Stephen 1986. How reliable are Soviet statistics on the kolkhoz markets? *Journal of Official Statistics* (Statistics Sweden) 2, 2: 181–91.

Shpolyanskaya, N. 1989. Tsena pyatogo milliona. (Report of round-table discussion organized by the journal.) *Sel'skoye Khozyaystvo Uzbekistana* 4: 4–11.

Sokolov, V. 1988. City residents in the cotton fields. *Current Digest of the Soviet Press* 40, 13: 11. Translated from *Literaturnaya Gazeta*, 15 March.

Sredneaziatskiy ekonomicheskiy rayon. 1972. Moscow: Nauka.

Starodubtsev, V. M. and Bogachev, V. P. 1983. Meliorativnaya otsenka stoka reki Syrdar'i. *Pochvovedeniye* 12: 90–101.

Statement by the Leaders of the Uzbek SSR, the Kazakh SSR, the Kirghiz SSR, the Tajik SSR, and the Turkmen SSR 1990. *FBIS Daily Report, Soviet Union,* 17 July: 86–7. Translated from *Sovetskaya Kirgiziya,* 24 June: 1.

Stebelsky, Ihor 1988. Food consumption patterns in the Soviet Union. In Joseph C. Brada and Karl-Eugen Wadekin (eds), *Socialist Agriculture in Transition: Organizational Response to Failing Performance,* pp. 98–109. Boulder, Colo.: Westview Press.

Strategiya vodoobespecheniya 1988. *Melioratsiya i Vodnaya Khozyaystvo* 7: 2–6.

Sychev, V., Kulakov, V. and Milovidov, A. 1989. Ekonomiko-demograficheskoye razvitiye strany i zadachi sotsial'nogo planirovaniya. *Planovoye Khozyaystvo* 4: 81–7.

Tadzhik SSR, Tsentral'noye Statisticheskoye Upravleniye and Goskomstat 1960–88 (annual volumes). *Narodnoye Khozyaystvo Tadzhikskoy SSR. Statisticheskiy Yezhegodnik.* Dushanbe: Irfon.

Tochenov, V. V. et al. (eds) 1983. *Atlas SSSR.* Moscow: Glavnoye Upravleniye Geogdezii i Kartografii SSSR.

Totskiy, I. 1988. Cotton grower's turn. *FBIS Daily Report, Soviet Union,* 19 September: 80–1. Translated from *Pravda,* 15 September: 2.

Turkmen SSR, Tsentral'noye Statisticheskoye Upravleniye and Goskomstat 1960–88 (annual volumes). *Narodnoye Khozyaystvo Turkmenskoy SSR. Statisticheskiy Yezhegodnik.* Ashkhabad: Turkmenistan.

Umarov, Kh. 1987. Intensivnoye ispol'zovaniye zemel'nykh resursov v usloviyakh rosta plotnosti naseleniya. *Izvestiya AN SSSR, Seriya Geograficheskaya* 6: 95–105.

Usmankhodzhayev, I. B. 1985. On the results of 1984 and the tasks of the collective farms, state farms and other enterprises of the repbulic's agro-industrial complexes in fulfilling plans and socialist pledges for 1985 in light of the decisions of the sixteenth plenary session of the Uzbekistan Communist Party Central Committee. *Current Digest of the Soviet Press* 37, 10: 1–4. Translated from *Pravda Vostoka,* 3 March.

Usmankhodzhayev, I. B. 1988. O zadachakh partiynoy organizatsii respubliki po vypolneniyu ukazaniy TSK KPSS o razvitii khlopkovodstva i ustraneniyu ser'yeznykh nedostatkov v etoy otrasli. *Kommunist Uzbekistana* 1: 4–25.

Usmanov, S. 1984. Puti povysheniya proizvoditel'nosti truda v sel'skom khozyaystve. *Sel'skoye Khozyaystvo Uzbekistana* 4: 2–4.

US Department of Agriculture, Foreign Agricultural Service 1977. *US Team Reports on Soviet Cotton Production and Trade.* Washington, DC: US Government Printing Office.

US Department of Agriculture, Foreign Agricultural Service 1989. *World Cotton Situation* (November). Washington, DC: US Government Printing Office.

USSR, CPSU 1988. At the CPSU Central Committee and USSR Council of Ministers. *FBIS Daily Report, Soviet Union,* 4 October: 56–9. Translated from *Pravda,* 30 September: 1–2.

USSR, Goskomstat—see USSR, Gosudarstvennyy Komitet po Statistike.

USSR, Gosudarstvennyy Komitet po Statistike 1986, 1987. *Narodnoye Khozyaystvo SSSR. Statisticheskiy Yezhegodnik.* Moscow: Finansy i Statistika.

USSR, Gosudarstvennyy Komitet po Statistike 1988a. *Kapital'noye Stroitel'stvo SSSR. Statisticheskiy Sbornik.* Moscow: Finansy i Statistika.

USSR, Gosudarstvenny Komitet po Statistike 1988b. *Kolkhozy SSSR. Kratkiy Statisticheskiy Sbornik.* Moscow: Finansy i Statistika.

USSR, Gosudarstvennyy Komitet po Statistike 1988c. *Naseleniye SSSR 1987. Statisticheskiy sbornik.* Moscow: Finansy i Statistika.

USSR, Gosudarstvennyy Komitet po Statistike 1988d. *Sel'skoye Khozyaystvo SSSR. Statisticheskiy Sbornik.* Moscow: Finansy i Statistika.

USSR, Gosudarstvennyy Komitet po Statistike 1988e. *SSSR i Zarubezhnyye Strany 1987. Statisticheskiy Sbornik.* Moscow: Finansy i Statistika.

USSR, Gosudarstvennyy Komitet po Statistike 1988f. *Trud v SSSR. Statisticheskiy Sbornik.* Moscow: Finansy i Statistika.

USSR, Gosudarstvennyy Komitet po Statistike 1989a. *Naseleniye SSSR 1988. Statisticheskiy Yezhegodnik.* Moscow: Finansy i Statistika.

USSR, Gosudarstvennyy Komitet po Statistike 1989b. O predvaritel'nykh itogakh vsesoyuznoy perepisi naseleniya 1989 goda. *Pravda*, 29 April: 2.

USSR, Gosudarstvennyy Komitet po Statistike 1989c. *Okhrana Okruzhayushchey Sredy i Ratsional'noye Ispol'zovaniye Prirodnykh Resursov v SSSR. Statisticheskiy Sbornik.* Moscow: Finansy i Statistika.

USSR, Gosudarstvennyy Komitet po Statistike 1989d. Posevnyye ploshchadi sel'skokhozyaystvennykh kul'tur po soyuznym respublikam. *Vestnik statistiki* 4: 73–6.

USSR, Gosudarstvennyy Komitet po Statistike 1989e. *Razvitiye agropromyshlennogo Proizvodstva v SSSR. Statisticheskiy Sbornik.* Moscow: Goskomstat.

USSR, Gosudarstvennyy Komitet po Statistike 1989f. *Sel'skokhozyaystvennoye Proizvodstvo v Lichnykh Podsobnykh Khozyaystvakh Naseleniya, Statisticheskiy Sbornik.* Moscow: Goskomstat.

USSR, Gosudarstvennyy Komitet po Statistike 1990. *Agropromyshlennyy Kompleks SSSR. Statisticheskiy Sbornik.* Moscow: Finansy i Statistika.

USSR, Gosudarstvennyy Komitet po Statistike, Byuro Sotsiologicheskikh Obsledovaniy 1990a. *Byudzhety Rabochikh, Sluzhashchikh i Kolkhoznikov v 1975–1988 gg. Sbornik Materialov po Dannym Byudzhetnykh Obsledovaniy.* Moscow: Goskomstat SSSR.

USSR, Gosudarstvennyy Komitet po Statistike, Byuro Sotsiologicheskikh Obsledovaniy 1990b. *Lichnoye Podsobnoye Khozyaystvo Naseleniva v 1988 Godu. Sbornik Materialov po Dannym Byudzhetnykh Obsledovaniy.* Moscow: Goskomstat SSSR.

USSR, Tsentral'noye Statisticheskoye Upravleniye 1958–85 (annual volumes). *Narodnoye Khozyaystvo SSSR. Statisticheskiy Yezhegodnik.* Moscow: Finansy i Statistika.

USSR, Tsentral'noye Statisticheskoye Upravleniye 1971. *Sel'skoye Khozyaystvo SSSR. Statisticheskiy Sbornik.* Moscow: Statistika.

USSR, Tsentral'noye Statisticheskoye Upravleniye 1987. *Sbornik Statisticheskikh Materialov (v pomoshch' agitatoru i propagandistu) 1986.* Moscow: Finansy i Statistika.

USSR, TsSU—see USSR, Tsentral'noye Statisticheskoye Upravleniye.

Uzbek SSR, Tsentral'noye Statisticheskoye Upravleniye and Goskomstat 1960–88 (annual volumes). *Narodnoye Khozyaystvo Uzbekskov SSR. Statisticheskiy Yezhegodnik.* Tashkent: Uzbekistan.

Valiyev, A. K. (ed.) 1988. *Nauchno-Tekhnicheskiy Progress i Sotsial'noye razsvitiye Sela v Uzbekistane.* Tashkent: Fan.

Vasil'yeva, E. K., Yeliseyeva, I. I., Kashina, O. N. and Laptev, V. I. 1985. *Dinamika Naseleniya SSSR. 1960–1988gg.* Moscow: Finansy i Statistika.

Voropayev, G. 1984. Sibirsko-Aral'skiy kanal i razvitiye proizvoditel'nykh sil prilegayushchikh territoriy. *Sel'skoye Khozyaystvo Uzbekistana* 2: 2–7.

Voropayev, G. V., Ismayylov, G. Kh., and Bostandzhoglo, A. A. 1988. Intensifikatsiya ispol'zovaniya vodno-zemel'nykh resursov rek basseyna Aral'skogo Morya s uchetom dostizheniy nauchno-tekhnicheskogo progressa. *Obshchestvennyye Nauki v Uzbekistane* 3: 5–14.

Yakubov, K. M. 1976. *Puti Povysheniya Proizvoditel'nosti Truda v Khlopkovodstve.* Tashkent: Fan.

Yazberdiyev, G. 1984. Uborku khlopka—na plechi mashin. *Sel'skoye khozyaystvo Turkmenistana* 8: 2–3.

Yunusov, B. 1983. Prodovol'stvennaya programma i effektivnost'ispol'-zovaniya trudovykh resursov. *Sel'skoye Khozyaystvo Uzbekistana* 11: 7–9.

Zyuzin, D. I. 1983. Prichiny nizkoy mobil'nosti korennogo naseleniya respublik Sredney Azii. *Sotsiologicheskoye Issledovaniye* 1: 109–13.

Zyuzin, D. 1985. Srednyaya Aziya—vazhneyshiy istochnik trudovykh resursov dlya narodnogo khozyaystva. In *Naseleniye Sredney Azii*, 88–95. Narodonaseleniye, V. 47. Moscow: Finansy i Statistika.

Work Force Composition, Patriarchy, and Social Change

Michael Paul Sacks

In any complex society economic development occurs unevenly. Some groups benefit more than others; some regions advance more quickly, while others may lag behind. This is clearly evident in Soviet society, and the way people are distributed across occupations is a particularly valuable measure of this inequality. Available data on the interrelationship between the regional, ethnic, and gender divisions in the work force provide important insights regarding the nature of inequality in Soviet Central Asia and the way in which work and family life may be suppressing out-migration of the indigenous ethnic groups of this region.

LESS DEVELOPMENT OR STIFLED DEVELOPMENT?

Soviet Central Asia is at a low level of economic development compared with other regions of the USSR. The more advanced position of particularly the Russian Republic is evident from a wide variety of measures (see Dellenbrant 1986: ch. 3; Clem 1988). The relationship between the Russian republic and Central Asia, however, cannot be characterized simply as an extension of tsarist colonialism.

Spechler (1979) labels the relationship "welfare colonialism" because of two related aspects of the situation. First, there are the absolute changes in Central Asia: "The long-term progress of Asian natives under Soviet rule—whether we measure in material, occupational or social (literacy, infant mortality) terms—has been impressive on any showing" (1979: 156). Secondly, income transfers from wealthier regions have bolstered the standard of living in Central Asia and assured that in this respect the region has not become relatively worse off (see also Dellenbrant 1986: 51).

Rywkin (1984) concurs with this view, although he stresses that this seemingly altruistic Soviet policy has ultimately resulted in a population that is "spiritually alienated" from the Soviet regime:

> [Soviet colonialism] is willing to go as far as to subsidize its dependencies instead of exploiting them. It is willing to offer inhabitants of these dependencies the equality of personal opportunities with the Russians, curtailing them only when state security is in question. Ideologically inspired, it is willing to strive for economic, social and educational equalization between the ruling and the ruled ethnic groups. (1979: 13–14)

Elsewhere, Rywkin (1984: 57) seems to stress far more the "politically opportunistic" aspects of policy aimed at accommodating the concerns of the "Muslim inhabitants in this politically sensitive geographic area."

Others see a more familiar form of colonialism. Dienes (1987) has attempted to explain why, despite the interrepublic redistribution of national income, a development gap persists. The investment in Central Asia may be aimed at the further exploitation of its cotton and other resources. Narrow development oriented towards export creates an economy that is increasingly dependent on the Russian Republic and not one which promotes self-sustaining economic growth. Based on 1966 economic data, Gillula finds that "major exports of less developed republics were commodities at or near the primary stage of production, while the imports of these republics were the output of a several-stage production process" (Dienes 1987: 650).

Dienes (1987: 133, 12) sees a "quasi-colonial economic structure" and views Central Asia as a "plantation province." Food and products of light industry must be imported from the RSFSR, while gas and especially cotton are exported. Particularly telling in Dienes's view is the contrast between the enormous quantity of cotton exported and the low proportion of cotton textiles manufactured, Central Asia and southern Kazakhstan accounting for only 6.5 percent of the Soviet total production (1987: 123, 125). Dienes (1987: 125) and Hodnett (1974: 67, 103) find similar criticism in Soviet monographs, although without the use of the label "colonialism."

Hodnett (1974: 95–6) argues that industrial development has been

> strictly cotton related: ginning, the textile industry, agricultural machine building, and so forth. . . . A great deal of the newer development that breaks away from cotton belongs either to one or another branch of extractive industry or to power production or both. . . . Thus, what appears to be happening is the creation of some new branches of primary industry which do not attack the cotton base of the regional economy and do not provide new large-scale employment opportunities.

Hodnett (1974: 103) concludes that the mechanization of cotton production has not been associated with a transformation of either agricultural production or the broader economy in such a way as to promote sustained economic development. The most likely future scenario, however, is the Soviet state's "attempt to find a structurally dependent pattern of development linking Central Asia to the rest of the Soviet Union, but also generating local employment" (Hodnett 1974: 105–6).

Rumer (1989) conceives of Soviet policy as far less benevolent. In *Soviet Central Asia: "A Tragic Experiment"* he documents the low level of Central Asian economic development, declining investment per capita and falling standard of living. Rumer (1989: 120) views this condition as the outcome of a "hidden conflict . . . between the decision makers in Moscow and the Central Asian republics." "[M]oscow is determined to preserve Central Asia as a peripheral region whose function is to provide mineral resources and raw materials for the advanced industrial complexes of European Russia" (p. 184). Insufficient investment, not overpopulation, has resulted in a shortage of jobs. Investments are geared to "one over-arching goal—cotton" (p. 40). Rumer quotes a 1987 article from the Soviet weekly, *Literaturnaya Gazeta* by Akhmed Ulmasov of the Tashkent Institute of Economics: "By being transformed into virtually one great cotton plantation, Uzbekistan embarked on a long, tragic experiment—to determine the capacity of monoculture to corrode not only agriculture, but also industry, education, health, and finally public morality" (quoted on p. 69). Labor-intensive industry needs to be fostered and dispersed among small and medium size towns, where it is most likely to promote better employment opportunities for the indigenous population (p. 120). "Reading between the lines" of Soviet sources reveals the exploitation of Central Asia by the authorities in Moscow (Rumer 1989: 59).

Not everyone may see the same writing that Rumer does, although the Soviet press continues to contain ample criticism of the dependence on and underpricing of cotton, the lack of textile factories, and the inadequate expansion of employment opportunities within the region. A Tashkent writer decries the "classic Third World features" of the Central Asian republics, where poverty has reached "60% in Tadzhikistan, over 46% in Uzbekistan, 40% in Kirgiziya and Turkmeniya"; he attributes the poverty to "mass unemployment and the meager wages of the bulk of the population working in the cotton fields" (Pulatov 1990: 7; see also Ziyadullayev 1987: 224–6; Pchelintsev and Ronkin 1988: 14–15; Ubayidullayeva 1988). The first secretary of the Communist Party of Uzbekistan attributes the cotton monoculture to the absence of regional economic autonomy and the absence of the influence of cotton workers and brigade leaders themselves (Karminov 1989: 1). Sharp criticism comes not only from the native population. A member of the Institute of Sociology of the USSR Academy of Science argues that the characterization of Central Asia and the Caucasus as labor-surplus regions and of other regions as short on labor derived from administration methods of the 1920s and 1930s: "This gave rise to the infringement upon the rights of the republics, thrusting upon them a direction and pace of economic development which issued from branches and departments and not the productive potential and nationality characteristics of the republics" (Morozova 1989: 75). Rejecting out-migration as a practical solution to unemployment, the author advocates investment in much needed social services and local industry accessible to the rural population (p. 78).

Gleason (1990) sees the problem in Central Asia as stemming from Brezhnev's failure to invest sufficiently in the mechanization of cotton harvesting. Huge sums were spent on the region, but this was for irrigation to expand cotton production and fulfill the production targets of central planners. Such plans provided distinct disincentive to produce other crops (p. 80–83).

Brezhnev strategy, as in other matters, focused on "political stability in the short and medium term at the expense of the long-term interests of Soviet economic development" (p. 85). Mechanizing production was put off because it would have displaced a large number of workers. Expanding labor intensive production allowed for the further absorption of population growth. Today, as a result, there is a far higher proportion of the rural population earning a very low income. Yet prices for this inefficiently produced cotton were, of course, higher than they would have been had mechanization been fully implemented. Thus, those in European Russia complain about the unfairness of continuing to subsidize Central Asian development, while Central Asians see the underpaid and ecologically destructive cotton monoculture as the source of their impoverishment.

But the native elite are not unaware of the danger now posed by a large-scale displacement of rural labor through either mechanization or even switching to a variety of other crops that require less labor. Gleason (1990, 92) argues that "cotton still has strong supporters." Moreover, he suggests that in the past "local leaders may have been more successful in promoting their own agenda than we—or Moscow leaders—have given them credit for" (p. 86). Developments in Kazakhstan, Latvia and Estonia showed that industrial development would not draw indigenous labor into the factories, but would bring an influx of Russians and Ukrainians. Gleason concludes:

> One can imagine that local elites felt that a delay of industrialization would keep Russians out while ensuring that indigenous birthrates would remain high and guarantee the long-term indigenous domination of the area. If it

was a conscious calculation, it seems to have achieved its objective (1990: 86).

OCCUPATIONAL DISTRIBUTION AS AN INDICATOR

Data on regional distribution of occupation provide an alternative source to evaluate the relationship between Central Asia and the Russian Republic. The number employed in detailed occupational categories is available for 1959 and 1970, but not as yet from either the 1979 or the 1989 census.

The shift of the work force out of agriculture (excluding the private subsidiary sector) occurred at a rapid rate in Central Asia. This shift is an important indicator of economic development (see Durand 1975). In 1959 under half the work force was employed in nonagrarian occupations (47.6 percent); by 1970 this reached nearly two-thirds (65.8 percent). The Russian Republic, already at a much higher level of industrial development, showed less of a shift (from 73.3 to 85.7 percent outside agriculture). By 1970 the percentage nonagrarian among males in Central Asia (73.9 percent) was about equal to the figure for the total work force of the Russian Republic in 1959. Although this is clearly a crude measure, the dramatic change in Central Asia does not seem consistent with the view that some form of aborted economic development was occurring.

Other aspects of the structure of the work force also suggest this. As already noted, Dienes suggests that one sign of a colonial structure is the absence of textile manufacturing commensurate with the level of cotton production. The relatively small textile industry is shown by the occupational data. In both 1959 and 1970 only about 2.5 percent of the work force were sewers or textile workers. This was only a fraction of a percentage point different from the Russian Republic (2.9 percent in 1959 and 2.7 percent in 1970). But would a work force in Central Asia highly skewed toward the manufacturing of clothing be a positive sign? Hodnett stresses the need for diversified development. Dienes (1987; 144) notes that, in debates regarding the kinds of industrial plants that should be constructed in rural areas, "Central Asian specialists differ on the extent to which such industrialization should concentrate on traditional wool, cotton, and food-processing branches or be extended to more modern industries."

Differences between the work force of the Russian Republic and that of Central Asia should reflect their relative levels of economic development. Overall differences can be assessed using a measure of differentiation. This shows the percentage of workers in Central Asia who would have to change occupations in order for them to be distributed across occupations in precisely the same way as workers in the Russian Republic. Ninety-four categories were used to calculate the measure. The measure dropped from 27 percent to 23 percent between 1959 and 1970. Agriculture accounted for nearly half these occupational differences. Eliminating agriculture produced a measure of 15 percent for both years, a seemingly low figure for a "metropolis"—colony comparison. Comparing the Ukrainian Republic and the Russian Republic in 1970 results in a measure of differentiation of 6 percent. This contrasts with the very sharp occupational differences between men and women. In 1970 in the Russian Republic, 54 per cent of females would have had to change occupations (excluding agriculture) in order for them to be distributed in the same way as males. The comparable figure for Central Asia was 44.5 percent.

Besides agriculture, the occupations accounting for the most significant differences between Central Asia and the Russian Republic were in machine construction, metallurgy, and engineering-technology. These are pivotal areas associated with large-scale industrial development. Here regional differences grew. In 1959 these occupations comprised 15.8 percent of workers in the Russian Republic as compared with 7.0 percent in Central Asia. In 1970 the figure rose by 7.7 points in the Russian Republic (to 23.5 percent), but by only 4.3 points in Central Asia (to 11.3 percent).

The lower initial figure for Central Asia points to the progress made by 1970. By 1970 nearly 18 percent of all males of Central Asia were in machine construction, metallurgy and engineering and technology. But it is also true that in Central Asia these occupations appear to be largely geared to the needs of cotton production. Seventy-five percent of the products of Uzbekistan's industry are reportedly connected with agriculture (Ziyadullayev 1984: 6, 28–9).

The third largest source of differences between Central Asia and the Russian Republic was primary and secondary education. Between 1959 and 1970 the percentage of the labor force teaching at this level grew from 2.5 to 4.0 in Central Asia; in the Russian Republic the corresponding figures were 1.9 and 2.4 percent. This is hardly surprising given the far younger age structure of the Central Asian population. All other differences in occupational distribution were far smaller than this, but, as is shown in the next section, are important for other reasons.

Rumer (1989; 55–7) contends that the system of territorial economic adminstration implemented by Khrushchev between 1957 and 1965 had very positive consequences for the industrial development of Central Asia. This included the growth of machine building and especially the chemical industry and nonferrous metallurgy. Though arguing that regional control over development was lost in the subsequent five-year plan, Rumer concedes that the positive consequences of the earlier policy gave sufficient momentum to sustain high industrial growth for some time. But Rumer emphasizes (p. 59) that "nonferrous metallurgy . . . is confined to the extraction and enrichment of metals—that is, the production of ore concentrates. The final processing and rolling (the manufacture of finished products) is generally performed outside the region." A Soviet scholar asserts, to the contrary, that

> in comparison with other republics, Uzbekistan SSR exceeds the average growth of industrial production due to the sped up growth of those branches which involve "higher stages" in the processing of raw materials. This includes machine construction, the chemical and petrochemical industry, as well as branches of light industry which turn out consumer products. (Ziyadullayev 1987: 220)

Hopefully, as yet unpublished occupational data from the censuses for 1979 and 1989 can provide a basis for evaluating the nature of regional differences in industrial development. But it does appear clear that changes in work force structure in recent decades were certainly not as great as in the 1960s. Rapawy and Heleniak (1987: 10–13) estimate that the percentage of the work force employed in agriculture in Central Asia fell from 39 to 35 percent between 1970 and 1985 as compared with a drop from 25 to 19 percent in the USSR as a whole. Central Asia appears to be further lagging behind in the development of its nonagrarian sector relative to the rest of the USSR, but the continuing decline in the share of agriculture despite very substantial growth in the employed population might be interpreted as a sign of the strength of the Central Asian economy. Over the same 15 years nonagrarian employment in Central Asia grew by 67 percent as compared with only a 31 percent growth in the USSR as a whole.

Another indicator, the change in average wages by branch of the economy, points to adverse developments in the labor force. Table 7.1 shows the ratio (times 100) of the wages in the republics of Central Asia to the wages of the Russian Republic for selected years and branches of the economy. In other words, for each year the wages of the Russian Republic were made equal to 100. Figures below 100 indicate that the republic had lower wages than the Russian Republic; above 100 means higher wages than the Russian Republic.

The pattern is strikingly clear. With the exception of Turkmeniya, the average wages among workers and employees in Central Asia (see column 1) had been at least 10 percent lower than in the Russian Republic at the beginning of the period. The discrepancy increased sharply by 1987; the indexed figures fell in all four republics. Declining wages signal a growth within the branch in the proportion of workers in lower-paid occupations and/or in lower skilled categories within

Table 7.1 Indexed wages of workers and employees by selected branches, 1975–87 (Russian Republic = 100 in each year).

	Year	Total	Industry	Agriculture	Transportation	Communication	Construction	Trade	Health	Education
Uzbekistan	1975	89	90	90	93	94	98	93	92	97
	1980	88	87	96	87	85	97	89	90	94
	1985	82	82	78	81	84	89	86	87	97
	1987	79	80	72	75	81	82	81	88	98
Kirgiziya	1975	88	92	86	90	97	90	91	90	98
	1980	83	89	80	82	91	85	86	88	94
	1985	81	86	76	77	83	81	84	86	94
	1987	79	86	69	76	82	78	82	85	96
Tadzhikistan	1975	89	87	82	92	92	94	96	99	106
	1980	82	83	79	79	83	85	89	94	101
	1985	78	80	68	74	80	81	85	93	104
	1987	77	79	64	70	77	78	83	87	105
Turkmeniya	1975	106	104	129	100	112	116	105	106	116
	1980	99	98	105	95	105	110	100	99	109
	1985	95	95	91	90	101	105	97	95	111
	1987	92	92	87	86	95	98	91	93	109

Source: Gosudarstvennyy Komitet SSSR po statistiki 1988: 20, 22–3.

occupations. In Central Asia this is surely associated with changes in the labor force composition: a larger proportion of females and a smaller proportion of Russian and other nonindigenous ethnicity. Low wages relative to the Russian Republic, however, cannot be attributed only to concentration in agrarian occupations like cotton picking. While the fall is very sharp in agriculture, it is evident across all branches with the exception of education.

More recent and more detailed occupational data may help to judge the degree to which Central Asia has been exploited to benefit other regions in the USSR or the extent to which its economic development has been crippled. Before terms such as "colonialism" or "neocolonialism" are applied to Soviet regional disparities, there is a need for more explicit definitions including empirical indicators. Comparative study, absent from the studies mentioned above, would also shed new light on both the theoretical and empirical issues. Substantial variation in development across regions, patterns of regional interdependency rather than self-sufficiency, and difficulties with providing sufficient opportunities for growing subpopulations are common in industrialized nations. The *maquiladora* industry tying major US corporations with impoverished Mexican labor (Tolan 1990), for example, suggests that the Soviet case should be evaluated in an even broader context. Looking at Soviet Central Asia in isolation may make it too easy to apply pejorative terms and confirm initial biases.

ETHNIC AND GENDER DIFFERENCES

The higher level of economic development of the Russian Republic has its counterpart in the more advantageous position of the Russian ethnic group within Central Asia. The change between 1959 and 1970 in Central Asia was associated with a large influx of Russians, and a wide range of sources show that these Russians were absorbed in the expanding industrial sectors far more than were the indigenous ethnic groups. These ethnic differences are reflected in the heavy urban concentration of the Russians in Central Asia. (Clem 1988).

Indirect measures are required to study detailed occupational differences among ethnic groups, for the published volumes of Soviet censuses after 1926 contain no tables on the ethnic composition of occupations (Sacks 1986). The approach used here is based on the assumption that Russians in Central Asia are not likely to be concentrated in occupations that are very different from those of workers in the Russian Republic. Males and female workers must be separated given the very substantial gender differences in employment.

Detailed occupational differences between regions can be detected by calculating for each occupation the ratio of workers in the Russian Republic to those in Central Asia. Such a ratio (multiplied by 100) shows the number of workers in Central Asia as a percentage of the number in the Russian Republic. In 1970 all nonagrarian workers in Central Asia taken together were equal to 8.4 percent of the number of nonagrarian workers in the Russian Republic (up from 6.4 percent in 1959). Given the low female labor force participation of the indigenous ethnic groups of Central Asia, it is not surprising that the figure for females was 7.0 percent as compared with 9.8 percent for males. An occupation has been defined as containing a particularly large number of males or females if the percentage for the category was twice or more these respective figures, that is, the number of women in Central Asia in 1970 equaled 14 percent or more of the number of women within the same occupation in the Russian Republic; the figure for men equaled or exceeded 19.6 percent.

Of the 93 nonagrarian occupational categories, there were only 2 in which females in Central Asia were found in particularly large numbers relative to the Russian Republic. Among males, however, there were 26 such categories. This points to the far greater similarity among women than among men when comparing Soviet regions (Sacks 1982: ch. 4). It is very significant that in only 2 of the 26 occupational categories was the total number of workers in Central Asia particularly large relative to the Russian Republic. Thus, this does not conflict with the findings in the previous section, for differences between the two regions were especially strong only for male workers.

Ethnic composition explains the gender differences. The female nonagrarian labor force of Central Asia has a far lower proportion of indigenous ethnicity than does the male labor force. Indeed, it is the low female labor force participation among the Turkic-Muslim groups that may explain the distinctive employment pattern of the males. The situation appears very comparable to what Youssef (1974) finds in the Middle East. Cultural restrictions exclude women from occupations that "involve public activity or presuppose contact with men." As a result, "occupations which in other countries become predominantly feminine from early industrialization onwards (such as the service occupations, domestic work, factory work, retail and clerical jobs) are in the Middle East staffed by men or foreign women" (1974: 37). In Central Asia, the "foreign women" are heavily of Russian ethnicity, but to a very substantial degree Turkic-Muslim men have made up for the low supply of indigenous women workers.

In Uzbekistan in 1926 restrictions on the work of indigenous females were particularly clear. Of the 56,000 indigenous women working outside of agriculture, 87 percent were textile making and sewing. These indigenous women were overwhelmingly in settings where they had very little contact with strangers. Sixty-one percent worked alone,

Table 7.2 Selected data on occupations with a large number of males in Central Asia relative to the Russian Republic in 1970.

Occupational category	CA/RSFSR ratio for males, 1970 (1)	Percent female in 1970 CA RSFSR (2)	Percent female in 1970 CA RSFSR (3)	Change in percent fem. 1959–70, CA (4)	Number of males, CA, 1970 (5)
Agronomists and zootechnicians	0.207	12.8%	46.1%	−2.7%	19,665
Physicians	0.197	60.0	77.4	−8.1	14,302
Dentists	0.240	52.1	83.1	−13.0	1,424
Pharmacists, assistant pharmacists	0.611	81.9	97.4	−6.7	911
Feldshers, midwives	0.417	87.1	96.5	1.1	14,230
Nursery directors and training personnel	0.468	98.0	99.7	−1.2	102
Other medical personnel	0.300	52.6	77.1	11.9	3,630
Primary and secondary schoolteachers	0.422	46.5	77.1	1.7	149,614
Heads of publishing houses and their divisions	0.214	52.3	78.4	1.6	1.87
Library heads	0.504	85.0	96.4	10.7	3,293
Telephone operators	0.397	88.4	97.4	−2.4	1,209
Store department managers	0.470	30.6	73.1	5.8	25,165
Heads of procurement and supply organizations	0.228	25.8	56.3	11.4	24,625
Bookkeepers, tally clerks	0.300	64.2	88.7	9.9	46,272
Statisticians	0.383	81.7	96.7	5.3	386
Cashiers	0.452	77.4	96.7	11.8	6,491
Accounting clerks	0.492	35.8	83.4	17.4	14,224
Typists and stenographers	0.600	96.0	99.3	−2.7	572
Secretaries and office personnel	0.325	87.4	96.5	13.7	4,035
Spinning machine operators	0.256	93.6	98.1	−0.8	472
Weavers	1.597	84.6	98.5	13.3	3,596
Embroiderers and other sewers	0.210	83.9	91.6	−0.6	1,216
Postal workers	0.409	46.3	89.8	4.0	8,351
salespersons	0.755	58.4	95.5	7.9	42,232
Cooks	0.485	63.1	92.7	−6.2	22,874
Communal and everyday service workers	0.201	63.9	82.6	13.8	112,785
Agriculture	0.214	57.9	51.8	5.6	958,571

Source: TsSU 1973b, tables 3, 7, 13, 14, 16, 19, 22, 29, 30, 32.

and an additional 21 percent worked only with other members of their family. Slavic females in Uzbekistan had far more diversified employment, and two-thirds worked in an office setting ("Sluzhashchiye") (TsSU Otd. Per. 1929: table IV).

Table 7.2 shows the occupations (excluding agriculture) where there were a particularly large number of Central Asian males in 1970 (defined as a Central Asia–Russian Republic ratio for males in excess of 0.196). Column 3 shows that in the Russian Republic these occupations were clearly "women's work." In 1970, 20 were over three-quarters female, and 14 were over 90 percent female. Only 3 occupations were over 90 percent female in Central Asia, and only 12 exceeded 75 percent female (column 2). With few exceptions, these were occupations which contained a far higher proportion of men in Central Asia than they did in the Russian Republic.

Column 4 shows an important additional dimension. In Central Asia since 1959, the percentage of women *increased* in 16 of the 23 categories; the increase was substantial (five or more points) in 11 categories. It is only in the medical field (physicians, dentists and pharmacists) and among cooks that males were clearly taking a greater share of the field.

While this method of analysis is surely imprecise and cannot show where all Turkic-Muslim men were employed, it does suggest that many of those outside of agriculture were in occupations that tended to be female-dominated in the Russian Republc and often characterized by a growing female presence.

WEAVERS, SALESPEOPLE, CLEANERS AND TEACHERS

A look at a few specific occupations shows the changes in occupational characteristics that are likely to influence gender composition.

The highest Central Asia–Russian Republic ratio by far (Table 7.2, column 1) was for weavers. Males outside of agriculture in Central Asia amounted to less than a tenth the number of such men in the Russian Republic, but the number of Central Asian male weavers vastly outnumbered those in the Russian Republic.

In 1926 in Uzbekistan weavers were nearly exclusively indigenous men (98.2 percent); these males either worked alone (64 percent) or headed firms employing only family members (30.5 percent) (TsSU Otd. Per. 1929: table IV). By 1939, a time when weaving was already a female-dominated field in the Russian Republic (8 percent male), males comprised the majority (58 percent) of Central Asian weavers. By 1970 only 15 percent of Central Asian weavers were men. Males were obviously leaving weaving as it changed from a rural handicraft involving skilled labor and a large degree of autonomy to urban factory employment. The number of male Central Asian weavers fell by nearly 30 percent between 1939 and 1970, while the total number of weavers grew from 8,828 to 23,317.[1]

The situation among salespeople shows a different trend. In 1926 in Uzbekistan 9 (0.6 percent) of 1,502 indigenous sales personnel were female; 184 (12.9 percent) of the 1,425 Slavs in sales were women (TsSU Otd. Per. 1929: table IV). As was true for weavers, only a majority of salespeople were males (again, 58 percent) in Central Asia by 1939. But in this occupation the number of Central Asian males grew nearly fourfold by 1970. Still the influx of females greatly exceeded the influx of males, and the occupation was 58 percent female by 1970. In the Russian Republic females comprised 61 percent in 1939, and by 1970 salespeople were almost exclusively female (95.5 percent). Here is a case where the occupation surely changed in character as more modern retail establishments expanded, but Turkic-Muslim males remained. Why?

This is a very broad category. It is impossible to know whether males and females are segregated into very different types of sales and levels of sales management. But there may be a strong financial incentive to tolerate female presence. Lubin (1984: 169), whose argument is more critically examined in the last section, stresses that the attraction of the Turkic-Muslim population to jobs in the service sector and light industry—areas that appear to be the "lowest-paid and least prestigious"—can be explained by the high income potential that escapes official statistics:

> Salespeople augment their incomes by selling items under the counter, by selling information on where to queue, by directly stealing goods or, as in the case of meat or food products, by directly manipulating the quality of the cut or the precision of the scale and increasing his or her own income accordingly. ... Thus, in 1979, a shoe salesman in Tashkent received a monthly salary of eighty roubles; by selling "left" (or "black") shoes, by fudging records or by selling shoes "under the counter", his net actual earnings were approximately 200 roubles per day—approximately 4,000 roubles per month (1984: 193)

A Soviet scholar, Perepelkin (1987: 87), argues that there is a strong interest among Uzbeks in occupations that afford "social interaction with the end users of the products produced and social interaction while working," and that this makes occupations in trade and everyday services, as well as work as railroad drivers and conductors, particularly attractive to them.

The category "communal and everyday service workers" is very diverse and includes janitors, launderers, cleaners and furnace workers. These occupations would appear to be far less attractive than sales as either sources of illicit income or rewarding social intercourse. Consequently there may have been less resistance to Turkic-Muslim women entering these jobs, many of which must have been at the lowest level of pay. In 1939 in Central Asia two-thirds of the jobs went to men; by 1970 the number of men had fallen sharply, and the jobs had become almost two-thirds female. The feminization had occurred earlier in the Russian Republic, so that even in 1939 the majority of these workers were women. (In 1926 in such comparable occupations as office cleaners, personal servants, and hospital attendants, 80 to 99 per cent were female among Slavs in Uzbekistan, but only 11 to 37 percent female among the indigenous ethnic groups—TsSU Otd. Per. 1929: table IV.)

Finally, teaching—the largest male category in Table 7.2—reflects the changes that took place with the transformation from exclusively male religious schools to a massive educational system for all children. In 1926 in Uzbekistan 9 percent of the teachers of indigenous ethnicity were female as compared with 60 percent among teachers of Slavic ethnicity (TsSU Otd. Per. 1929: table IV). In Central Asia as a whole in 1939 three-quarters of the 81,000 primary and secondary schoolteachers were men (as compared with only 40 percent in the Russian Republic). By 1970 there were nearly 280,000 such teachers; the number of males had grown, but by then they comprised just under the majority (47 percent) as compared with less than a quarter in the Russian Republic.

All four areas show growing female presence. In some cases this may have been a result of men moving into new areas that were segregated from females. Female entry appears less important than social change which altered the character of the occupations and made the work inherently less desirable for men. Given their presence in occupations which in the Russian Republic were dominated by women and their limited alternative employment opportunities, men of Turkic-Muslim ethnicity increasingly found themselves in occupations similar to those of Russian women who had migrated to Central Asia. In recent years Turkic-Muslim women may be competing for these typically "female" occupations.

RECENT DATA ON ETHNIC COMPOSITION

Some confirmation of this picture of ethnic and gender occupational distribution comes from 1979 census data revealed by Perepelkin (1987). The figures in Table 7.3 show the extent to which Uzbek males and Russian males in the city of Tashkent are over- (in excess of 1.0) or underrepresented (less than 1.0) in selected occupations. They are equal to the ratio for each ethnic group of the proportion of those in the occupation to the proportion in the work force as a whole. (The actual proportion of the work force consisting of Uzbek males and Russian males is omitted from the article, as are any data on females.) Excessive overrepresentation (over 2.0) of Uzbek males occurs precisely in areas where Russian men are most underrepresented: among workers in trade and public dining, drivers and conductors, food industry workers, and to a lesser degree among textile workers. In the Russian Republic in 1970 these areas varied between 83 and 96 percent female, with the exception of drivers, who were 56 percent female. (The category for drivers in the census, however, included drivers of trams, buses, and trains.)

Table 7.3 Russian and Uzbek male representation in selected branches of the economy of Tashkent in 1979.

	Uzbeks	Russians
Workers in machine construction and metallurgy	0.79	1.39
Textile workers	1.35	0.68
Sewers	1.19	0.48
Leather and fur workers	1.19	1.00
Food industry workers	2.03	0.31
Construction workers	0.92	1.00
Machine operators	0.25	1.41
Concrete and asphalt workers	1.24	0.69
Workers in railroad transport	1.43	0.86
Locomotive operators and their assistants	0.80	1.34
Drivers and conductors	2.38	0.15
Workers in trade and public dining	2.38	0.08

Source: Perepelkin, 1987: 85.

The underrepresentation of Uzbeks is greatest in areas of machine construction and metallurgy as well as the operation of machinery (locomotive operators and their assistants; machine operators in construction)—areas of high Russian male concentration. Thus, areas which were shown above to be key indicators of the greater industrial development of the Russian Republic relative to Central Asia are precisely the same as the occupations in which Russians—or at least Russian males—within Central Asia tend to be concentrated. There is a clear interconnection between regional, intraregional, and ethnic differences.

The proportion of indigenous ethnicity within the work force of Central Asia, however, is rising significantly. This is shown by data on the representation of the titular ethnic groups within their respective republics. Summing the number of titular group ethnicity within each of the four Central Asia republics (that is, the number of Uzbeks in Uzbekistan, Tadzhiks in Tadzhikistan, and so on) shows that these titular ethnic groups comprised only 39 percent of workers and employees

Table 7.4 Titular ethnic group representation among workers and employees by branch of the economy.

Republic	Total	Industry	Agriculture	Transport and communic.	Construction	Trade and public dining	Housing and everyday sevices	Health physical culture	Education	Art and culture	Science	Government
(a) Percentage of titular ethnicity in 1987												
Uzbekistan	61	53	76	55	50	66	55	64	69	63	39	57
Kirgizia	41	25	69	35	26	34	30	46	43	46	27	42
Tadzhikistan	54	48	63	57	48	61	56	50	58	56	31	51
Turkmeniya	59	53	81	48	54	65	53	62	67	70	48	51
(b) Percentage change in titular ethnic group representation between 1977 and 1987												
Uzbekistan	10	15	9	11	14	3	3	12	8	2	8	11
Kirgizia	8	10	6	7	10	11	11	14	−1	0	7	7
Tadzhikistan	9	13	1	12	12	9	10	17	7	1	8	9
Turkmeniya	14	19	23	8	13	18	12	14	11	6	13	6

Source: Gosudarstvennyy Komitet SSSR 1988: 20, 22–3.

in 1967, 47 percent by 1977, and 57 percent by 1987. In the period between 1977 and 1987 the titular ethnic group members within each republic comprised 84 percent of all additional workers and employees in Central Asia (TsSU 1969: 551, 1978: 381; Gosudarstvennyy Komitet 1988: 20, 22–3).

Table 7.4 shows that, while titular ethnic group representation is highest in agriculture (top half of table), between 1977 and 1987 large increases have occurred in widely diverse areas (bottom half of table). The rise in industry is particularly sharp. This has occurred despite the fact that there has been only a slight increase in the proportion of all titular ethnic group workers and employees found within industry. For example, in 1977 in Uzbekistan 16 percent of all Uzbeks were in industry; by 1987 this rose only three points. (This compares with about one-quarter of other ethnic groups in Uzbekistan being in the branch of industry at both points in time.) But industry became sharply more Uzbek in ethnic composition, rising from 38 percent to 53 percent Uzbek over the same period. The number of Uzbeks in industry doubled as compared to about a 9 percent rise among non-Uzbeks (Gosudarstvennyy Komitet 1988: 24, 52).

These figures, unfortunately, say nothing about the occupations within each branch that the Uzbeks are employed in. Wage declines by branch (Table 7.1) suggest that this is very important. Moreover, to some degree the rise in the titular ethnic group is a sign not only of Russian out-migration or lesser growth but also of the out-migration of indigenous ethnic groups other than the titular group (Anderson and Silver 1989). But these data certainly do point to a complex and not yet clear picture of indigenous ethnic group entry into the modern sectors of the economy.

The enormous increase in workers of indigenous ethnicity is likely to reduce the segregation of males and females. This can be understood by looking at geographical distribution of employment.

THE GEOGRAPHY OF GENDER DIFFERENCES

The geographic separation of ethnic groups appears to mitigate problems that might arise as a consequence of Turkic-Muslim male concentration in areas overlapping Russian women.

Soviet census data for 1970 show how the workers in broad branches of the economy are distributed between urban and rural areas. These data also show differences by gender and, thereby, make it possible to draw inferences regarding ethnic divisions.

Tables 7.5 and 7.6 include two categories which are likely to have a large concentration of Turkic-Muslim men. Data are shown separately for the Russian Republic and for Central Asia. Females were between 72 and 79 percent of these workers in the Russian Republic (see Table 7.5). In the area of education, science, and public health, those employed in rural areas were somewhat more likely to be women than those in urban areas. The pattern in Central Asia was strikingly different. The percentage female was far lower with the one exception among those employed in education, science, and public health in *urban* areas. In rural areas males predominated in both categories; the percentage female was over 20 points higher in urban areas.

Table 7.5 Percentage female by branch for urban and rural areas of Central Asia and the Russian Republic, 1970.

Branch group	Central Asia		RSFSR	
	Urban	Rural	Urban	Rural
Trade, public catering, material–technical supplies and marketing, and procurement	55.4	33.1	79.0	72.3
Education, science, and public health	68.9	48.3	72.0	74.5

Source: TsSU 1973a: Tables 13, 16, 23, 24, 26.

Table 7.6 shows the same data percentaged differently. Females of the Russian Republic in these branches of the economy were about as likely to be in cities as were the females in the same jobs in Central Asia. But sharp regional difference exists among males. The Central Asian males were far more concentrated in rural areas. In education, science, and public health the majority of the males of Central Asia worked in rural areas; three-quarters of the men of the Russian Republic worked in urban areas.[2]

Table 7.6 Percentage urban by branch for males and females in Central Asia and the Russian Republic, 1970.

Branch group	Central Asia		RSFSR	
	Males	Females	Males	Females
Trade, public catering, material–technical supplies and marketing, and procurement	58.2	77.7	71.2	78.1
Education, science, and public health	47.0	67.8	75.7	73.2

Source: TsSU 1973a: tables 13, 16, 23, 24, 26.

This all reflects the concentration of Russians in the cities of Central Asia and the indigenous ethnic groups in rural areas. Russian females employed in the service sector of the economy (as well as in the work force as a whole) were located overwhelmingly in the cities; in the countryside the restrictions on the participation of indigenous women meant that these jobs were filled by Turkic-Muslim men. Geographic segregation formed a barrier to gender and ethnic integration. But this barrier is eroding both with the growing female proportion in many of the occupations and the increasing numbers of Turkic-Muslims in the cities of Central Asia. Moreover, there are clear signs of an increasing number of Turkic-Muslim females in the nonagrarian work force (Sacks 1982: 80–2, 1988: 98–9).

PROBLEMS OF PRESERVING PATRIARCHY

It is ironic that Turkic-Muslim males, likely to uphold highly patriarchal values, are perhaps most vulnerable to female competition in the labor force. This vulnerability appears to be the outcome of growing contact between the traditional rural Central Asian society and the Russian-dominated urban industrial society. The past isolation of rural Central Asian society may explain the strength of the patriarchal culture, the high fertility of the region (see van de Walle and Knodel 1980), and the reason men are so heavily represented in occupations which are "women's work" in the more modernized regions of the USSR.

This isolation has declined as a result of rising educational attainment, increased nonagrarian labor force participation, increased exposure to mass media, and a moderate rise in migration and travel to cities. Fertility levels are falling. Less isolation may have heightened the perception of Turkic-Muslim men that patriarchal privileges are being challenged. Their response may be to bolster the geographic barrier that served so well in the past. This conclusion is suggested by recent feminist research on the dynamics of patriarchy.

In *Patriarchy and Socialist Revolution in China*, Stacey (1983) traces the persistence of the patriarchal family in China despite revolutionary change. Critical here is Stacey's analysis of patriarchy as a powerful and independent force. Early policies of the Chinese Communist Party, especially the distribution of land to peasant households, actually resuscitated faltering patriarchal structures. The relatively autonomous peasant family economy gained strength. Problems arose when revolution and patriarchy were no longer compatible. By the 1950s "Chinese patriarchy reached the limit of its capacity to serve as a revolutionary force, and became instead, the constraining context in which strategies for socialist development were destined to proceed" (1983: 203). The initiation of the Great Leap Forward threatened the economic interdependence of family members that was at the root of Chinese patriarchy. This family system then served to impede or halt further change in rural China.

A key point here is that the family may shape economic arrangements. This is illustrated by Stacey's interpretation of the source of large, complex family structures in Taiwan in the 1960s. Myron Cohen argued that these structures were the outcome of the high labor input required for tobacco growing and other aspects of employment opportunities in the region "which made the economic flexibility and interdependence of the large, joint-family household an economically sensible pattern" (Stacey 1983: 23). Stacey suggests that the reverse may have been the case: "Chinese patriarchs from all social classes were predisposed to take advantage of whatever economic conditions facilitated large, extended family life. They even may have attempted to generate such conditions themselves because that was the family structure most conducive to the realization of the Confucian patriarchal ideals" (1983: 26).

Walby (1987) traces change in several industries in England to show the independent influence of patriarchal forces, sometimes supporting the dominant capitalist interests but other times opposing them. This feminist approach again reverses the typical relationship between variables. Family responsibilities have not been the key impediment to women's labor force participation, but rather "women's position in the family is largely determined by their position in paid work" (p. 70). Under capitalism "the primary mechanism that ensures that women will serve their husbands is their exclusion from paid work on the same terms as men. Patriarchal relations within waged work are crucial in preventing women from entering that work as freely as men, and are reinforced by patriarchal state policies" (Walby 1987: 54).

In Central Asia during the 1920s violent and effective resistance to the attacks on patriarchy compelled the Soviet state to retreat from its attempt to bring about radical change in women's status (Massell 1974). Continued very high fertility and restrictions on women's education and employment are some indicators of the persistence of patriarchal values (see Jones and Grupp 1987: 180, 217). The institutional structure of the cotton economy may have helped to preserve the integrity of the traditional culture. Even during the tsarist period landownership tended to remain with the indigenous population. In Uzbekistan in 1926 less than 2 percent of those in agriculture who hired workers were of Slavic ethnicity (TsSU Otd. Per. 1929: table IV). Commercial exploitation was via "an intermediate social stratum that performed the role of linking native growers (many of them tenant farmers for native landowners) and Russian consumers. Essentially the same social formation exists today operating within a bureaucratic rather than commercial framework" (Hodnett 1974: 66).

But based on an extensive review of the evidence, Jones and Grupp (1987: 253) conclude that the region has clearly undergone substantial social change:

> The cultural heritage of Islam has made some of the USSR's Muslim regions more resistant to the erosion of traditional family values associated with

modernization. Isalamic heritage has not, however, rendered those groups immune to the effects of socioeconomic development. Exposure to mass communications, education, urban life, modern occupations, and the modern bureaucracy has produced dramatic change in family patterns and gender roles.

This chapter points to another type of change: the rising female representation in occupations where Turkic-Muslim males are working. The patriarchal response to these changes since World War II may be to attempt even more staunchly to preserve a threatened rural way of life and family structure. Recent widespread incidents of female suicides by self-immolation in Central Asia may be connected to such a patri-archal backlash (Tashbayeva and Savurov 1989: 46). It was reported in Tadzhikistan in 1988 that the most suicides occurred, not in rural areas where patriarchy was least disturbed, but rather in the more economically developed urban areas (Zhukov 1988: 16–17). There is an ongoing dispute as to whether these suicides are the result of new pressures brought on by conditions of female employment during the Soviet era or are the outcome of the oppressiveness of traditional culture and the female passivity it encourages (Rumer 1989: 121–2; Tashbayeva and Savurov 1989: 46–7). Patriarchal pressures on women may stem from both factors. Very low rates of out-migration despite the declining employment opportunities in rural Central Asia may be a strategy for sustaining patriarchy. Perhaps even the growth of rural industrial production may be resisted because of the opportunities it provides for female work and autonomy.

THE DILEMMA FOR TURKIC-MUSLIM MEN

Nancy Lubin (1984) argues that the indigenous ethnic groups of Central Asia are content with their position within the work force despite a stratification system that appears to put Russians at a distinct advantage within the modern sectors of the economy:

> the indigenous nationalities do not regard themselves as a deprived social class in Central Asia, but rather as economically and culturally superior to the Slavs in their midst. They have not been excluded from the modern "Russian world" as much as they have resisted Soviet efforts to integrate them fully into that world. (1984: 231)

Lubin stresses the persistence of the traditional culture among the Turkic-Muslims which leads to a preference for a rural way of life and occupations which are quite different from those that are highly valued by the Russians.

The concern with preserving threatened patriarchal privilege may explain why traditions persist. Burdened with large families, concentrated predominantly in the difficult unskilled aspects of field work and exposure to chemical pollutants, women of indigenous ethnicity have far less to be content about than do the men. "Hence comes their attitude toward this profession, and striving to protect their own children, especially girls, from this kind of hardship. To a certain extent this results in the growth of the education attainment and cultural demands of the rural population" (Tashbayeva and Savurov 1989: 71–2). The men are likely to be far more ambivalent about the situation that Lubin indicates. Furthermore, the situation is hardly a static one. As a result of exposure to modern society and growing female employment, the men are likely to be increasingly attracted to the same jobs as men elsewhere in the USSR. Among rural Uzbek families in a recent study it was common to find children studying in higher educational institutions in Central Asia. The men saw the most prestigious occupations as lawyer, accountant, and engineer, while women list teacher, medical worker, worker in preschool institution, bookkeeper, sewer, and salesperson—very familiar stereotypes (Tashbayeva and Savurov 1989: 42, 72).

In discussing the situation of ethnic groups in the developing nations of Asia and Africa, Horowitz (1985: 112) finds support for Lubin's position. Disparaging the occupations and work habits of another ethnic group serves to defend the status of one's own ethnic group. Economic advancement may be less significant than "counter-vailing values like sociability, resignation, and the fulfillment of religious obligation." But Horowitz implies that upward social mobility erodes such values: "economic aspirations among peasants and urban workers in the developing world are generally quite modest, certainly far more modest than those of their middle-class counterparts. Moderate aspiration levels tend to inhibit resentment of the more advantaged ethnic strangers with whom they deal" (p. 120). Higher educational attainment and exposure to mass media and urban living are raising the aspirations of Central Asian youth and their parents. Declining fertility levels are also likely to stem from heightened aspirations.

Treiman (1977) directly attacks the view that occupational prestige is a product of culture and argues instead that prestige is derived from characteristics inherent in structure of work. He shows a remarkable consistency in occupational prestige ratings across nations and ethnic groups, and even over time. He asserts that this uniformity is due to the "differential control over scarce and valued resources" afforded by specialized occupational roles (1977: 12):

There are two main reasons why roughly the same complement of occupational roles is found in all complex societies. First, all societies, of any level of complexity, face the same *functional imperatives*, these same needs that must be met if the society is to survive. Second, complexity carries its own *organizational imperatives*; some social roles require others and some depend upon others—for example, complex organization is not possible without specialized managerial and clerical roles. (1977: 7–8, italics in original)

Treiman does find differences among societies in the way in which occupations are evaluated. This is especially likely to occur for blue-collar occupations and when comparing societies at very different stages of industrial development. Industrial development brings into existence blue-collar jobs that did not exist before and also may substantially change characteristics of other jobs: "new occupational roles are created and old ones are transformed to accomplish the new set of highly specialized and finely differentiated tasks characteristic of the modern mass-factory system" (1977: 130). This was noted above, for example, with regard to textile manufacturing.

And finally, Olzak and Nagel (1986) in reviewing a large body of research conclude that ethnic group job competition and conflict are associated with urban growth, expansion of industrial and service sector jobs, and the development of peripheral areas. They also see such competition emerging from increased involvement of the state in ethnic affairs: "policy changes (such as instituting ethnic and racial civil rights laws, designating official lands, or implementing language rules) increase ethnic awareness and the likelihood of ethnic movements" (1986: 3–4). It is remarkable how well Soviet Central Asia fits all these conditions. This clearly suggests reasons not only for conflict between Russians and the indigenous ethnic groups but also for the conflict among indigenous groups common today.

Overall, the literature indicates that differences in the culture of Russians and Turkic-Muslims cannot under present conditions prevent job competition.

CONCLUSION

Patriarchal forces may be behind what Lubin (1984: 209) sees as a "preference for a rural way of life" that is shaping occupational choices. But Turkic-Muslim males are increasingly likely to have a positive evaluation of the very jobs in the modern industrial economy that do not provide the best defense against threats to patriarchal privileges. Far from the consensus which Lubin sees within the society, Turkic-Muslim men appear to be subject to divergent pressures quite different from those of Turkic-Muslim women.

An economy dominated by cotton and other agricultural production was more compatible with patriarchy than one which has witnessed the sharp growth in occupations outside agriculture. As in the case for China described by Stacey (1983), the further achievement of social change may be impeded by an emerging clash between the objectives of the state and the concerns of patriarchal forces. But the social divisions may not be drawn solely along ethnic lines. A growing modernized sector of the indigenous population in Central Asia exists and is advocating increased industrial production and greater econmic self-sufficiency, and the female population may have increasing ability to resist subordination. Occupational data from the 1989 census would provide an important indicator of the outcome of these social forces. The persistence of high fertility and the resistance to leaving the countryside may be important indicators of the vitality of patriarchy in conflict with modern economic development.

Central Asian republics clearly continue to lag behind most other Soviet regions in terms of their level of economic development. Making the situation more difficult to reverse is rapid population growth combined with a persisting rural concentration of indigenous ethnic groups despite the mechanization of agriculture. Russians and other nonindigenous ethnic groups have predominated in the modern sector of the economy. Turkic-Muslims have been far more concentrated in agriculture. Consequently, in important ways regional economic differences are directly reflected in ethnic employment differences with Central Asia. Significant change is shown by titular ethnic group growth in nearly all branches of the economy. The slow shift out of agriculture and into heavy industry marks the continuity of past inequality. The present reflects past patterns of stratification, but the pressures for change have probably never been greater.

NOTES

1. The figures for 1939 throughout this section are based on TsSU (1962, 1963a, 1963b, 1963c, 1963d: tables 42, 43, 47, 48).

2. These figures are also confirmed by data on white-collar work force ("sluzhashchiye") as a whole. In 1970 in rural Central Asia 34 percent were female; in urban areas 57 percent were female. In the Russian Republic 61 percent were women in both urban and rural areas. In Central Asia 78 percent of the female white-collar workers were in urban areas as opposed to 58 percent of the male workers. In the Russian Republic 81 percent of males and of females were in urban areas (TsSU 1973a: table 11).

REFERENCES

Anderson, Barbara and Silver Brian D. 1989. Demographic sources of the changing ethnic composition of the Soviet Union. *Population and Development Review* 15, 4 (December): 609–56.

Clem, Ralph S. 1988. The ethnic factor in contemporary Soviet society. In Michael Paul Sacks and Jerry G. Pankhurst (eds) *Understanding Soviet Society*, pp. 3–30. Boston, Mass. Unwin Hyman.

Dellenbrant, Jan Ake 1986. *The Soviet Regional Dilemma: Planning, People, and Natural Resources*. Armonk, NY: M. E. Sharpe.

Dienes, Leslie 1987. *Soviet Asia: Economic Development and National Policy Choices*. Boulder, Colo.: Westview Press.

Durand, John D. 1975. *The Labor Force in Economic Development: A Comparison of International Census Data, 1946–1966*. Princeton N.J.: Princeton University Press.

Gleason, Gregory. 1990. Marketization and migration: the politics of cotton in Central Asia. *Journal of Soviet Nationalities*, 1, 2 (Summer): 66–98.

Gosudarstvennyy Komitet SSSR no Statistike 1988. *Trud v SSSR, Statisticheskiy Sbornik*. Moscow: Finansy i Statistika.

Hodnett, Grey, 1974. Technology and social change in Soviet Central Asia: the politics of cotton growing. In Henry W. Morton and Rudolf L. Tokes (eds), *Soviet Politics and Society in the 1970's*, pp. 60–117. New York: Free Press.

Horowitz, Donald L. 1985. *Ethnic Groups in Conflict*. Berkeley, Calif.: University of California Press.

Jones, Ellen and Grupp, Fred W. 1987. *Modernization, Value Change and Fertility in the Soviet Union*. Cambridge: Cambridge University Press.

Karminov, I. A. 1989. Oriyentiry obnovleniya: na voprosy gazet ''Pravda Vostoka'' i ''Sovet Uzbekistana'' otvechaet pervyy sekretar' TsK Kompartii Uzbekistana. *Pravda Vostoka*, 27 September, pp 1–2.

Lubin, Nancy 1984. *Labour and Nationality in Central Asia*. Princeton, NJ: Princeton University Press.

Massell, Gregory J. 1974. *The Surrogate Proletariat: Moslem Women and Revolutionary Change in Soviet Central Asia, 1919–1929*. Princeton, NJ: Princeton University Press.

Morozova, Galina Fedorovna 1989. Trudoizbytochna li Srednyaya Aziya? *Sotsiologicheskiye Issledovaniya* 6: 74–9.

Olzak, Susan and Nagel, Joane (eds) 1986. *Competitive Ethnic Relations*. Orlando, Fla: Academic Press.

Pchelintsev, O. S and Ronkin, G. S. 1988. Novyye oriyentiry sotsial'nogo planirovaniya v regionakh. *Sotsiologicheskiye Issledovaniya* 2: 13–20.

Perepelkin, L. S. 1987. K voprosu ob etnokul'turnykh faktorakh trudovoy deyatel'nosti rabotnika sovremennoy promyshlennosti (nekotoryye itogi ekspertnogo oprosa rabochikh Uzbekistana). *Sovetskaya Etnografiya* 2: 83–8.

Pulatov, Timur. 1990. Is democracy a burden on the poor? How to involve the destitute millions in Renewal? *Moscow News*, 20–7 May: 7.

Rapawy, Stephen and Helniak Timothy E. 1987. Annual average employment of the socialized sector in Soviet Central Asia, 1970–1985. Research note from the Soviet Branch, Center for International Research, Bureau of the Census, Washington, DC.

Rumer, Boris Z. 1989. *Soviet Central Asia: "A Tragic Experiment"*. Boston: Unwin Hyman.

Rywkin, Michael 1982. *Moscow's Muslim Challenge*. Armonk, NY: M. E. Sharpe.

Rywkin, Michael 1984. National symbiosis: vitality, religion, identity, allegiance. In Yaacov Ro'i (ed.) *The USSR and the Muslim World*, pp. 3–15. London: Allen & Unwin.

Sacks, Michael Paul 1982. *Work and Equality in Soviet Society: The Division of Labor by Age, Gender and Nationality*. New York: Praeger.

Sacks, Michael Paul 1986. Occupation and work force data in Russian and Soviet censuses. In Ralph S. Clem, (ed.) *Research Guide to the Russian and Soviet Censuses*, pp. 98–112. Ithaca, NY: Cornell University Press.

Sacks, Michael Paul 1988. Shifting strata: ethnicity, gender and work in Soviet Central Asia. In Terry L. Thompson and Richard Sheldon (eds), *Soviet Society and Culture: Essays in Honor of Vera S. Dunham* pp. 87–105. Boulder, Colo: Westview Press.

Spechler, Martin 1979. Regional developments in the USSR, 1958–78. In *Soviet Economy in a Time of Change*, Vol. 1, pp. 141–63. Washington, DC.: US Government Printing Office.

Stacey, Judith 1983. *Patriarchy and Socialist Revolution in China*. Berkeley, Calif.: University of California Press.

Tashbayeva, T. Kh. and Savurov M. D. 1989. *Novoye i Traditsionnoye v Bytu Sel'skoy Sem'i Uzbekov*. Tashkent: Izdatel'stvo "Fan" Uzbekskoy SSR.

Tolan, Sandy 1990. The border boom: hope and heartbreak. *New York Times Magazine*, 1 July 16–40.

Treiman, Donald J. 1977. *Occupational Prestige in Comparative Perspective*. New York: Academic Press.

Tsentral'noye Statisticheskoye Upravleniye, Otdel Perepisi (TsSU Otd. Per.) 1929. *Vsesoyuznoy Perepisi Naseleniya 1926 Goda*. Vol. 32: *Uzbekskaya SSSR: Zanyatiya*. Moscow: Izdaniye TsSU Souza SSSR.

Tsentral'noye statisticheskoye Upravleniye pri Sovete Ministrov SSSR (TsSU). 1962. *Vsesoyuznoy Perepisi Naseleniya 1959 Goda: Uzbekskaya SSR*. Moscow: Gosstatizdat.

TsSu. 1963a. *Vsesoyuznoy Perepisi Naseleniya 1959 Goda: Kirgizskaya SSR*. Moscow: Gosstatizdat.

TsSu. 1963b. *Vsesoyuznoy Perepisi Naseleniya 1959 Goda: RSFSR*. Moscow: Gosstatizdat.

TsSu. 1963c. *Vsesoyuznoy Perepisi Naseleniya 1959 Goda: Tadzhikskaya SSR*. Moscow: Gosstatizdat.

TsSu. 1963d. *Vsesoyuznoy Perepisi Naseleniya 1959 Goda: Turkmenskaya SSR*. Moscow: Gosstatizdat.

TsSu. 1963a. *Vsesoyuznoy Perepisi Naseleniya 1959 Goda: SSR* Moscow: Gosstatizdat.

TsSu. 1969. *Narodnoe Khozyaistvo SSSR v 1968 g.: Statisticheskiy Yezhegodnik*. Moscow: Statistika.

TsSu. 1973a. *Vsesoyuznoy Perepisi Naseleniya 1970 Goda*. Vol. 5: *Raspredeleniye naseleniya SSSR po Obshchestvennyn Gruppam*. Moscow: Statistika.

TsSu. 1973b. *Vsesoyuznoy Perepisi Naseleniya 1970 Goda*. Vol. 6: *Raspredeleniye Naseleniya SSSR po Zanyatiyam*. Moscow: Statistika.

TsSu. 1978. *Narodnoe Khozyaistvo SSSR v 1977 g.: Statisticheskiy Yezhegodnik.* . Moscow: Statistika.

Ubayidullayeva, R. A. 1988. Problemy obespecheniya ratsional'noy zanyatosti naseleniya v usloviyakh uskoreniya sotsial'no-ekonomicheskovo razvitiya. *Obshchestvennyye Nauki v Uzbekistane* 7: 5–12.

Youssef, Nadia Haggag 1974. *Women and Work in Developing Societies*. Population Monograph Series, No. 15. Berkeley, Calif.: Institute of International Studies, University of California.

van de Walle, Etienne and Knodel John 1980. Europe's fertility transition: new evidence and lessons for today's developing world. *Population Bulletin* 34 (February).

Walby, Sylvia 1986. *Patriarchy at Work: Patriarchal and Capitalist Relations in Employment*. Minneapolis, Minn.: University of Minneapolis Press.

Zhukov, V. 1988. Rising above the mass of falsehoods. Translated from *Kommunist Tadzhikistana*, 11 September; 2 *Current Digest of the Soviet Press*, 40, 44: 16–17.

Ziyadullayev, S. K. 1984. *Industriya Sovetskogo Uzbekistana*. Tashkent: ''Uzbekistan''.

Ziyadullayev, S. K. 1987. *Sovershenstvovaniye Struktury Regional'nogo Promyshlennogo Proiuzvodstvo: na Materialakh Uzbekskoy SSR*. Tashkent: Izdatel'stvo ''Fan'' Uzbekskoy SSR.

Part IV

Social Geography

Chapter 8 ──────────────────────────────────

The Demography of Soviet Central Asia and its Future Development*

Ozod Baba-Mirzayevich Ata-Mirzayev and
Abdukhakim Abdukhamidovich Kayumov

At present, unique demographic conditions relative to the remainder of the USSR, as well the world, are developing in Soviet Central Asia. Historically demographic processes are not totally universal. In accord with the general theory of modernization, all populations pass through a transition from high, uncontrolled fertility and mortality to low, controlled fertility and mortality. Specific features of the demographic transition, however, such as its essence and temporary lag and final reproductive norms, vary widely among populations. From this perspective, the study of demographic processes in Soviet Central Asia as a special demographic region is particularly interesting.

Demographic development is a complex sociodemographic and geographic phenomenon. It is determined by many interrelated processes, affecting population reproduction, territorial and social mobility, urbanization, work force participation and so on. One of the most important characteristics of the demography of Soviet Central Asia is the rapid absolute and relative growth of its population. The current population of the republics of Soviet Central Asia (Uzbek SSR, Kirgiz SSR, Tadzhik SSR, and Turkmen SSR) is more than 33 million people, which is more than 11 percent of the Soviet population. The absolute annual growth of the population of Soviet Central Asia is currently about 900,000, which is more than 30 percent of the total Soviet population growth. During the 1980s the population of the region grew about 800,000 per year or about 2.6 percent per year, which exceeded by almost three times the Soviet rate (Table 8.1).

* Translated by Robert A. Lewis.

Table 8.1 Population change in Soviet Central Asia, 1913–87.

	Population (000)								Average annual growth rate (%)						
	1913	1926	1939	1950	1959	1970	1979	1987	1913–26	1926–39	1939–50	1950–9	1959–70	1970–9	1979–87
Uzbekskaya SSR	4,334	4,621	6,347	6,194	8,119	11,799	15,931	19,026	0.50	2.50	–0.20	3.05	3.45	3.00	2.70
Kirgizskaya SSR	864	1,002	1,458	1,716	2,066	2,934	3,529	4,143	1.15	2.90	1.50	2.10	3.25	2.10	2.05
Tadzhikskaya SSR	1,034	1,032	1,485	1,509	1,981	2,900	3,801	4,807	0.00	2.85	0.15	3.05	3.55	3.05	3.00
Turkmenskaya SSR	1,042	998	1,252	1,197	1,516	2,159	2,759	3,361	–0.30	1.75	–0.40	2.65	3.25	2.75	2.50
Central Asia	7,274	7,653	10,542	10,616	13,682	19,792	25,480	31,337	0.40	2.50	0.05	2.85	3.40	2.85	2.60
USSR	159,153	147,028	190,678	178,547	208,827	241,720	262,436	281,689	–0.60	2.00	–0.60	1.75	1.35	0.90	0.90

Sources: Goskomstat SSSR, Narodnoye Khozyaystvo SSSR Za 70 Let 1987: 377;
Goskomstat SSSR, Naseleniye SSSR, 1987 1988a: 8–15.

A progressive increase in the growth of the Central Asian population has occurred. During the Soviet period, in particular from 1926 to 1987, its population grew from 7.6 to 31.3 million, or more than four times. In the past 25 years alone it doubled. Population decline, however, occurred during World War II. It was not until 1950 that the prewar population was once again achieved (Table 8.1). The most rapid rate of growth of the Central Asian population occurred between 1959 and 1970, when it grew 3.4 percent per year. Subsequently, a decline in the rate of growth has occurred, and the decline marks the beginning stage of the demographic transition in the region, a universal, global population trend. This beginning stage of demographic transition will last about 50 years (1960–2010).

The high current population growth rates of the region are being maintained mainly by the high level of fertility and natural increase. At present, about 1.2 million children are born each year, or about 300,000 more than in 1980. Crude birth rates in the republic of Soviet Central Asia have stabilized in the past decade at between 34 and 38 per thousand, which is twice as high as the Soviet average. A high proportion of the indigenous female population in the child-bearing ages is an important demographic feature of the Central Asian population, which results in the high level of natural increase of the population. In Central Asia, the proportion of women in the high-fertility ages is twice the Soviet average. In the USSR as a whole, the period of relatively high fertility for women begins at age 20 and ends toward age 30, whereas in the Central Asian republics high fertility continues to age 40. Moreover, the fertility of the Central Asian women in the older ages is considerably higher than the Soviet average. At present in Soviet Central Asia about a half of the children are born to women with three or more children, as opposed to 25 percent as an average for the USSR.

As a result, the population of the Central Asian republics is characterized by high total fertility rates. In 1986–7, this rate was 5.7 in Tadzhik SSR, 4.8 in Turkmen SSR, 4.6 in Uzbek SSR, and 4.2 in Kirgiz SSR, as compared to the Soviet average of 2.5 (Table 8.2). Age-specific maternal birth rates for the republics of Central Asia reveal the general trends of present and future fertility. These trends can be summarized as: (a) some increase in the fertility of women younger than 20 (except in Turkmen SSR), (b) a concentration of births in high-fertility age groups (20–24, 25–29, and 30–34), (c) a noticeable decline in fertility among women in the older ages, beginning with the 35–39 age group.

The Central Asian republics are characterized by low crude death rates, largely as a result of the young age structure. In the past decade, crude death rates in Central Asia generally ranged from 6 to 7 per thousand, in contrast to the Soviet average of 9 to 10. Infant mortality, however, is particularly high in Central Asia. In 1986, reported infant mortality in Kirgiz SSR was 42.4 per thousand; in Tadzhik SSR, 47.8; in

Table 8.2 Age-specific birth rates: Soviet Central Asia (per thousand).

	Uzbek SSR			Kirgiz SSR			Tadzhik SSR			Turkmen SSR			USSR		
	1965–6	1978–9	1986–7	1965–6	1978–9	1986–7	1965–6	1978–9	1986–7	1965–6	1978–9	1986–7	1965–6	1978–9	1986–7
0–20	30.20	35.40	37.70	28.00	41.20	39.20	30.90	37.90	38.40	32.70	21.80	20.70	25.50	39.40	44.40
20–24	252.80	277.30	295.50	283.30	252.80	268.10	224.50	316.30	319.90	282.60	256.20	239.00	159.60	174.60	192.60
25–29	270.20	281.70	278.10	225.10	202.80	258.00	244.40	299.90	317.30	290.90	289.30	311.00	136.00	125.60	145.90
30–34	238.10	210.70	180.60	223.70	187.30	151.70	240.90	243.50	229.90	251.20	229.60	204.80	97.00	72.10	79.20
35–39	181.30	134.70	98.40	134.60	122.90	85.60	194.50	177.20	150.80	200.10	151.20	119.60	50.60	31.90	33.80
40–44	99.20	66.90	36.80	66.20	61.50	35.10	110.30	54.30	69.30	109.90	88.10	51.90	19.10	11.70	8.70
45–49	41.00	12.50	5.10	25.80	13.50	6.10	52.20	24.90	12.90	40.30	18.30	7.10	4.40	1.60	0.70
15–49	165.30	149.60	160.60	137.20	128.80	139.50	166.20	168.90	186.70	176.60	151.30	156.80	70.80	69.90	79.70
Total fertility rate	5.50	5.10	4.60	4.60	4.50	4.20	5.40	6.00	5.70	6.00	5.30	4.80	2.50	2.30	2.50

Source: Goskomstat SSSR, *Naseleniye SSSR, 1987* 1988b: 209–14.

Table 8.3 Birth, death, and natural increase rates: Soviet Central Asia (per thousand).

	Crude birth rate				Crude death rate				Natural increase			
	1940	1960	1970	1986	1940	1960	1970	1986	1940	1960	1970	1986
Uzbek SSR	33.80	39.80	33.60	37.80	13.20	6.00	5.50	7.00	20.60	33.80	28.10	30.80
Kirgiz SSR	33.00	36.90	30.50	32.60	16.30	6.10	7.40	7.10	16.70	30.80	23.10	25.50
Tadzhik SSR	30.60	33.50	34.80	42.00	14.10	6.10	6.40	6.80	16.50	28.40	28.40	35.20
Turkmen SSR	36.90	42.40	35.20	36.90	19.50	6.50	6.60	8.40	17.40	35.90	28.60	28.50
USSR	31.20	24.90	17.40	20.00	18.00	7.10	8.20	9.80	13.20	17.80	9.20	10.20

Source: Goskomstat SSSR, *Narodnoye Khozyaystvo SSSR za 70 Let* 1987.

Uzbek SSR, 46.2; and in Turkmen SSR, 58.2. Among the rural population infant mortality is somewhat higher. The high infant mortality has resulted from the relatively poorly developed public health services, especially for maternal and child health, and the deterioration of the ecology of the region.

High crude birth rates and high crude death rates in the republics of Central Asia result in high natural increase, almost three times the average rate for the USSR. Natural increase rates vary from 25.5 per thousand in Kirgiz SSR to 35.2 in Tadzhik SSR (Table 8.3).

High fertility has resulted in a considerable population potential, which is the outstanding feature of the demography of Central Asia. The result is a young age structure, with a predominance of children and young people. At present (1987), almost 70 percent of the population is under 30, 40 percent under age 15, and 16 percent under age 5. The corresponding percentages for the USSR are 49, 26, and 9. With respect to the growth of the population of Soviet Central Asia, the considerable absolute and relative growth in the number of women in the reproductive years (15–49), who currently comprise 45.6 percent of the female population, is particularly important.

A rather high marriage rate and a particularly low divorce rate among the indigenous population, especially women, are characteristic of Central Asia. For example, in 1985 per thousand women over age 16, 690 were married in Kirgiz SSR, 647 in Tadzhik SSR, and 638 in Uzbek SSR, but in the USSR as a whole 588 (Table 8.4). The divorce rate in Central Asia is increasing, but it is only about half the Soviet average. Compared to the average Soviet rate of 3.4 per thousand, in Turkmen SSR it is 1.4; in Uzbek and Tadzhik SSRs, 1.5 and in Kirgiz SSR, 1.9 (Table 8.5). Per thousand married couples in 1984–5, there were 8.5 divorces in Kirgiz SSR, 8.4 in Tadzhik SSR, 8.1 in Turkmenskaya SSR, and 7.4 in Uzbek SSR, in contrast to the USSR average of 14.1 (Table 8.6).

Another demographic characteristic of Central Asia is the low level of urbanization of the indigenous population. Only about 20 percent of the indigenous population lives in urban settlements. The urban population of Central Asia at present totals 12.8 million people or 41

Table 8.4   Marriage rates: Soviet Central Asia (per thousand, age 16 and over).

	1979		1985	
	Men	Women	Men	Women
Uzbek SSR	660	613	684	638
Kirgiz SSR	670	699	677	690
Tadzhik SSR	602	623	640	647
Turkmen SSR	653	610	659	588
USSR	707	580	730	588

Source: Goskomstat SSSR, *Naseleniye SSSR, 1987* 1988b: 174–89.

Table 8.5   Marriage and divorce rates: Soviet Central Asia (per thousand population).

| | 1960 | | 1987 | |
	Marriages	Divorces	Marriages	Divorces
Uzbek SSR	13.40	0.30	9.80	1.50
Kirgiz SSR	14.30	0.30	9.60	1.90
Tadzhik SSR	13.10	0.40	9.50	1.50
Turkmen SSR	14.50	0.50	9.20	1.40
USSR	12.10	1.30	9.80	3.40

Source: Goskomstat SSSR, *Naseleniye SSSR, 1987* 1988b: 190–7.

Table 8.6   Divorce rates: Soviet Central Asia (per thousand married couples).

	1958–1959	1969–1970	1978–1979	1984–1985
Uzbek SSR	1.40	5.90	8.10	7.40
Kirgiz SSR	1.40	6.20	10.10	8.50
Tadzhik SSR	1.50	5.90	8.60	8.40
Turkmen SSR	2.20	6.60	8.60	8.10
USSR	5.30	11.50	15.20	14.10

Source: Goskomstat SSSR, *Naseleniye SSSR, 1987* 1988b: 208.

percent of the region's population. During the Soviet period, the urban population has increased rapidly, almost tenfold (Table 8.7). Between 1959 and 1987 it increased by more than 8 million people, or 2.7 times.

The level of urbanization of Central Asia increased from 18 to 41 percent between 1926 and 1987 (Table 8.8). In recent decades there has been a large absolute increase in the urban population, but the level of urbanization has changed slowly because of rapid growth of the rural population. Between 1959 and 1987, the rural population increased by 9.6 million people, whereas the rural population of the USSR as a whole decreased by 13.2 million. The rate of growth of the rural population of Central Asia continues to be rapid (Table 8.9), which dampens urbanization.

The urban network of Central Asia consists primarily of small urban settlements. Of the 180 cities and the 248 urban settlements, only 33 have a population over 50,000, and only 10 have a population over 200,000.

The urban settlements of Central Asia are multinational and contain almost all the nationalities of the USSR (more than 100). In general, the indigenous population (Uzbeks, Tadzhiks, Kirgiz, Turkmen, Karakalpaks, and Kazakhs) predominates and accounts for about 80 percent of the population. Because the indigenous population has significantly higher natural increase than the European and other nationalities, and indigenous out-migration is almost nonexistent, the indigenous population continues to increase its share of the population. This share will continue to increase, because in the past decade migration into the

Table 8.7 Urban population change in Soviet Central Asia (thousands).

| | Urban population | | | | | | | | Average annual growth rate (%) | | | | | | |
	1913	1926	1939	1950	1959	1970	1979	1987	1913–26	1926–39	1939–50	1950–9	1959–70	1970–9	1979–87
Uzbek SSR	1,060	1,012	1,470	1,917	2,729	4,322	6,348	7,974	−0.35	2.90	2.45	4.00	4.25	4.35	2.90
Kirgiz SSR	106	122	270	475	696	1,098	1,366	1,646	1.10	6.30	5.25	4.35	4.25	2.45	2.35
Tadzhik SSR	95	106	249	391	646	1,077	1,325	1,603	0.85	6.80	4.20	5.75	4.75	2.35	2.40
Turkmen SSR	117	137	416	433	700	1,034	1,323	1,601	1.25	8.90	0.35	5.50	3.60	2.75	2.40
Central Asia	1,378	1,377	2,405	3,216	4,771	7,531	10,362	12,824	0.00	4.35	2.65	4.50	4.25	3.60	2.70
USSR	28,452	26,314	60,409	69,414	99,978	135,991	163,586	186,001	−0.60	6.60	1.25	4.15	2.85	2.05	1.60

Source: Goskomstat SSSR, *Naseleniye SSSR, 1987* 1988b: 8–15.

Table 8.8   Urbanization in Soviet Central Asia (percent).

	1913	1926	1939	1950	1959	1970	1979	1987
Uzbek SSR	24	22	23	31	34	37	41	42
Kirgiz SSR	12	12	19	28	34	37	39	40
Tadzhik SSR	9	10	17	26	33	37	35	33
Turkmen SSR	11	14	33	36	46	48	48	48
Central Asia	19	18	23	30	35	38	41	41
USSR	18	18	32	39	48	56	62	66

Source: Goskomstat SSSR, *Naseleniye SSSR, 1987* 1988b: 8–15.

region from other parts of the USSR has almost ceased. In recent years Central Asia has experienced net out-migration.

The share of the Central Asian population relative to the total Soviet population has increased. In 1959 it was 4.5 percent, and in 1979, 7.4 percent. The share of the indigenous Central Asian population will also continue to grow primarily as a result of the high fertility. The high fertility of the indigenous population is related to the following objective, interrelated demographic, socioeconomic, and ethnic factors:

Demographic factors

(a) The dominance of rural settlement among the indigenous population.
(b) High net reproduction rates.
(c) High fertility throughout the reproductive ages.
(d) A young age structure.
(e) High marriage and low divorce rates.
(f) Low rates of geographic mobility.
(g) Low rates of female participation in the work force, as a result of lack of jobs and a poorly developed occupational structure.

Social factors

(h) Low rates of social mobility, a high peasant share in the social structure.
(i) Inadequate development of the social infrastructure, in general, and of the medical services, in particular, especially in rural areas.
(j) A low level of sanitation and medical education among the rural population.
(k) Social approval of high fertility.
(l) The low level of education and professional training of the population, particularly for the women.

Table 8.9 Rural population change in Soviet Central Asia, 1913-87.

	Rural population (thousands)								Average annual change (%)						
	1913	1926	1939	1950	1959	1970	1979	1987	1913-26	1926-39	1939-50	1950-9	1959-70	1970-9	1979-87
Uzbek SSR	3,274	3,609	4,877	4,277	5,390	7,477	9,043	11,052	0.75	2.35	-1.10	2.60	3.00	2.15	2.55
Kirgiz SSR	758	880	1,188	1,241	1,370	1,836	2,163	2,497	1.15	2.35	0.40	1.10	2.70	1.85	1.80
Tadzhik SSR	939	926	1,236	1,118	1,335	1,823	2,476	3,204	-0.10	2.25	-0.85	2.00	2.85	3.45	3.30
Turkmen SSR	925	861	836	764	816	1,125	1,436	1,760	-0.55	-0.20	-0.80	0.75	2.95	2.75	2.60
Central Asia	5,896	6,276	8,137	7,400	8,911	12,261	15,118	18,513	0.50	2.00	-0.80	2.10	2.95	2.35	2.60
USSR	130,701	120,714	130,269	109,133	108,849	105,729	98,850	95,688	-0.60	0.60	-1.50	0.00	-0.25	-0.70	-0.40

Source: Goskomstat SSSR, *Naseleniye SSSR, 1987* 1988b: 8-15.

(m) The desire for many children for social security in old age.
(n) The communal economy that large families sustain.
(o) The relatively small expenditures on children in large families and low consumer expectations.
(p) Governmental financial encouragement of high fertility.

Ethnic factors

(q) A national tradition of high fertility.
(r) The extended family.
(s) The strong influence of the older generation on high fertility.
(t) The dependent status of women.
(u) The preference for sons.

These interrelated factors will determine the future reproductive behaviour of the indigenous population and the dominant features of the demography of Soviet Central Asia.

REFERENCES

Goskomstat SSSR 1987. *Narodnoye Khozyaystvo SSSR za 70 Let*. Moscow: Finansy i Statistika.
Goskomstat SSSR 1988a. *Narodnoye Khozyaystvo SSSR za 70 Let*. Moscow: Finansy i Statistika.
Goskomstat SSSR 1988b. *Naseleniye SSSR, 1987*. Moscow: Finansy i Statistika.

Chapter 9

Demographic Trends in Soviet Central Asia and Southern Kazakhstan

Richard H. Rowland

An understanding of any region requires a knowledge of its population characteristics. Indeed, the increased attention placed on Soviet Central Asia in recent years has probably been largely due to its rapid population growth and increasing shares of the total population of the USSR, the Soviet work force, and the military-age population.

The purpose of this chapter is to describe and analyze population patterns in Soviet Central Asia in the postwar period. Post-1979 trends during the 1980s will be compared with trends in the intercensal periods of 1959–70 and 1970–9, as well as with those of 1951–9. Specific topics to be investigated will include overall population growth; the components of population growth: fertility, mortality, migration, and age composition; the share which Central Asia has comprised of the total Soviet population; and the geographical distribution and redistribution of population within Central Asia, including urbanization and urban–rural population change. An examination of the question of the future indigenous out-migration to local cities and to Slavic areas to the north will also be considered. Demographic aspects of nationality will also be touched upon. However, they are discussed in greater detail in the chapter on nationality by Robert Kaiser.

DATA SOURCES, REGIONAL DEFINITION, AND PROCEDURES

This study will be based chiefly on data from the postwar Soviet censuses of 1959, 1970, and 1979, as well as limited published data from the 1989 census and recently published data on regional population

Map 9.1   Soviet economic regions, 1961.

estimates in 1951 (USSR 1962–3, 1972–4, 1984, 1988a: 16–33; O predvaritel'nykh . . . 1989: 2). Interregional comparisons will be based on the 19 economic regions of 1961, the comparable set of regions utilized in other studies by the author (Map 9.1) (Lewis et al. 1976; Lewis and Rowland 1979). To maintain comparability with other chapters in this volume, population patterns in southern Kazakhstan will be investigated, although this discussion will be limited because of the lack of data.[1]

POPULATION GROWTH AND SHARE OF USSR POPULATION

The population of Central Asia more than tripled between 1951 and 1989 and has grown much more rapidly than that of the USSR as a whole, both overall and in every period since 1951 (Table 9.1). In fact, its growth rate has been more than double the national average in the overall 1951–89 period and the intercensal periods of 1959–90, 1970–9, and 1979–89. The peak intercensal annual growth rate, 3.4 percent, was reached between 1959 and 1970. The growth rate has currently slowed somewhat to 2.5 percent per year between 1979 and 1989.

Given the relatively rapid growth of Central Asia, its share of the Soviet population has consistently increased (Table 9.1). In 1951 it had only 6.0 percent of the national population, buy by 1989 this proportion had nearly doubled to 11.5 percent.

Table 9.1 Total population change in Central Asia, southern Kazakhstan, and the USSR, 1951–89.

Year	Central Asia	Southern Kazakhstan	USSR	Percent of USSR population in: Central Asia	Southern Kazakhstan
1951	10,977,000	2,464,000	181,603,000	6.0	1.4
1959	13,667,813	3,212,864	208,826,650	6.5	1.5
1970	19,790,716	4,625,358	241,720,134	8.2	1.9
1979	25,480,009	5,495,229	262,436,227	9.7	2.1
1989	32,843,000	6,363,000	286,717,000	11.5	2.2

Year	Average annual percentage change in population Central Asia	Southern Kazakhstan	USSR	Percentage point change in percent of USSR population Total Central Asia	Southern Kazakhstan	Average annual Central Asia	Southern Kazakhstan
1951–9	2.8	3.3	1.7	0.5	0.1	0.06	0.01
1959–70	3.4	3.3	2.3	1.7	0.4	0.15	0.04
1970–9	2.8	1.9	1.0	1.5	0.2	0.17	0.02
1979–89	2.5	1.5	0.9	1.8	0.1	0.18	0.01
1951–89	2.9	2.5	1.2	5.5	0.8	0.14	0.02

Sources: USSR 1962–3 (for 1959), 1972–4 (for 1970), 1984a (for 1979), 1988a: 16–33 (for 1951); O predvaritel'nykh . . . 1989: 2 (for 1989).

The pace of its share increase in recent years has continued to be *unprecedented* in the Soviet context. During the overall 1951–89 period and the 1951–9, 1959–70, 1970–9, and 1979–89 periods, average annual share increases were 0.14, 0.06, 0.15, 0.17, and 0.18 percentage points per year, respectively (Table 9.1). In any intercensal period between 1897 and 1989 the closest intercensal increase by another economic region was only 0.08 (Kazakhstan, 1959–70) (Lewis and Rowland 1979: 59; Rowland and Lewis 1982: 77). Between 1979 and 1989, the next highest increase was only 0.02 (East Siberia and Kazakhstan) (Rowland 1989: 642).

Furthermore, Central Asia has accounted for a relatively large and increasing share of the absolute population growth of the USSR in recent years. The Central Asian share of the total population growth of the USSR from 1951 to 1989 was 20.8 percent, and that share has consistently increased: 1951–9, 9.9 percent; 1959–70, 18.6 percent; 1970–9, 27.5 percent; and 1979–89, 30.3 percent. Thus, in recent years, whereas Central Asia has accounted for roughly 11 percent of the Soviet population, it has accounted for roughly 25 to 30 percent of the Soviet population growth. In the mid-to-late 1980s, including 1989, Central Asia also became the most populous economic region in the Soviet Union (Rowland 1989: 642).[2]

Some international perspectives further highlight the increasing demographic importance of Central Asia. With a population of 32,843,000 in 1989, Central Asia is larger than most countries of the world and was surpassed in population by only approximately two dozen (25) of the roughly 14 dozen countries of the world. Its growth rate of 2.5 percent per year during 1979–89 places it well above the world (1.8 percent) and underdeveloped (2.1 percent) averages in 1989 (Population Reference Bureau 1989).

The population of southern Kazakhstan has also grown more rapidly than the national average overall and in each period since 1951, but its growth rate has sharply declined and has been appreciably below that of Central Asia during 1970–9 and 1979–89 (Table 9.1). Indeed, southern Kazakhstan's 1979–89 growth rate was less than one-half of its 1951–9 and 1959–70 rates. The relative shift of the Soviet population to southern Kazakhstan has thus been less dramatic than that toward Central Asia, and, in fact, hardly any shift at all occurred to southern Kazakhstan between 1979 and 1989. In 1989, southern Kazakhstan contained roughly 6 million people or roughly 2 percent of the Soviet population and 38.5 percent of the population of the Kazakh SSR.

In 1989, the population of Central Asia and southern Kazakhstan was nearly three times that of 1951, and contained over 39 million people or nearly 14 percent of the Soviet population.

FERTILITY, MORTALITY, AGE COMPOSITION, AND MIGRATION

Of the three immediate factors in the growth of the population of any area (fertility, mortality, and migration), the chief factors in the rapid growth of Central Asia in recent decades have been a relatively high fertility rate coupled with a relatively low mortality rate. This has resulted in the highest rate of natural increase of any Soviet economic region (Rowland 1986: 177; USSR 1988a, 160–9). Such a situation signifies that Central Asia has, like underdeveloped countries, recently been in the middle or explosive phase of the demographic transition. This theory is based on Western experience and maintains that with economic development and modernization societies pass from an initial stage of high fertility and mortality to a stage of low fertility and mortality, which results in a transition from slow to rapid to slow natural increase.

Crude birth, death, and natural increase rates for Central Asia for 1951 through 1988 are shown in Table 9.2 and reflect rates for both the indigenous and nonindigenous populations. They reveal that the crude birth and natural increase rates of Central Asia have generally been almost double and roughly triple the USSR average in recent years, respectively. The relatively high fertility of Central Asia is also demonstrated by total fertility rates. These rates are age-adjusted in that they remove the influence of age composition and are essentially projections of how many children a woman will have as she goes through the reproductive years, assuming a continuation of this annual rate. As can be seen in Table 9.3, in recent years total fertility rates have been roughly 4 to 6 children in the four Central Asian republics, rates which have typically been double to triple the Soviet or RSFSR average. The high fertility rate of Central Asia chiefly reflects the relatively low socio-economic status of the highly rural indigenous nationalities of the region.

However, as can also be seen in Table 9.3, the total fertility rate is generally declining in Central Asia. As the Russian presence is on the wane, as will be discussed later in this chapter, this decline cannot be

Table 9.2 Crude birth, death, and natural increase rates of Central Asia, southern Kazakhstan, and the USSR, 1951–88.

	Central Asia			USSR		
	CBR	CDR	CRNI	CBR	CDR	CRNI
1951	32.7	8.0	24.7	27.0	9.7	17.3
1959	35.8	6.1	29.7	25.0	7.6	17.4
1970	33.5	6.0	27.5	17.4	8.2	9.2
1979	34.4	7.4	27.0	18.2	10.1	8.1
1983	35.2	7.6	27.6	19.8	10.4	9.4
1984	36.1	7.6	28.5	19.6	10.8	8.8
1988	35.4	7.0	28.4	18.8	10.1	8.7

	Southern Kazakhstan			USSR		
	CBR	CDR	CRNI	CBR	CDR	CRNI
1970	26.2	6.2	20.0	17.4	8.2	9.2
1980	26.1	7.9	18.2	18.3	10.3	8.0
1985	27.4	7.8	19.6	19.4	10.6	8.8
1986	28.1	7.3	20.8	20.0	9.8	10.2

Sources: for Central Asia and USSR—USSR 1975: 69–83 (for 1951–9), 1988a: 8–15, 110–43 (for 1970–84), 1988b: 173, 1989d: 26 (for 1988) for southern Kazakhstan and USSR—USSR 1988a: 150–69; and *Narodnoye Khozyaystvo SSSR*, various issues.

Table 9.3 Total fertility rates by Central Asian Republic, RSFSR, and the USSR, 1958–87.

TFR	Uzbek SSR	Kirgiz SSR	Tadzhik SSR	Turkmen SSR	RSFSR	USSR
1958–9	4.9	4.2	3.8	5.0	2.6	2.8
1965–6	5.5	4.6	5.4	6.0	2.1	2.4
1969–70	5.7	4.9	5.9	6.0	2.0	2.4
1975–6	5.7	4.9	6.3	5.7	2.0	2.4
1978–9	5.1	4.5	6.0	5.3	1.9	2.3
1982–3	4.6	4.1	5.5	4.7	2.0	2.3
1986–7	4.6	4.2	5.7	4.8	2.2	2.5

Source: USSR 1988a: 209–14.

explained by an increased Russian presence, but instead by a real fertility decline among the indigenous nationalities. The fact that the reported crude birth rate of Central Asia has been increased from 1970 to 1979 and from 1979 to 1988 is chiefly due to age composition, as well as to the improved recording of vital events.

The crude death rate of Central Asia is relatively low by Soviet standards, but like that of the USSR has generally been increasing in recent years from the 1970s to the 1980s. The low level of the crude death rate would appear to be chiefly due to the high fertility and, as will be seen, young age composition of Central Asia. However, reported mortality rates increased even on an age-adjusted basis from the 1970s to the 1980s. In particular, life expectancy levels in the mid-to-late 1980s were lower than those for the USSR as a whole and for those around 1970, although they were higher than those for the late 1970s (USSR 1987: 408–9, 1989a: 31). In addition, and somewhat similarly, infant mortality rates in the mid-to-late 1980s were higher than those for the USSR as a whole and for those in 1970, but were lower than those for the mid-1970s (USSR 1988a: 344–6, 1989a: 29). For example, in 1988, three of the four republics had life expectancy levels below the union-wide level of 69.5 years. Levels ranged from 65.9 for Turkmeniya to 69.7 for Tadzhikistan, while Uzbekistan, the most populous republic, had a life expectancy level below the national average (68.7 years). In addition, in 1988, life expectancy levels for the four republics were generally surpassed by most other republics, and only the anomalously low level for earthquake-ridden Armenia (62.3) was lower than that for Turkmeniya. In 1988, the infant mortality situation for Central Asia was even poorer. Not only did all four republics have infant mortality rates above the national average of 24.7 per thousand, but they constituted the four highest rates for any of the 15 republics (they ranged from 36.8 for Kirgiziya to 53.3 for Turkmeniya) (USSR 1989a: 29, 31). However, a part of the mortality increase has been due to improvements in reporting (Anderson and Silver, 1989a: 260).

The combination of high fertility and low mortality has resulted in one of the most distorted age structures of any society in the world in recent years. In particular, the rural population of Central Asia has had an extraordinarily low percentage comprised by the working-age population. Indeed, in 1970, only 33.9 percent of the rural population was comprised of the 20–59 or "working age" cohort. This share was the lowest for any of the 19 economic regions and well below the all-union average of 46.0 percent. Few countries, apparently, have ever had such a low working-age share.[3] Conversely, Central Asia had an extraordinarily high young dependent share. In 1970, more than half (57.1 percent) of the rural population was comprised of the 0–19 age cohort; in contrast, the corresponding all-union share was only 40.8 percent. By 1987, these age structure patterns were not quite as extreme as they were in 1970, as the rural 20–59 and 0–19 shares were now 37.7 and 56.4 percent, respectively, as compared to corresponding Soviet averages of 46.6 and 37.3 percent (USSR 1988a: 48–95).

These age imbalances have had and will have major implications for the future of Central Asia and the USSR. Most notable, as the large young dependent population ages, the working-age population has increased and will increase rapidly. Indeed, based on age data from the 1970 and 1979 censuses, as well as age estimates for 1987, the working-age population of Central Asia as a whole nearly doubled from 7.4 million to 13.3 million between 1970 and 1987, and increased by more than 3 million or by nearly one-third (31.1 percent) from 1979 to 1987 (USSR 1988a: 48–95). The rural working-age population of Central Asia, in particular, increased by 75.2 percent or more than 3 million from 1970 to 1987 and accounted for more than half (50.9 percent) of the total Central Asian working-age increase during 1970–87. Meanwhile, the more industrialized northern regions experienced either declines or slow increases. Overall, the corresponding USSR changes for 1970–87 were an increase of only 25.0 percent for the total working-age population and a *decline* of 3.9 percent for the rural working-age population. Central Asia alone accounted for about 20 percent (19.6) of the total working-age population increase of the USSR between 1970 and 1987 and for more than one-fourth (25.7 percent) during 1979 and 1987. This topic and its implications will be discussed further later in this chapter.

Migration has played only a small role in the growth of the population of Soviet Central Asia, and it appears also that the influence of net in-migration, while never great, is now nonexistent. In a previous study, the author determined that the rate of net migration (the difference between the number of in-migrants and out-migrants) for Central Asia declined from 1959–70 to 1970–9. Based on data used for that study, it is also possible to conclude that the share of the total population growth of Central Asia due to net in-migration also declined from 1959–70 (9.9 percent) to 1970–9 (1.5 percent) (Rowland 1982: 559).

Since 1979, Central Asia has become a region of *net out-migration* (Table 9.4).[4] Between 1979 and 1989, Central Asia experienced an overall net *out-migration* of more than –800,000 people. Average annual net migration represented –2.9 per 1,000 midpoint population, which in turn represented a decline of 3.8 points from 0.9 percent in 1970–9.[5] The turnaround has also been striking in absolute terms as the amount of net migration for Central Asia during 1979–89 (–844,222) was more than *one million* less than that of 1970–9 (Table 9.4). Furthermore, there was also virtually no republic variation within Central Asia from this 1979–89 average as all four republics had net out-migration (Table 9.4). From a broader perspective, during the 1980s specifically here 1979–87, only the four northern republics (the RSFSR and the three Baltic republics) had net in-migration, while the remaining eleven, more southern republics (including the four of Central Asia) had net out-migration (Rowland 1988: 812). As will be discussed in greater detail later, net out-migration from Central Asia probably greatly reflects a net out-migration of Russian rather than indigenous peoples.

Table 9.4   Net migration for Central Asia and southern Kazakhstan, 1970–89.

Republic	Total absolute		Average annual absolute	
	1970–9	1979–89	1970–9	1979–89
Uzbek	242,092	–505,217	26,899	–50,522
Kirgiz	–75,256	–156,413	–8,362	–15,641
Tadzhik	26,436	–100,672	2,937	–10,067
Turkmen	–1,007	–81,920	–112	–8,192
Central Asia	192,265	–844,222	21,363	–84,422
Southern Kazakhstan	–25,181	–279,080	–2,798	–27,908

Republic	Average annual net migration		Point change 1970–9 to 1979–89
	1970–9	1979–89	
	Per 1000 population of:		
	July 1974	January 1984	
Uzbek	2.0	–2.9	–4.9
Kirgiz	–2.6	–4.0	–1.4
Tadzhik	0.9	–2.3	–3.2
Turkmen	–0.0	–2.6	–2.6
Central Asia	0.9	–2.9	–3.8
Southern Kazakhstan	–0.5	–4.7	–4.2

Sources: Natural increase, total population, and thus net migration data come from USSR 1988a: 8–15, 110–26, 1988b: 173, 1989: 26; see notes 4 and 6 for an explanation of estimating procedures.

Fortunately, data have been made available in recent years to determine fertility, mortality, and migration patterns for the units comprising southern Kazakhstan (Tables 9.2 and 9.4). They suggest that the slower growth of southern Kazakhstan as compared to Central Asia has been due to its lower fertility and natural increase as well as a higher rate of net out-migration. In particular, data on births, deaths, and natural increase are available for oblast-level units of Kazakhstan as well as the USSR as a whole for 1970, 1980, 1985, and 1986. As can be seen in Table 9.2, athough southern Kazakhstan, like Central Asia, has had crude birth and natural increase rates above and crude deaths below the national average, its crude birth and natural increase rates have been lower than those for Central Asia. These data in turn can be used to estimate net migration for southern Kazakhstan during 1970–9 and 1979–89 (USSR 1988a: 110–69).[6] As can be seen in Table 9.4, southern Kazakhstan already had net out-migration during 1970–9 (–0.5 per

1,000), and the rate deepened during 1979–89 to –4.7 per 1,000, a net out-migration rate which was roughly equal to that of Central Asia during the 1980s.

POPULATION DISTRIBUTION, REDISTRIBUTION, AND GROWTH

In 1989 the population density of Central Asia was twice the Soviet average (25.7 vs 12.8 people per square kilometer). However, this masks the considerable variations that exist within the region. In terms of population concentrations and voids, the distribution of population within Central Asia is, in fact, probably one of the most uneven of any of the Soviet economic regions. Indeed, roughly one-half of the area has a rural density of less than one person per square kilometer, while some areas have the highest rural densities in the USSR, and Central Asia is the *only* economic region which contains areas with rural densities of more than 200 people per square kilometer (USSR 1986: 130–1).

Most of the population of Central Asia is concentrated in the river valleys and along canal systems (Map 9.2). The bulk of the population is concentrated in the eastern half of the region in the Zeravashan and upper Syrdar'ya river valleys. Areas with rural densities of more than 200 people per square kilometer are found in these valleys, especially: (a) the Fergana Valley of the upper Syrdar'ya (b) the Leninabad and Tashkent areas downstream on the Syrdar'ya, including the Chirchik River, a major branch of the Syrdar'ya; and (c) the Samarkand and Bukhara areas of the Zeravshan River valley.

The western part of Central Asia is generally very sparsely populated, particularly in the vast Kyzyl Kum and Kara Kum deserts. However, dense concentrations of settlement exist in the areas of the lower Amudar'ya River, especially Khorezm Oblast and the Ashkhabad and Mary areas of the Kara Kum Canal Zone of the Turkmen SSR. On the other hand, the extreme eastern part of Central Asia is very sparsely populated, due to very mountainous conditions.

Southern Kazakhstan also reveals an oasis type of settlement pattern, although the densities are less than those of Central Asia. Unlike Central Asia, no area has rural densities of 200 or more people per square kilometer (USSR 1986a: 130–1). Major areas of settlement exist in river valleys or branches thereof of the Syrdar'ya (especially Chimkent and Kyzl-Orda), Ili (Alma-Ata), and Karamal (Taldy-Kurgan) rivers. Much like Central Asia, however, vast areas are very sparsely populated, particularly the extension of the Kyzyl Kum Desert south-west of the Syrdar'ya, the deserts north and east of the Syrdar'ya (and

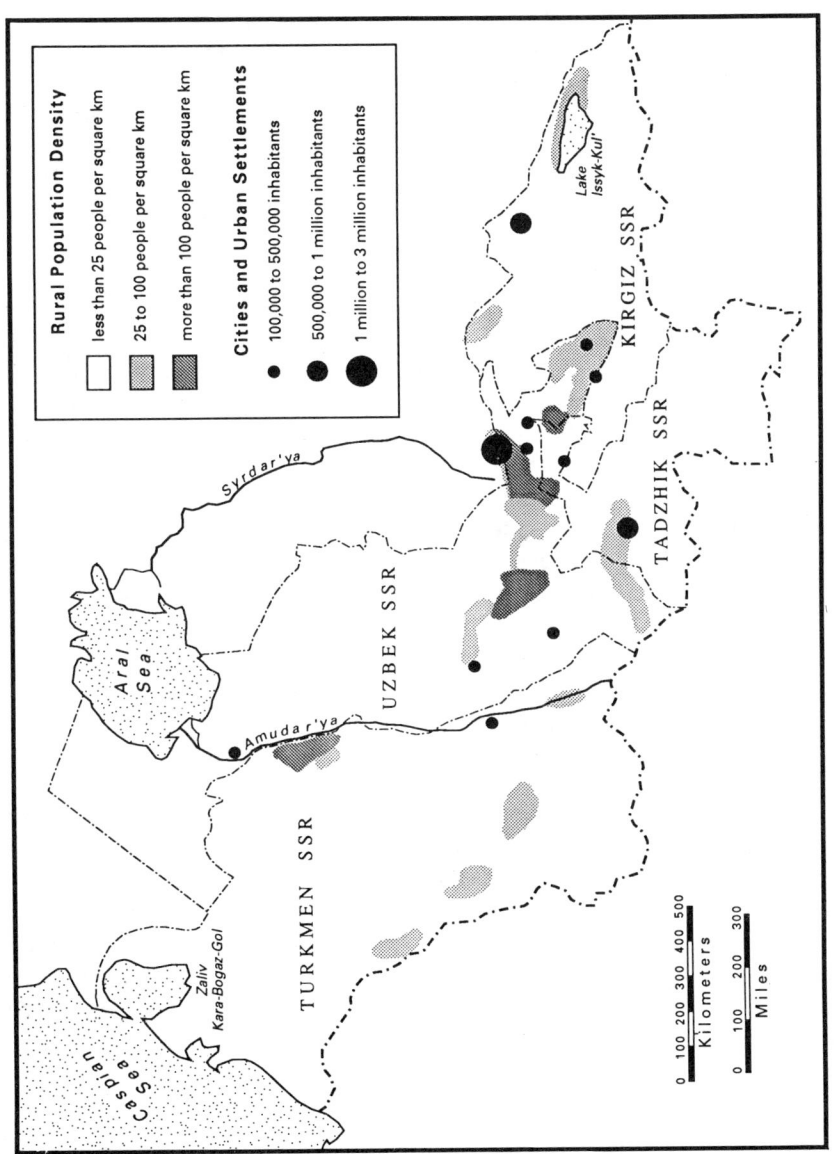

Map 9.2 Rural population density, 1979.

south and west of Lake Balkhash), and the mountainous areas in the extreme east and southeast.

On a republic basis, in Central Asia the highest densities in 1989 were found in Uzbekistan (44.5) and Tadzhikistan (35.7). Kirgiziya had an intermediate density (21.6), while Turkmeniya had by far the lowest density of the four republics (7.2), reflecting the fact that the vast majority of the republic is covered by the Kara Kum Desert.

On an oblast basis, the population of Central Asia is relatively highly concentrated in oblasts extending from Bukhara eastward through Tashkent to Andizhan—namely, from the western Zeravshan River valley through the eastern Syrdar'ya River valley. In 1989, virtually one-half (49.9 percent) of the population of Central Asia was comprised of Bukhara, Samarkand, Syrdar'ya, Tashkent (including Tashkent City), Leninabad, Namangan, Andizhan, and Fergana oblasts, which contained only roughly one-fifth (19.8 percent) of the area of Central Asia. Of course, since some of these oblasts contain extensive sparsely populated desert areas (especially Bukhara and Samarkand), the degree of concentration is even greater. Furthermore, in 1989, three of the four most densely populated units (oblasts or equivalent units) of Central Asia were all located in the Fergana Valley of the upper Syrdar'ya River: Andizhan, which is the most densely populated unit in the USSR (411.4 people per square kilometer), Fergana (303.2), and Namangan (186.7). The fourth most densely populated unit, Tashkent City and Oblast combined (271.5), is located on the Chirchik River. On the other hand, the most sparsely populated units in 1989 had very low densities: Gorno-Badakhshan Autonomous Oblast (2.5 people per square kilometer) and rayons of republic subordination of Turkmeniya (3.8). The former includes the Pamir Mountains, the highest in the USSR, while the latter are covered by the Kara Kum Desert.

The lesser degree of settlement in southern Kazakhstan is reflected by the fact that densities by political units are very low in comparison to those of Central Asia. In 1989, the overall density of southern Kazakhstan (8.9 people per square kilometer) was only about one-third that of Central Asia (25.7 people per square kilometer), and the Uzbek SSR alone had a population roughly triple that of southern Kazakhstan on an area less than two-thirds that of southern Kazakhstan. Only two units of southern Kazakhstan had densities in excess of 10 people per square kilometer: Alma-Ata City and Oblast combined (20.1) and Chimkent Oblast (15.7) and these, of course, were well below the highest densities of Central Asia.

The percentage population distribution by republic of Central Asia for each census year and 1989 is shown in Table 9.5. The chief pattern to emerge is the *considerable consistency* of the distribution of the population. In each year, the decided majority of the population (roughly three-fifths or 60.6 percent in 1989) has been located in the Uzbek SSR.

The remaining two-fifths has been roughly divided equally by the other three republics, each of which has had no less than 10 percent of the population in any year and no more than 17 percent. In 1989, the Tadzhik SSR had the second highest share (15.6 percent), while the Turkmen SSR had the lowest share (10.8 percent).

Table 9.5   Total population growth, distribution, and redistribution by Central Asian SSR, 1951–89.

	Uzbek	Kirgiz	Tadzhik	Turkmen
(a) *Population in thouands*				
1951	6,434	1,764	1,554	1,225
1959	8,119	2,066	1,981	1,516
1970	11,799	2,934	2,900	2,159
1979	15,391	3,529	3,801	2,759
1989	19,906	4,291	5,112	3,534
(b) *Average annual percentage change*				
1951–9	2.9	2.0	3.0	2.7
1959–70	3.4	3.2	3.4	3.2
1970–9	2.9	2.0	3.0	2.7
1979–89	2.5	2.0	2.9	2.5
1951–89	3.0	2.3	3.1	2.8
(c) *Percentage distribution*				
1951	58.6	16.1	14.2	11.2
1959	59.3	15.1	14.5	11.1
1970	59.6	14.8	14.7	10.9
1979	60.4	13.9	14.9	10.8
1989	60.6	13.1	15.6	10.8
(d) *Percentage point change*				
1951–9	0.7	–1.0	0.3	–0.1
1959–70	0.3	–0.3	0.2	–0.2
1970–9	0.8	–0.9	0.2	–0.1
1979–89	0.2	–0.8	0.7	0.0
1951–89	2.0	–3.0	1.4	–0.4

Sources: USSR 1988a: 8–15 (for 1951–79); O predvaritel' nykh . . . 1989: 2 (for 1989).

This consistency reflects the fact that relatively little redistribution or growth rate variations have occurred by republic (Table 9.5). This, in turn, mirrors the existence of little republic variation in crude birth, death, and natural increase rates (Table 9.6).

In recent years, the only notable deviation has been the relatively slow growth of Kirgiziya as compared to the other three republics. Between 1979 and 1989 Kirgiziya had an average annual rate of change from 0.5 to 0.9 percentage points below the other republics. This was probably due to the fact that it had by far the highest Russian share of any of the four republics (in 1979, 25.9 percent, vs Uzbek, 10.8; Tadzhik, 10.4; and Turkmen, 12.6; and in 1989, 21.5, 8.3, 7.6, and 9.5 respectively) (Anderson and Silver 1989b: 628; USSR 1989b: 3–113). Kirgiziya has also had the lowest crude birth and natural increase rates and highest rate of

Table 9.6 Crude birth, death, and natural increase rates by Central Asian republics, 1951–88.

	Crude birth rate				Crude death rate				Crude rate of natural increase			
	Uzb.	Kirg.	Tad.	Turk.	Uzb.	Kirg.	Tad.	Turk.	Uzb.	Kirg.	Tad.	Turk.
1951	32.2	33.3	30.5	37.6	7.6	7.9	7.8	10.3	24.6	25.4	22.7	27.3
1959	37.1	33.6	30.3	39.3	6.1	6.2	5.5	6.9	31.0	27.4	24.8	32.4
1970	33.6	30.5	34.8	35.2	5.5	7.4	6.4	6.6	28.1	23.1	28.4	28.6
1979	34.4	30.1	37.8	34.9	7.0	8.3	7.7	7.7	27.4	21.8	30.1	27.2
1983	35.3	31.4	38.3	35.1	7.5	7.9	7.6	8.4	27.8	23.5	30.7	26.7
1984	36.2	32.1	39.8	35.2	7.4	8.3	7.4	8.2	28.8	23.8	32.4	27.0
1988	35.1	31.2	40.0	36.0	6.8	7.4	7.0	7.8	28.3	23.8	33.0	28.2

Sources: USSR 1975: 70–83 (for 1951–9), 1988a: 127–43 (for 1970–84), 1989: 26 (for 1988).

net out-migration of any of the four republics during the 1980s (Tables 9.4, 9.6).

It is possible to extend these relationships to explain further the relatively slow growth of southern Kazakhstan. Namely, such growth may also be accounted for by a relatively high Russian presence. In 1979, Russians comprised 32.9 percent of these six units combined in comparison to only 13.0 percent for Central Asia.

URBANIZATION AND URBAN–RURAL POPULATION CHANGE

Another major facet of the population geography of any region is urbanization or the distribution by urban–rural residence. Urbanization refers to the percentage of the total population residing in urban centers (level of urbanization) and changes therein. Urban and rural population change refers to the change in the urban and rural populations separately.

Central Asia has been and still is a decidedly rural region. In 1989, only slightly less than two-fifths (39.6 percent) of the population resided in urban centers, a level which was less than two-thirds of the Soviet average of 65.8 percent. In fact, the Central Asian level has been below the Soviet level and increasingly so. The gap has widened from 9.5 percentage points in 1951 to 13.0 in 1959, 21.6 in 1979, and 26.2 points in 1989, based on official definitions (Table 9.7). In 1989, the level of urbanization of Central Asia was the *lowest* of any of the 19 economic regions, and Central Asia was, along with Moldavia, the only economic region to have a level of less than 50 percent (Rowland 1989: 653).

Within Central Asia, all republics and most other political units are still predominantly rural. In terms of republics, the levels of urbanization in 1989 ranged from 45.4 percent for Turkmeniya to only 32.6 for

Table 9.7 Urbanization and urban and rural population change for Central Asia, southern Kazakhstan, and the USSR, 1951–89.

	Urban population[1]		Rural population		Level of urbanization			Central Asia as percentage of Soviet total	
	Central Asia	Southern Kazakhstan	Central Asia	Southern Kazakhstan	Central Asia	Southern Kazakhstan	USSR	Urban	Rural
1951	3,371,000	906,000	7,606,000	1,558,000	30.7	36.8	40.2	4.6	7.0
1959	4,771,762	1,341,533	8,896,051	1,871,331	34.9	41.8	47.9	4.8	8.2
1970	7,530,000	2,187,919	12,260,716	2,437,439	38.0	47.3	56.3	5.5	11.6
1979	10,362,284	2,736,251	15,117,725	2,758,978	40.7	49.8	62.3	6.3	15.3
1989	13,017,000	3,338,000	19,826,000	3,025,000	39.6	52.5	65.8	6.9	20.2

Average annual percentage change

	Urban population			Rural population			Average annual percentage point change in level of urbanization[2]		
	Central Asia	Southern Kazakhstan	USSR	Central Asia	Southern Kazakhstan	USSR	Central Asia	Southern Kazakhstan	USSR
1951–9	4.4	4.9	3.9	2.0	2.3	0.0	0.53	0.63	0.96
1959–70	4.2	4.4	2.8	2.9	2.4	-0.3	0.28	0.50	0.76
1970–9	3.6	2.5	2.0	2.3	1.4	-0.8	0.30	0.28	0.67
1979–89	2.3	2.0	1.4	2.7	1.0	-0.1	-0.11	0.27	0.35
1951–89	3.6	3.4	2.5	2.5	1.7	-0.3	0.23	0.41	0.67

[1] Based on official urban definitions.
[2] Absolute change in level of urbanization divided by number of years in period.
Sources: USSR 1962–3 (for 1959), 1972–4 (for 1970), 1984a (for 1979), 1988a: 16–33 (for 1951); O predvaritel'nykh ... 1989: 2 (for 1989).

Tadzhikistan (the levels for Uzbekistan and Kirgiziya were 40.7 and 38.2 percent, respectively). In fact, no subrepublic political unit (excluding those involving capital cities) had a majority (50.0 or more percent) in urban centers.

The *pace* of urbanization in Central Asia has also been well below the Soviet average (Table 9.7). During 1951–89 as a whole and 1959–70 and 1970–9 in particular, the average annual percentage point change in the level of urbanization for Central Asia was always less than one-half the Soviet average; and during each of these periods, the Central Asian urbanization pace was the slowest of any of the 19 economic regions (Rowland and Lewis 1982: 84; USSR 1988a: 16–33; Rowland 1989: 653).

Moreover, the urbanization pace of Central Asia for 1979–89 was especially extraordinary, since it *declined* by –0.11 percentage points per year, which means there was a ruralization of the population. In fact, it was the only economic region to experience a decline in the level of urbanization in any of the subperiods during 1951–89. Urbanization patterns in recent years for Central Asia have been distorted slightly by boundary and population changes for the city of Samarkand, although adjustments for these changes do not fundamentally change the above conclusions (Rowland 1989: 654–65).[7]

The prime immediate reason for the low urbanization level and pace of Central Asia is the fact that, despite rapid urban growth, it also has relatively rapid rural growth. This dampens the level of urbanization and changes therein. Central Asia has certainly experienced relatively rapid urban growth (Table 9.7). During 1951–89 as a whole and 1951–9, 1959–70, 1970–9, and 1979–89 individually, its urban growth rate was above the national average. In fact, Central Asia has generally ranked at the top or near the top among the 19 economic regions in this regard (Rowland and Lewis 1982: 84; Rowland 1989: 650–66). Largely inconsequential deviations from these conclusions resulted from the Samarkand situation.[8] In 1989, Central Asia had roughly one-sixteenth (6.9 percent) of the Soviet urban population, its highest share ever.

In comparison to the USSR as a whole, the rapid urban growth has been especially due to natural increase rather than migration-reclassification (Table 9.8). Natural increase accounted for the slight majority of the growth of the urban population of Central Asia (55.8 percent) during 1970–9, but for *most* (91.0 percent) of the urban growth during 1979–89. Corresponding figures for the urban population of the USSR as a whole were much less (43.3 and 57.9 percent, respectively) (USSR 1988d: 8–15, 110–26; Rowland 1989: 637).

But, despite the rapid urban growth of Central Asia, it is even more notable for its relatively rapid rural growth. This offsets much of the urban growth and leads to the relatively low and slow urbanization of Central Asia, a pattern which is similar to that of underdeveloped countries.

Table 9.8 Urban and rural fertility, mortality, natural increase and net migration-reclassification for Central Asia, 1970–89.

	Urban			Rural		
	1970–9	1979–89	Change	1970–9	1979–89	Change
Average annual per 1,000 midpoint population (July 1974 and January 1984):						
Births	26.9	27.4	0.5	38.6	41.0	2.4
Deaths	7.0	7.1	0.1	6.9	7.6	0.7
Natural increase	19.9	20.3	0.4	31.7	33.4	1.7
Net migration-reclassification	15.7	2.0	–13.7	–8.6	–5.6	3.0
Percent of change due to:						
Natural increase	55.8	91.0	35.2	137.1	120.4	–16.7
Net migration-reclassification	44.2	9.0	–35.2	–37.1	–20.4	16.7

Source: USSR 1988a: 8–15, 110–26; data on births, deaths, and natural increase were available for 1979–88 and not 1979–89; data for 1979–88 were inflated by a ratio of 1.11 to provide 1979–89 estimates.

The rate of rural population change in Central Asia has been above the national average and usually appreciably so (Table 9.7). As a result, Central Asia has had an increasing share of the rural population of the USSR. In 1989, it had the largest regional rural population and accounted for more than one-fifth (20.2 percent) of the rural population of the USSR, in comparison to shares of roughly one-ninth of the total population and one-sixteenth of the urban population in 1989 and one-ninth of the rural population in 1970 (Tables 9.1 and 9.7).

Furthermore, Central Asia has had the most rapid rate of rural growth of any of the 10 economic regions in recent years. Based on official definitions, it had the most rapid rural growth of any region during 1951–89 overall and, more specifically, during 1959–70, 1970–9, and 1979–89 (Rowland and Lewis 1982: 74; USSR 1988a: 16–33; Rowland 1989: 659). Indeed, it is also one of the very few Soviet regions remaining with rural population increase, and its 1979–89 rate of rural growth (2.7 percent per year) exceeded that of 1970–9 and easily exceeded that of the second ranking region (the Far East: 1.0) during 1979–89. The rate for Central Asia was higher than its 1970–9 rate, partly because of the Samarkand situation. Based on the Samarkand adjustment, the rural rates of increase for 1970–9 and 1979–89 were roughly *equal* (2.4 and 2.6 percent per year), and thus they have neither subsided nor increased. However, the rural rate of change for the USSR did *increase* from 1970–9 to 1979–89 from –0.8 percent per year to –0.1 (Table 9.7).

Although southern Kazakhstan follows many of the general urban–rural trends of Central Asia, some differences are evident (Table 9.7). Like Central Asia, southern Kazakhstan has a relatively low level of urbanization by Soviet standards. However, the level and pace of urbanization have been higher in southern Kazakhstan than in Central Asia. Whereas the Central Asian level of urbanization declined between 1979 and 1989, the pace of urbanization in southern Kazakhstan virtually equaled the Soviet pace, and, furthermore, the level came to exceed 50 percent (52.5 percent in 1989) (Table 9.7). The more rapid urbanization of southern Kazakhstan is chiefly due to the fact that its rural rate of change is well below that of Central Asia. Between 1979 and 1989, as during 1951–89 as a whole, the urban growth rate of southern Kazakhstan continued to be nearly as high as that of Central Asia. However, a major departure has occurred with respect to rural change. Whereas the rate of rural change of southern Kazakhstan was higher than that of Central Asia during the 1951–9 period, during the 1959–70, 1970–9, and 1979–89 periods it was decidedly lower. Indeed, between 1979 and 1989 one of the five units of southern Kazakhstan actually experienced absolute rural population decline (Taldy-Kurgan Oblast). This differential rural change may be due in part to nationality differences. Namely, in 1970 (the last year of available urban–rural nationality data by oblast) Russians comprised a much larger share of the rural population of southern Kazakhstan than of the rural population of Central Asia (18.1 vs 4.0 percent, respectively). Thus, the considerable slowing of rural population change in southern Kazakhstan may reflect the general waning presence of Russians.

SUMMARY AND CONCLUSIONS

In summary, the population geography of Soviet Central Asia evinces a number of striking features. (a) In recent decades it has been characterized by very high fertility, relatively low mortality, and rapid population growth to the point where it is now *the* most populous and rapidly growing Soviet region. (b) Population distribution is extremely uneven, with very high rural densities coupled with vast expanses of uninhabited areas typical of oasis areas. (c) The level of urbanization is relatively low and has been increasing at a very slow rate overall, because the relatively rapid urban growth is accompanied by even more rapid rural growth.

Post-1979 population trends do, however, suggest a slight slowing in many of the major Central Asian demographic trends, even though population change continues to be very dynamic. In particular, changes in the total and urban growth rates, and age-standardized total fertility rates have slowed during the 1980s relative to the 1970s, and the level of urbanization has actually declined. In addition, Central Asia has changed from a region of net in-migration to one of net out-migration, and, as will be seen, net in-migration to urban centers has also declined.

To a certain extent southern Kazakhstan shares many of these general patterns of Central Asia. Most notable are an oasis-type settlement, and relatively rapid population growth and lower urbanization by Soviet standards. However, a number of differences are also evident. The highest rural densities in southern Kazakhstan are decidedly lower than those of Central Asia, as are the rates of total and rural population growth.

IMPLICATIONS AND FUTURE PROSPECTS

The relatively rapid growth of the population, particularly the rural population, of Central Asia has given rise to what is commonly known as the "Central Asian problem", and the possible solutions to the increasing labor surpluses in these areas.

A considerable literature has evolved discussing the fate of Central Asia. The arguments basically center around whether or not a sufficient amount of investment and jobs will be placed in Central Asia to accommodate a rapidly growing indigenous work force, and thus whether or not their out-migration to other regions of the USSR will occur. The different scholarly viewpoints on this problem have been discussed in detail elsewhere (for example, see Feshbach 1979: 656–709; Lewis and Rowland 1979: 412–24). Arguments for such a migration, which have been propounded especially by the author and colleagues, are based on the universal experience, which suggests that rapid population growth and labor surpluses in less developed regions increase population pressures and give rise to a migration to local cities and eventually to more developed regions, which have a demand for unskilled and semiskilled workers. In addition, such a migration typically involves ethnic intermixing and tensions in the destination regions (for example, Mexicans to California, Blacks to northern cities of the United States, Puerto Ricans to New York City, peoples of the mediterranean basin to northwest Europe, Indians and Pakistanis to Great Britain). Arguments against such an out-migration include indigenous reluctance to leave the ethnic homeland for a more severe environment further north, although such migrations have been common elsewhere in the world.

That labor surpluses and unemployment are occurring in rural Central Asia has been suggested in a number of studies (for example, see Lewis and Rowland 1979: 412–13; Lubin 1984: 102–5; Porket 1989: 106–7, 176–7; Soviet unemployment . . . 1989: 6–7). Major manifestations include declines in arable land per collective farmer and number of man-days worked per able-bodied farmer, and, as elswehere, the surplus is being aggravated by increased mechanization (Lubin 1984: 102–5). This topic is discussed in greater detail in the chapter on agriculture by Peter Craumer.

The central question is what responses are being made to this problem, both by the government in terms of investment policies and by the rural populace in terms of possible out-migration? In very simple hypothetical terms, both mechanisms are for the most part ultimately mutually exclusive in that, if one is insufficient, the other will have to occur.

Capital investment

An investigation of capital investment trends per worker and per capita by republic is possible through the 1980s. In a recent study, the author investigated such trends for 1970–9 and 1979–86 by union republic (Rowland 1988: 821–7). The number of workers was estimated by published data on the average annual number of rabochiye, sluzhashchiye, and kolkhozniki. Although per capita and per worker investment continued to increase, a deteriorating Central Asian situation was, nonetheless, evident (Rowland 1988: 824–85). First, average annual percentage increases in capital investment per average annual worker and per capita in Central Asia during 1970–9 and 1979–86 were always below the national average. Furthermore, capital investment per average annual worker in Central Asia changed from being 9.2 percent *above* the national average in 1970, to 5.0 percent *below* in 1979 and 19.5 percent *below* in 1986. Capital investment per capita was below the all-union average in *each* of these years, and a deepening and widening gap was again noticeable (21.6 percent lower in 1970, to 34.1 percent in 1979, to 44.0 percent below the national average in 1986).

More recently published data on capital investment in 1980 and 1988, as well as age data for 1989 and 1987 and population data for 1989, allow for an update of these trends (USSR 1988a: 48–95, 1989a: 55; O predvaritel'nykh ... 1989: 2). They further suggest a relatively deteriorating situation for Central Asia. Whereas per capita investment for the USSR increased by nearly one-third from 575.0 rubles in 1980 per 1979 population to 761.0 rubles in 1988 per 1989 population, that for Central Asia barely increased from 384.6 to 393.3 rubles. In addition, whereas investment per population aged 20–59 for the USSR as a whole also increased by roughly one-third from 1,089.1 rubles in 1980 per 1979 population to 1,446.7 rubles in 1988 per 1987 population aged 20–59, that for Central Asia barely increased from 965.3 to 970.5 rubles per popula-tion aged 20–59. These trends were due to the fact that, although the amount of capital investment in Central Asia increased more rapidly than the national average, the total population and work force of Central Asia increased *much* more rapidly than the corresponding national averages.

An especially deteriorating situation was evident for Uzbekistan, the chief republic of Central Asia, which actually had *declining* per worker and per capita capital investment levels between 1979 and 1986. In addition, between 1984 and 1986, the amount of capital investment in Uzbekistan in particular *decreased* in absolute terms by 2.3 percent. Meanwhile, between 1984 and 1986 absolute capital investment for the USSR as a whole increased by 11.5 percent (Rowland 1988: 823–6).

There are other indications of insufficient economic attention towards Central Asia. Published planned increases in industrial and agricultural production for the 1981–6 and 1986–90 five-year plans also do not indicate a strong reorientation toward Central Asia. Whatever absolute industrial and agricultural gains are experienced by Central Asian republics will probably be offset by a relatively rapidly growing population and number of workers (*Sovetskaya Rossiya* 1981: 3–6; *Osnovnyye* . . . 1985: 18. 44. 70–84). It appears that the most recent five-year plan of 1986–90 and Gorbachev's policies particularly emphasize redevelopment and increased efficiency in the European USSR, which would appear to give relatively low priority to Central Asia (Shabad 1986: 1–16). It is also interesting to note that the Soviets are abandoning one specific facet of investment that would have benefited Central Asia; namely, the diversion of rivers from north to south (Micklin 1986: 287–329).

It should be kept in mind that the above economic data are somewhat crude in that they do not indicate investment into specific branches of industry, such as those that are labor intensive or capital intensive. Obviously, substantial investment in labor-intensive textiles would be especially beneficial for Central Asia, given its large cotton and labor supply. More detailed aspects of the economy and investment diversions are discussed in the chapter by Liebowitz.

Because it does not appear that sufficient investment will be given to Central Asia, one could conclude that indigenous out-migration will accelerate. Accordingly, it seems appropriate to investigate the current extent of such migration.

Migration

Although there is some evidence that the traditionally immobile population of Central Asia has become more mobile, the indigenous population is still relatively immobile by Soviet standards, and relatively little rural out-migration has yet to occur. This conclusion is supported by data through the 1979 census (Shabad 1979: 456, 1985: 109–53; Kozlov 1982: 100; Rowland 1982; Karakhanov 1983: 227–8; Mullyadzhanov 1983: 214; *Naseleniye SSSR* 1983: 39).

Rural-to-urban migration can be estimated for 1970–9 and 1979–89 based on recently published annual births, deaths, and natural increase for the urban and rural population for January 1970 through January 1988. Results indicate that there has been a reduced rural-to-urban net migration-reclassification (Table 9.8). In particular, average annual net migration-reclassification for the urban population declined by 13.7 points from 15.7 during 1970–9 per 1,000 July 1974 population to only 2.0 per 1,000 January 1984 population during 1979–89, although this was partly due to the Samarkand situation. However, even if the Samarkand situation of 130,000 is adjusted for (that is, by adding 13,000 per year to the average annual rural-to-urban net migration-reclassification category during 1979–89), the rate of net migration-reclassification still declined for the urban population by 12.6 points (15.7 to 3.1). Reduced rural-to-urban net migration may also reflect the apparent reduced Russian in-migration and/or increased Russian out-migration from Central Asia, since Russians have resided chiefly in urban centers. In contrast, the net migration-reclassification rate for the rural population increased by 3.0 points from –8.6 to –5.6 per 1,000 from 1970–9 to 1979–89 (even with the Samarkand adjustment it would increase by 2.2 points from –8.6 to –6.4).

Other recently available sources further testify to the relatively low degree of rural-to-urban migration and mobility in Central Asia (Tarasova 1985: 58; *Razvitiye* . . . 1988: 50).

In addition, data from the "micro-census" or sample census of 1985 also do not suggest much rural-to-urban migration. On 1 January 1985 a sample micro-census was conducted in the USSR based on a sample of more than 13 million people or roughly 5 percent of the Soviet population (*Vestnik Statistiki* 1986; no. 6: 53). As in the 1979 census, migrants are defined as those who have resided in their permanent place of residence *not* from birth. In nontabular form, "active migration was observed in [the seven republics of] the RSFSR, Ukraine, Belorussia, Kazakhstan, and [three] Baltic republics, where the share of the population that changed their permanent place of residence ranged from 41 percent to 61 percent" (*Vestnik Statistiki* 1986, no. 6: 55). This implies that the migration rate for the other eight republics, including the four of Central Asia, was less than 41 percent and thus relatively low. According to the 1979 census, the comparable "migration rate" for the seven "active" republics ranged from 43.7 percent (Ukraine) to 62.3 percent (Estonia), while the rates for the remaining eight republics were below 40 percent. The Central Asian republics ranged from 22.3 percent for the Uzbek SSR to 35.9 percent for the Kirgiz SSR.

With regard to the Central Asian issue, a particularly valuable type of 1985 data is rural–urban migration data by republic. Central Asian republics have relatively low values, reflecting comparatively low indigenous rural-to-urban migration. For the USSR as a whole, 40.4 percent of the migrants were rural-to-urban migrants, but the Central Asian republics were all well below this average, with the four lowest republic values (Uzbek, 26.7 percent; Kirgiz, 29.6; Tadzhik, 24.5; and Turkmen, 31.8). (*Vestnik statistiki* 1986, no. 8: 80). These figures are consistent with the slow rate of urbanization in Central Asia through the mid-to-late 1980s.

With regard to the migration of indigenous peoples of Central Asia to the north, this also does not yet appear to be occurring to a significant degree. According to the 1979 census, more than 98 percent of the four SSR-status indigenous nationalities lived in the Central Asian republics. Data from the 1989 census further suggest little change in this pattern, as 97.4 percent of these four nationalities still resided in Central Asia. In addition, although the population of the four major Central Asian nationalities residing in the RSFSR nearly doubled from 1979 to 1989, their total population in the RSFSR in 1989 was only roughly 250,000, which accounted for only about one percent (0.9) of their combined total population and for only a minuscule 0.2 percent of the total population of the RSFSR (USSR 1989b: 3–113).

However, as noted earlier, since 1979, Central Asia has changed from a region of net in-migration to one of *net out-migration* (Table 9.4). Although this might provide support for the thesis of out-migration from Central Asia, at this stage it probably reflects nonindigenous migration trends, especially by Russians. Namely, it appears that this turnaround chiefly reflects a net out-migration of Russians (Anderson and Silver 1989b: 641). Indeed, data from the 1989 census indicate that the Russian population of Central Asia actually declined slightly between 1979 and 1989 from 3,321,620 to 3,289,829, marking the first time such a decline has occurred in an intercensal period (Lewis et al. 1976; Rowland 1982). Thus, Russian net out-migration was roughly equal to Russian natural increase, and there has been reduced Russian migration to Central Asia probably because of labor deficits in the Russian areas further north and the increased education of the rapidly growing indigenous Turkic-Muslims in Central Asia. This makes it increasingly possible for them to fill many higher-level jobs, which Russians previously moved in to fill given the former relative backwardness of the indigenous peoples (Zyuzin 1983: 115; Rywkin 1984: 88–91).

A number of factors may explain the continued low indigenous out-migration. First, it may be that urban–rural and regional wage differentials in Central Asia and the USSR may simply not yet be great enough to generate such a migration (Lewis 1983: 93). Other impediments may include ethnic discrimination and the different cultural environment of the Russian-influenced cities and industrial activities, as well as the continued strength and attraction of the rural family structure. These topics are discussed in greater detail in other chapters in this volume.

Interesting perspectives on reasons for the low mobility of the indigenous populations of Central Asia have also been provided in two recent surveys for 1978–80 and 1979–81 (Zyuzin 1983: 109–17, 1985: 88–95). In general, they suggest that the chief factor limiting out-migration from rural areas was the current higher standard of living in the villages as compared to cities. Specific factors noted included: higher real family incomes due to income from private plots and lower costs of living; more housing space in rural areas; conditions more favorable for raising large families, including more housing; and nationality tradition in that older children stay to help with young and with aging parents.

However, one of the two studies also indicates a certain receptiveness on the part of the indigenous population toward moving to the RSFSR. For example, 70 percent of the surveyed (graduates of secondary schools and working youths) said that they would live in regions outside of Central Asia, and 40 percent expressed a very strong desire to do so (Zyuzin 1985: 93). Indeed, in recent years, there has been increased receptiveness for migration out of Central Asia by indigenous nationalities in the form of response to "public appeals" (Zyuzin 1985: 93).

In summary, this section has attempted to provide an up-to-date assessment of the actual extent of two possible major responses to the "Central Asian problem" since 1979; namely, the degree of investment in and rural out-migration from Central Asia. As noted at the outset, it is possible to hypothesize simply that the two mechanisms are somewhat and ultimately mutually exclusive in that, if one does not occur, the other will. It is thus, perhaps, paradoxical that results from this study suggest that through the mid-to-late 1980s *neither* mechanism appears to be in operation to any great extent. Soviet regional investment plans show no dramatic redirection toward Central Asia. Indeed, per capital and worker investment in Central Asia continues to increase more slowly than the Soviet average, or even to decline by some indicators. North-to-south water transfer programs have also apparently been dropped. In addition, although the mobility of the indigenous nationalities is increasing, it is still low by Soviet standards, and there does not yet appear to be any strong indigenous rural-to-urban or northward migration. However, surveys do suggest an increasingly indigenous desire to move to the RSFSR.

Of course, it may still be too early to test the thesis of indigenous out-migration from Central Asia, which was partly predicated on projected regional work force changes up to 1990. Based on 1970 regional age data, it was projected that the working age population would be increasing substantially in Central Asia, while actually declining or increasing slowly (in the absence of migration) in northern regions (Lewis 1983: 80). This projection was borne out by subsequent trends discussed earlier for 1970–87. Continued low investment per worker and per capita in Central Asia could further exacerbate labor surpluses, thus promoting a northward migration and increased ethnic intermixing and tensions.

On the other hand, such a northward migration could be thwarted somewhat if another policy associated with increasing efficiency and productivity is carried out, namely, the reduction of redundant labor in the northern "labor-deficit" areas (Kostakov 1987: 78–89). If this policy were realized, unemployment might then exist among a certain share of the ethnic Russian work force, and thus excess Russian labor could, in turn, help alleviate labor deficits in the RSFSR, thus obviating a south-to-north Turkic-Muslim migration. The continued lack of appreciable investment in Central Asia, plus the reduction of job opportunities for them outside of Central Asia, could further amplify labor surpluses and dissatisfaction in Central Asia.

Migration and more detailed nationality data from the 1989 census will shed further light on the status of indigenous out-migration. Indeed, like the 1985 micro-census (and unlike the 1979 census), 1989 census migration data will apparently include a question on urban or rural origin (*Vestnik Statistiki* 1986, no. 3: 19).

Finally, with respect to the theme of integration, the topic of this chapter, population, would generally support the viewpoint that little integration is occurring. Most important, despite Central Asia's increasing demographic importance in the USSR, little indigenous rural out-migration to and integration with Russian-influenced cities within and in areas outside Central Asia has yet to occur. In addition, central Asia does not appear to be increasingly integrated from an economic perspective, as investment directives from Moscow appear to be waning overall and in particular paying relatively less rather than the needed more attention to Central Asia. However, it is still possible that these situations will eventually change and that with increased indigenous rural-to-urban and interrepublic migration, and increased investment in Central Asia, the indigenous population of the region will become more integrated with the mainstream of Soviet society.

To repeat, it is possible to foresee an indigenous rural out-migration due to increasing labor surpluses of Central Asia and possible labor shortages in areas further north. To date, such an out-migration has apparently been delayed chiefly due to the fact that wage differentials between rural Central Asia, on the one hand, and local and distant cities, on the other, have not yet been great enough. In addition, the strength of the family and the relatively familiar and more attractive cultural and natural environment have also impede such an out-migration.

However, it is difficult to foresee an eternal continuation of these circumstances in the face of increasing labor surpluses and insufficient investment. Under such surplus conditions wage differentials should eventually increase and give rise to out-migration. Of course, it is also possible that the indigenous population would be willing to accept docilely a deteriorating standard of living, or that the government could respond by changing its current policy and devoting more funds to Central Asia either to create more jobs or to support a massive unemployed welfare population.

However, these scenarios may not occur. It is doubtful that any fully fledged Marxist Soviet regime would tolerate the existence of a region of massive unemployment, because it goes against the tenets of Marxist ideology. On the other hand, there are indications that with perestroyka some unemployment may be tolerated, and also the degree of Marxist preoccupation under the current Gorbachev regime is certainly lessening. In addition, Soviet regional policies, at least current ones, obviously show no major inclination to increase funds for Central Asia to the necessary levels. Nonetheless, few if any populations in the history of the world have been content to remain in an area of declining living standards, regardless of the economic-ideological orientation of the government.

It should also be noted that birth control policies would have only a negligible influence on reducing population pressures. Even if there was an adopted birth control policy which was immediately effective, this would have an influence only on new entrants to the work force in the early 2000s. The immediate concern is the necessity of accommodating those born in the 1960s and 1970s and entering the work force in the late 1980s or 1990s. In addition, if anything, the Soviet policy toward fertility is pronatalist, not antinatalist.

In short, Central Asia could be described as a demographic cauldron, and because the temperature will probably not be turned down by either necessary funding or birth-control measures, there may be only one realistic alternative: out-migration.

NOTES

1. Southern Kazakhstan is operationally defined here as the combination of the current units of Kyzl-Orda Oblast, Chimkent Oblast, Dzhambul Oblast, Alma-Ata City and Oblast, and Taldy-Kurgan Oblast. These units generally have much in common with Soviet Central Asia. In particular: (a) they either border on or are very close to Central Asia; (b) in 1979, the Muslim Kazakhs outnumbered Russians in all units, except for Alma-Ata City; (c) they contain areas of irrigated agriculture and thus oasis-type settlement; and (d) they are relatively lowly urbanized. It is also possible to investigate these units combined

on a comparable basis since 1951, in that the only major change to occur has been former Alma-Ata Oblast being subdivided into Alma-Ata Oblast and Taldy-Kurgan Oblast. Thus, the current six units of today are in aggregate comparable to the five units in 1951. Recently abolished Mangyshlak Oblast, which borders on the Turkmen SSR, has not been included, because it contains little agriculture, irrigated or otherwise, and is highly urbanized. Indeed, in 1987, only one-ninth (10.7 percent) of the population was rural, and nearly one-half (49.2 percent) of the population resided in the oil city of Shevchenko.

2. In 1985, Central Asia became *the most populous* of the 19 economic regions of 1961 for the first time in the time span beginning in 1897, the year of the first Russian census, as its population surpassed that of the Center or Moscow area (29,629,000 vs 28,818,000) (for appropriate census data since 1897, see Lewis and Rowland 1979: 442; Rowland 1989: 642–3). In addition, with regard to *current* economic regions, in 1986, Central Asia also surpassed the Center, which, unlike the 1961 regions, contains Orel Oblast (30,456,000 vs 29,795,000).

3 Indeed, in recent years, only a small number of countries have had comparable proportions; for example, the 20–59 cohort accounted for only 32.3 percent of the rural population of Zimbabwe in 1982 and 33.3 percent of the rural population of Kenya in 1979 (see United Nations 1984: 222–9).

4 Net migration for 1970–9 and 1979–89 has been estimated on the basis of published annual absolute natural increase data for all years in each period. Such data are explicitly available for 1970–9 and 1979–88 in *Naseleniye SSSR, 1987* (USSR 1988a: 110–26). Natural increase for 1988–9 was estimated on the basis of natural increase rates for 1988 multiplied by the midyear population of 1988, which in turn was estimated by averaging the January 1988 and January 1989 populations (USSR 1988b: 173, 1989a: 26; O predvaritel'nykh . . . 1989: 2).

5. The Central Asian net migration rate of 0.9 per 1,000 or 0.1 percent for 1970–9 in Table 9.4 used here differs slightly from that of 0.0 percent for 1970–9 mentioned in a previous study (Rowland, 1982), because data in Table 9.4 are based on more complete annual natural increase from *Naseleniye SSSR, 1987* (see sources, Table 9.4).

6. In order to estimate net migration for 1970–9, 1970 and 1980 natural increase for the units of southern Kazakhstan were summed. This sum for 1970 and 1980 was then taken as a percentage of the sum of 1970 and 1980 natural increase for Kazakhstan as a whole. The percentage was then multiplied, absolute natural increase for Kazakhstan as a whole during 1970–9 (USSR 1988a: 116). In short, southern Kazakhstan's percentage share of Kazakhstan's natural increase for the sum of 1970 and 1980 was multiplied by Kazakhstan's natural increase for 1970–9 to estimate its natural increase and thus net migration for 1970–9. For 1979–89, natural increase and net migration for Kazakhstan as a whole could be estimated as they were for Central Asia (see note 4). A ratio, 3.30, was then derived between natural increase for Kazakhstan as a whole during 1979–89 to that of the sum of that for 1980, 1985, and 1986. Natural increase for southern Kazakhstan in 1980, 1985, and 1986 was then multiplied by 3.30 to estimate its natural increase and ultimately net migration for 1979–89.

7. Samarkand's reported population declined from 515,000 in 1984 to 371,000 in 1985, and, in fact, between 1984 and 1985, "deurbanization" occurred for Central Asia and Uzbekistan as the level of urbanization for both units declined (the former from 41.2 to 40.9 percent and the latter from 42.3 to 41.9

percent) (see USSR 1984b: 12–17, 22, 1985: 14–19, 24). It appears that Samarkand has returned to its pre-1979 boundaries. Between 1978 and 1979 its reported population skyrocketed from 318,000 to 476,000 due to the annexation of surrounding rural areas. Thus, the annexed rural areas have reverted to the rural population (see Shabad 1986a: 65). Subsequently published preliminary results from the 1989 census indicate populations for Samarkand of 346,000 in 1979 and 366,000 in 1989 (O predvaritel'nykh . . . 1989: 2). It is also interesting to note that, unlike for other cities which have had boundary changes, the Soviets never did provide pre-1979 populations for Samarkand in 1979 boundaries, apparently because, it now can be concluded, such an enlargement was only temporary. If, for comparability purposes, 130,000 people are subtracted from the urban population of Central Asia in 1979 and added to the rural population, the 1979 level of urbanization for Central Asia would be 40.2 percent instead of 40.7.

Despite these adjustments, Central Asia still emerges as the least urbanized Soviet economic region in 1979 and 1989, and the region with the slowest urbanization during 1970–9 and 1979–89. With respect to urbanization change, the average annual percentage point change for each period would be 0.24 and –0.06 points per year, respectively, instead of changes of 0.30 and –0.11 as shown in Table 9.7. In short, this also signifies a declining rate of urbanization increase from 1970–9 and 1979–89, and the slowest urbanization pace in the USSR despite the adjustments in 1979 for Samarkand and the fact that the urbanization change for 1979–89 would be slightly higher with this adjustment (–0.06 vs –0.11). It should also be noted that the USSR as a whole experienced a slowing urbanization process during the 1980s (Table 9.7).

8. If an adjustment is again made for Samarkand (namely, subtracting 130,000 people from the urban population of Central Asia in 1979), its urban growth for 1979–89 turns out to be 2.4 percent per year rather than 2.3. Conversely, based on this adjustment, the 1970–9 Central Asian urban growth rate would be 3.4 percent per year rather than 3.6. Nonetheless, even with the adjustment, the Central Asian urban growth rate in the 1980s was still somewhat lower than during the 1970s.

REFERENCES

Anderson, Barbara A. and Silver, Brian D. 1989a. The changing shape of Soviet mortality, 1958–1985: an evaluation of old and new evidence. *Population and Development Review* 43: 243–65.

Anderson, Barbara A. and Silver, Brian D. 1989b. Demographic sources of the changing ethnic composition of the Soviet Union. *Population and Development Review* 15: 609–56.

Feshbach, Murray 1979. Prospects for massive out-migration from Central Asia during the next decade. In US Congress, Joint Economic Committee, *Soviet Economy in a Time of Change*, pp. 656–709. Washington DC: Joint Committee Print.

Karakhanov, M. K. 1983. *Nekapitalisticheskiy Put' Razvitiya i Problemy Narodon-aseleniya*. Tashkent: Izdatel'stvo "Fan" Uzbekskoy SSR.

Kostakov, V. 1987. Zanyatost': defitsit ili izbytok? *Kommunist* 2: 78–89.

Kozlov, V. I. 1982. *Natsional'nosti SSSR: Etnodemograficheskiy Obzor.* Moscow. Finansy i Statistika.

Lewis, Robert A. 1983. Regional manpower resources and resource development in the USSR: 1970–90. In Robert G. Jensen, Theodore Shabad, and Arthur K. Wright (eds), *Soviet Natural Resources in the World Economy,* pp. 72–97. Chicago: University Press of Chicago.

Lewis, Robert A. and Rowland, Richard H. 1979. *Population Redistribution in the USSR: Its Impact on Society, 1897–1977.* New York: Praeger.

Lewis, Robert A., Rowland, Richard H. and Clem, Ralph S. 1976. *Nationality and Population Change in Russia and the USSR: An Evaluation of Census Data, 1897–1970.* New York: Praeger.

Lubin, Nancy 1984. *Labour and Nationality in Soviet Central Asia.* Princeton, NJ: Princeton University Press.

Micklin, Philip 1986. The status of the Soviet Union's north–south water projects before their abandonment. *Soviet Geography* 27: 287–329.

Mullyadzhanov, I. R. 1983. *Demograficheskoye Razvitiye Uzbekskoy SSR.* Tashkent: Uzbekistan.

Naseleniye SSSR: spravochnik 1983. Moscow: Izdatel'stvo Politicheskoy Literatury.

O predvaritel'nykh itogakh vsesoyuznoy perepisi naseleniya 1989 goda 1989. *Pravda,* 29 April: 2.

Osnovnyye Napravleniya Ekonomicheskogo i Sotsial'nogo Razvitiya SSSR na 1986–1990 Gody i na period do 2000 Goda 1985. Moscow: Izdatel'stvo Pravda.

Population Reference Bureau 1989. *World Population Data Sheet, 1989.* Washington DC: Population Reference Bureau.

Porket, J. L. 1989. *Work, Employment and Unemployment in the Soviet Union.* New York: St Martin's Press.

Razvitiye Narodonaseleniya i Problemy Trudovykh Resursov Respublik Sredney Azii 1988. Tashkent: Izdatel'stvo "Fan" Uzbekskoy SSR.

Rowland, Richard H. 1982. Regional migration an ethnic Russian population change in the USSR (1959–79). *Soviet Geography: Review and Translation* 23: 557–83.

Rowland, Richard H. 1988. Union republic migration trends in the USSR during the 1980s. *Soviet Geography* 29: 809–29.

Rowland, Richard H. 1989. National and regional population trends in the USSR. 1979–89: Preliminary results from the 1989 census. *Soviet Geography* 30: 635–69.

Rowland, Richard H. and Lewis, Robert A. 1982. Regional population growth and redistribution in the USSR, 1970–9. *Canadian Studies in Population* 9: 71–93.

Rywkin, Michael 1984. The impact of socio-economic change and demographic growth on nationality identity and socialization. *Central Asian Survey* 3: 79–88.

Shabad, Theodore 1979. News Notes. *Soviet Geography: Review and Translation* 20: 440–56.

Shabad, Theodore 1985. Population trends of Soviet cities, 1970–84. *Soviet Geography* 26: 109–53.

Shabad, Theodore 1986a. News Notes. *Soviet Geography* 27: 65.

Shabad, Theodore 1986b. Geographic aspects of the new Soviet five-year plan, 1986–90. *Soviet Geography* 27: 1–16.

Sovetskaya Rossiya 1981. 5 March: 3–6.

Soviet unemployment: facing the facts 1989. *Current Digest of the Soviet Press* 41: 6–7.

Tarasova, N. V. 1985. Problema povysheniya migratsionnoy aktivnosti naseleniya. In *Sovremennyye Problemy Migratsii*, pp. 47–63. Moscow; Mysl'.

United Nations, Department of International Economic and Social Affairs, 1984. *Demographic Yearbook, 1984*. New York: United Nations.

USSR, Glavnoye Upravleniye Geodezii i Kartografii pri Sovete Ministrov SSSR. 1986. *Atlas SSSR*. Moscow.

USSR, Gosudarstvennyy Komitet SSSR po Statistike (Goskomstat) 1987. *Narodnoye Khozyaystvo SSSR za 70 Let*. Moscow: Finansy i Statistika.

USSR, Goskomstat 1988a. *Naseleniye SSSR, 1987*. Moscow: Finansy i Statistika.

USSR, Goskomstat 1988b. *SSSR v Tsifrakh v 1897 g*. Moscow: Finansy i Statistika.

USSR, Goskomstat 1989a. *Narodnoye Khozyaystvo SSSR v 1988 g*. Moscow: Finansy i Statistika.

USSR, Goskomstat 1989b. *Natsional'nyy Sostav Naseleniya i Ekonomicheskoye i Sotsial'noye Razvitiye Soyuznykh i Avtonomnykh Respublik, Avtonomnykh Oblastey i okrugov*. Moscow: Informatsionno-izdatel'skiy Tsentr.

USSR, Tsentral'noye Statisticheskoye Upravleniye (TsSU) SSSR 1962–3. *Itogi Vsesoyuznoy Perepisi Naseleniya 1959 Goda*. Moscow: Gosstatizdat.

USSR, TsSU 1972–4. *Itogi Vsesoyuznoy Perepisi Naseleniya 1970 Goda*. Moscow: Statistika.

USSR, TsSU 1975. *Naseleniye SSSR, 1973*. Moscow: Izdatel'stvo "Statistika".

USSR, TsSU 1980. *Narodnoye Khozyaystvo SSSR v 1979 g*. Moscow: Statistika.

USSR, TsSU 1984a. *Chislennost' i Sostav Naseleniya (po dannym vsesoyuznoy perepisi naseleniya 1979 Goda)*. Moscow: Finansy i Statistika.

USSR, TsSU 1984b. *Narodnoye Khozyaystvo SSSR v 1983 g*. Moscow: Statistika.

USSR, TsSU 1985. *Narodnoye Khozyaystvo SSSR v 1984 g*. Moscow: Finansy i Statistika.

USSR, TsSU 1986b. *Narodnoye Khozyaystvo SSSR v 1985 g*. Moscow: Finansy i Statistika.

Vestnik Statistiki 1986. Nos 3, 6, 8.

Zyuzin, D. 1983. Prichiny nizkoy mobil'nosti korennogo naseleniya respublik Sredney Azii. *Sotsiologicheskiye issledovaniya* 1: 109–17.

Zyuzin, D. 1985. Srednyaya Aziya—vazhneyshiy istochnik trudovykh resursov dlya narodnogo khozyaystva. In *Naseleniye Sredney Azii*. Moscow: Finansy i Statistika.

Chapter 10 ————————————————————————————

Social Mobilization in Soviet Central Asia

Robert J. Kaiser

INTRODUCTION

The purpose of this chapter is to examine the social mobilization of the indigenous nations of Soviet Central Asia and Kazahkhstan,[1] and the extent to which this has resulted in a "Sovietization" of the population. Social mobilization is defined as "the process in which major clusters of old social, economic and psychological commitments are eroded and broken and people become available for new patterns of socialization and behaviour" (Eisenstadt 1966: 41–2). This process may be viewed as resulting in the social integration of a nation's population into the modernized sectors of Soviet society.

This process is, of course, intimately connected with the broader process of modernization. The latter is a complex, interdependent process involving radical social, economic, political, and cultural change. Modernity, the ill-defined and perhaps unattainable objective of this process, is most frequently viewed as the antithesis of traditional society. Thus, it is said that modernization is the transition from "traditional" to "modern" life.[2] This definition of the process is much too simplistic, since, as Goldscheider (1971: 97) has pointed out, "elements of modernity may be observed within traditional societies and traditional patterns persist within modern social systems." A more accurate depiction of modernization is as "the process by which historically evolved institutions are adapted to the rapidly changing functions that reflect the unprecedented increase in man's knowledge, permitting control over his environment, that accompanied the scientific revolution" (Black 1966: 7, italics added). From this definition, Black (1966: 48–9) goes on to note that while there is a convergence or "universalization" of functions performed across modernizing societies, "however great the institutional change the very fact of adaptation means that many features of the traditional institutions survive."

It has sometimes been said that modernization is Westernization, or Europeanization, and there is certainly some truth to this equation, particularly in the former colonies of the Third World. However, the socialization that occurs with modernization is of a more universal nature and is not simply an aping of what has occurred in Western Europe. And, since modernization is a process of adaptation, it is also true that local, indigenous sociocultural forms influence the new social, economic, and political institutions which emerge with development.

The view of modernization as a process of adaptation is applicable in the USSR generally, and is of particular relevance in the case of Soviet Central Asia. Modernization in the Soviet Union has been equated with Sovietization, and at a certain level of analysis this is undoubtedly true.[3] However, misinterpreting modernization as the transformation from a traditional society to modernity has led some analysts, such as Carrere d'Encausse (1981: 250), to proclaim that "in the Moslem regions the attempt to transform human thought has run into an almost impenetrable socio-cultural situation." A closer examination of change in the region provides clear evidence that, while some aspects of traditional life have survived and continue to influence the structure of society, extensive adaptation to a more modern sociocultural milieu has also occurred. This has resulted in the reorientation of the indigenous nations of Soviet Central Asia to a way of life more compatible with that obtaining in other parts of the Soviet Union. At this level, one may speak of Sovietization.

This is at the same time not to deny that traditional sociocultural patterns and institutions continue both to exist and to influence the process of modernization itself. For example, in explaining the slow transition to low fertility in the region, Jones and Grupp (1987: 183) found that

> The later onset of fertility decline in the Muslim areas of the USSR is due in part to the low stage of socioeconomic development in these regions on the eve of the Revolution. But this is only part of the explanation. An additional, and perhaps more crucial factor, is the persistence of traditional sex roles and family systems in these areas that reinforce cultural traditions assigning women to a maternal role. Muslim traditionality, however, has not meant that these minorities are immune to the effects of social and economic change.

Social mobilization in Soviet Central Asia as elsewhere is a process with both its universal and unique aspects; the process itself and the consequences for the individuals and nations involved cannot be assessed without this adaptive framework in mind.

MEASURING SOCIAL MOBILIZATION

To measure the extent to which social mobilization has occurred, we focus on a number of socioeconomic trends, and examine surveys designed to measure attitudinal changes associated with social mobilization. Measurement of this process is difficult in a region such as Soviet Central Asia, because at present there is a great degree of overlap between traditional and modern society, and each influences the population in particular spheres of life. Given this fluid situation, we attempt to answer two general questions in this chapter. First, is the indigenous population, or more precisely the younger generation, undergoing a basic reorientation in the social and economic spheres that are associated with modernization? Second, is there evidence of a convergence of aspirations between younger members of the indigenous nations, on the one hand, and those of the more modernized nations, on the other?

To answer these questions, we focus on education, Russian bilingualism, and changes in the occupational structure within Soviet Central Asia. Education has been cited as the preeminent vehicle for sociocultural change (Inkeles, 1966: 146). It has been stated that "modernization can be achieved only by improving and extending education" (Anderson 1966: 69). Education serves as the major means through which social values are restructured to conform with the needs of modernizing society. If successful, this "socialization" results in the sociocultural, economic, and political integration (that is, "Sovietization") of the population, in that it prepares the population to participate more fully in the modernized sectors of society while at the same time assisting them in the adaptation to the new norms prevalent in such a society.

A second measure used in the examination of social mobilization is the increasing Russian fluency levels among the indigenous nations. In the Soviet context, learning the Russian language is behavior associated with social mobility. This is particularly true in Soviet Central Asia, where the urban and industrial sector has been dominated by Russians, who entered the region with socioeconomic development. Learning the state's lingua franca is akin to education itself, since it signifies one's willingness to participate in the modernization process. However, this relationship holds only so long as the Russian nation and language are accepted (however grudgingly) as dominant. During the past few years, the status of Russians in the non-Russian periphery has been increasingly challenged, and as this has occurred the importance of the Russian language itself has been undermined.

The third major aspect of social mobilization examined is changes in the occupational structure. A transition toward greater diversification and specialization of the work force is at the heart of modernization. At the same time, evidence of upward mobility in the workplace indicates that a convergence in the valuation of occupations is occurring, that the younger generation of Central Asians have employment aspirations similar to those of other, more modernized nations in the USSR. Obviously, higher educational attainment serves as a precursor of upward mobility in the workplace. If the latter is not occurring at a rate sufficient to satisfy the aspirations of the educated indigenous elite, we expect to find evidence of growing relative deprivation, defined as the "actors' perception of discrepancy between their value expectations and their value capabilities" (Gurr 1970: 24).

In all three sections, we assess changes in the status of women in Soviet Central Asia. Equalization in educational attainment, Russian bilingualism, and occupation structure signify that movement away from the traditional male-dominated society in the region is occurring. This transition is not only important in the sense that it leads to the more complete social mobilization of the indigenous populations; it is also an important factor in the demographic transition to low fertility in the region (Rowland, this volume).

EDUCATION AND SOCIAL MOBILIZATION

At the time of its formation, the Soviet Union was a multinational, multi-homeland state with vast disparities in the level of social and economic development. Central Asia, the last region conquered by the Russian Empire, contained a population that was primarily illiterate, with a feudal–patriarchal socioeconomic system in which most of the social action was prescribed. The major task in the early years of Soviet power was to educate the Turkic-Muslim nationalities in Central Asia in an effort to create an indigenous proletarian element loyal to the socialist state. Education was seen as holding the potential not only for the secularization of society in Central Asia, but also for the political socialization of the Central Asian nations to an alien Marxist–Leninist ideology. For these reasons, an all-out effort to improve educational attainment in the region was launched (Makarova 1987: 136–9).

The region has experienced a rapid rise in the level of educational attainment during the Soviet era. In 1926, only about 7 percent of the Central Asian men and less than one percent of the women were literate, compared to about one-half of Russian males and one-third of Russian females (Arutyunyan and Bromley 1986: 48). By 1985, universal literacy had been achieved, and over two-thirds of the population aged 10 years and older had a higher or secondary education (Statisticheskiye materialy 1986: 67).

While none of the indigenous nations of Soviet Central Asia has yet achieved equality of educational attainment with Russians, all but the Turkmen experienced a faster rate of increase than the dominant nation between 1959 and 1985 (Table 10.1).[4] Differences between the Central Asian nations and Russians continue to exist, primarily due to: (a) the relatively low level of educational attainment among the older generation in Central Asia, (b) the relatively high number of children in the 10–15 age cohort, who are still mainly in primary school, and (c) the relatively large percentage of the population residing in rural areas, which in general has a lower level of educational attainment than the urban population.[5]

Table 10.1 Educational level by nation, 1959–85 (per 1,000 aged 10 and older with higher or secondary [complete and incomplete] education, and percentage point change).

Nation	1959	1970	1979	1985	% point change
Russian	378	508	661	709	33.1
Uzbek	311	412	615	683	37.2
Kazakh	268	390	592	678	41.0
Kirgiz	299	400	590	666	36.7
Tadzhik	299	390	565	643	34.4
Turkmen	363	430	597	677	31.4
Karakalpak	252	384	593	n.a.	(34.1)

Sources: Zinchenko 1984: 160; (data for 1959–79) Statisticheskiye materialy 1986: 67 (data for 1985).

An examination of the higher education rates by age indicates that the younger members of the Central Asian nations had achieved parity with Russians by 1970 (Figures 10.1, 10.2). Education rates for the 20–29-year-olds are particularly striking, in that only the rural Tadzhik rate fell below the Russian rate in this age cohort. Thus, over time it appears that all of the indigenous nations of Soviet Central Asia are attaining higher rates of education than the dominant Russians, and that according to this measure at least, social mobilization is increasing dramatically in this region of the country. By the 1989–90 school year, the proportion of Uzbeks in higher educational institutions ("vuzy") was equal to the proportion of Uzbeks in the total population of Uzbekistan, and for the Kazakhs, Kirgiz, Tadzhiks, and Turkmen the respective proportion of indigenes in each republic's "vuzy" exceeded the nation's proportion in the total population (Goskomstat SSSR 1990a: 1).

Changes in the composition of students in higher education indicate that Central Asians are becoming more oriented towards urban, industrial careers. In 1979, Uzbek students enrolled in polytechnic institutes comprised almost one-third of all Uzbek students in universities; their percentage of the total student body in polytechnic institutes in

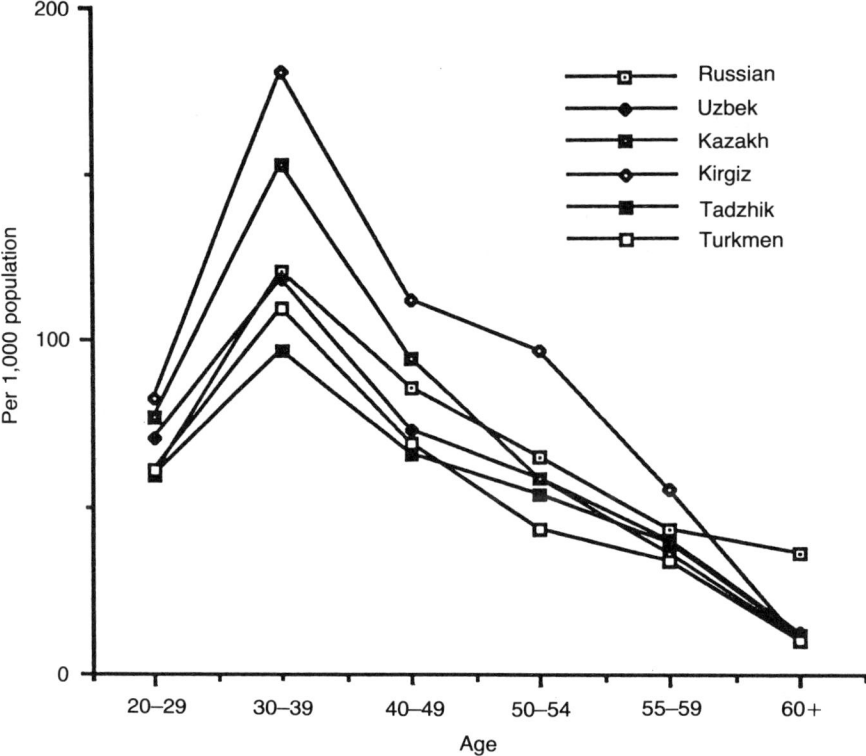

Figure 10.1 Urban rate of completed higher education by nationality and age, 1970.
Source: TsSU SSSR 1973, vol. 4: 549–62.

Uzbekistan increased dramatically between 1970 and 1979, from 41 percent to 53 percent (Arutyunyan and Bromley 1986: 87). This indicates that young Central Asians are no longer willing to pursue traditional careers for which they have a secure niche, but rather are increasingly oriented towards high-status positions in the modernized sectors of Soviet Central Asia that have until now been dominated by Russians and other nonindigenes. In this way, social mobilization among the indigenous nations of Soviet Central Asia is intensifying the level of international competition for high-status positions between indigenes, on the one hand, and Russians and other socially mobilized nonindigenes, on the other.

Status of women

In a region marked by a traditional, male-dominated society such as Soviet Central Asia, changes in the status of women are indicative of the

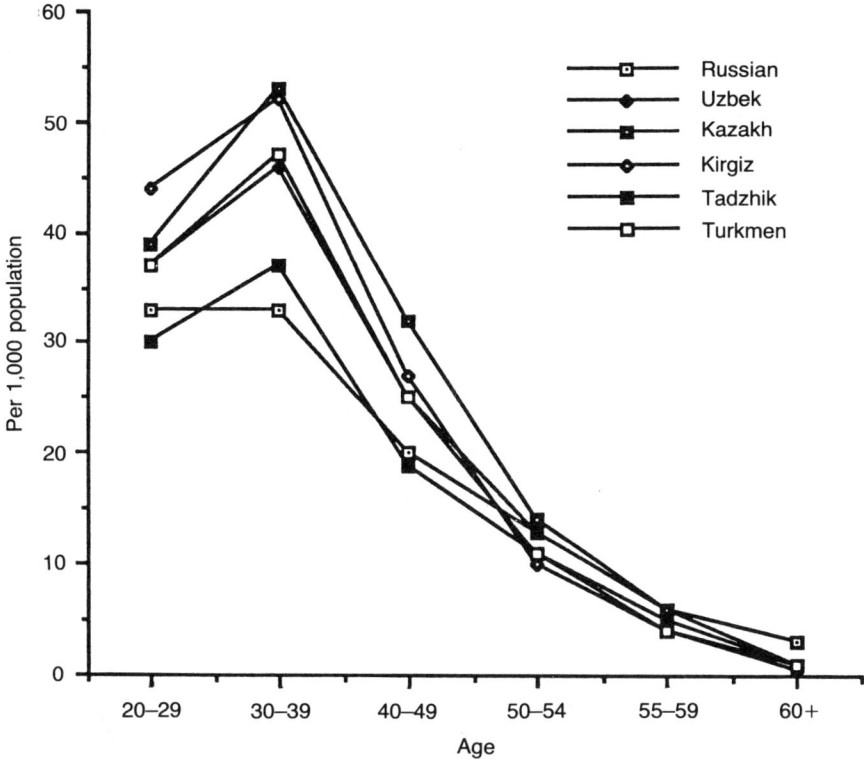

Figure 10.2 Rural rate of completed higher education by nationality and age, 1970.
Source: TsSU SSSR 1973, vol. 4: 549–62.

broader process of social mobilization. Improvement in the position of women in the social sphere is also important in an economic sense, since it would represent a much greater development of human resources in the region. Finally, the prospect for a demographic transition to low birth rates in Soviet Central Asia is closely tied to the improved status of women. Of particular significance is the equalization between men and women in the level of educational attainment (Jones and Grupp 1987: 166–8).

Unfortunately, education data for the 1979 census by sex are provided only for administrative units, and not for national communities. However, the 1985 Soviet pre-census, a survey of 5 percent of the population, does provide this information for nations (Statisticheskiye materialy 1986: 67), which can be compared with earlier census results.

In order to examine the question of equalization between the sexes, a number of measures may be used. Following Jones and Grupp (1987: 172), we utilize a female–male ratio, rather than a measure such as percentage point change, since we are most interested in the comparative level of educational attainment between the sexes. According to this measure, a clear process of equalization in educational attainment between indigenous men and women is occurring in Central Asia. By 1985, the female–male ratios for the nations of Soviet Central Asia were comparable with Russians in the urban setting, and exceeded the Russian ratio in rural areas. Indeed, of the national communities listed in Table 10.2, only the Russian women experienced growing inequality compared to the educational attainment of Russian men.

Table 10.2 Equalization in educational attainment by sex, 1959–85 (female–male ratios of the population aged 10 and older with higher or secondary [complete and incomplete] education).[1]

Nation	Urban			Rural		
	1959	1970	1985	1959	1970	1985
Uzbek	66.3	75.9	93.5	62.0	76.4	89.6
Kazakh	61.9	80.1	95.6	51.9	71.1	85.6
Kirgiz	67.9	85.6	99.4	55.2	71.4	86.6
Tadzhik	54.7	67.0	88.0	65.4	75.7	88.7
Turkmen	62.0	73.0	94.0	75.2	82.3	91.6
Russian	96.6	93.1	94.0	83.4	84.0	80.9

[1] Ratio = female rate ÷ male rate × 100. A ratio of 100 would indicate equality.
Sources: Tsentral 'noye Statisticheskoye Upravleniye SSSR 1973, vol. 4: 393–403 (data for 1959–70); Statisticheskiye materialy 1986: 67 (data for 1985).

Although the level of educational attainment has increased for men and women in both urban and rural areas, it is clear that the female rate of education continues to lag behind that of men in Soviet Central Asia. The gap between men and women is larger in rural areas except for Tadzhiks, and this may be due both to the stronger hold of a traditional patriarchal value system among the rural population and also to the influence of more "European" values in the urban areas of Soviet Central Asia, which have a large number of more "modernized" European nations, and particularly Russians, living in them.

As with the general level of educational attainment, equalization between men and women was occurring in the number of indigenes who went on to a higher education between 1959 and 1970. However, the ratios in the latter category, a more stringent measure of the status of women, are on the whole much lower among the indigenous nations of Soviet Central Asia (Table 10.3). The Russian nation was much more equalized at this level of education in 1970, indicating that the process of social mobilization had proceeded much further than was the case for the Central Asian nations.

Table 10.3 Equalization in educational attainment by sex, 1959–70 (female–male ratios of the population aged 10 and older with higher [complete and incomplete] education).

Nation	Urban			Rural		
	1959	1970	% point change	1959	1970	% point change
Uzbek	41.9	55.0	13.1	11.8	20.0	8.2
Kazakh	42.9	58.7	15.8	22.7	37.8	15.1
Kirgiz	48.5	61.4	12.9	20.8	34.2	13.4
Tadzhik	32.2	39.1	6.9	11.1	13.9	2.8
Turkmen	23.8	37.2	13.4	8.7	15.0	6.3
Russian	84.2	84.7	0.5	86.7	90.5	3.8

Source: Tsentral'noye Statisticheskoye Upravleniye SSSR 1973, vol. 4: 393–403.

Unfortunately, data for higher education by sex and nationality were not published in the 1979 census and have not yet been made available for the census of 1989. However, according to surveys of higher educational attainment in Soviet Central Asia in the early 1980s, "in universities (vuzy) and technicums the number of female Turkmen, Tadzhiks, Uzbeks and Kirgiz increased by several times during the past twenty years, although their proportion among students of their nationalities still comprises less than forty percent" (Arutyunyan and Bromley 1986: 86). Thus, while there is substantial evidence of equalization between men and women in this measure of social mobilization, equality between the sexes in Soviet Central Asia has yet to be achieved. This gap in educational attainment indicates that the traditional male-dominated society has retained some degree of relevance in contemporary Soviet Central Asia, and this will continue to have repercussions for the status of women throughout the social, economic, and political spheres of life in the region.

Conclusions

Trends in education suggest that an educated, socially mobile, indigenous elite exists in Soviet Central Asia, and that the size of this elite is growing rapidly. This signifies that the younger members of the indigenous nations of Soviet Central Asia are becoming increasingly prepared to participate more fully in the modernized sectors of society. Movement toward equalization in educational attainment between men and women is an important trend in the region in both social and economic terms. First, it provides evidence that the strength of the patriarchal family value system is eroding, particularly in urban areas. Second, increasing educational attainment among Central Asian women has a striking impact on the number of children they expect to have. Surveys have shown that "the lower educated women had an average expected family size that was three children, or 80 per cent higher, than

their college-educated counterparts" (Jones and Grupp 1987: 166).
Third, the equalization between men and women in educational attainment means that the potential development of "human capital" in the region is more fully realized. This is positive for the USSR in the sense that the human population is a resource which may lead to great economic development of the country. However, two other contemporary trends in the region have led to increasing problems for the Soviet Union. Rapid population growth in Soviet Central Asia and the relative lack of investment capable of absorbing the growing cadre of young educated indigenes have served to increase the relative deprivation experienced by upwardly mobile Central Asians whose aspirations are growing much faster than their ability to attain high-status positions. The ability of the governmental leadership in Moscow to manage this "national problem" not to mention those confronted in other parts of the country, is increasingly in doubt.

RUSSIAN BILINGUALISM AS SOCIOCULTURAL ADAPTATION

For most national communities in the Soviet Union and elsewhere in the developed world, rising education among the rural population has corresponded with increasing urbanization. This correlation is clearly weaker among the indigenous nations of Soviet Central Asia, and particularly among the female members of these nations. Several reasons for the relative lack of urbanization among the more educated segments of the rural population in this region of the country may be cited. Rural incomes are relatively high; early marriages and large families impede migration; housing shortages in the cities have become a chronic problem; the hold of parents continues to be stronger in the region; and traditional attitudes toward the seclusion of women have not been completely eliminated in the countryside (Zyuzin 1983). While all these factors undoubtedly play a role, an additional impediment to greater rural-to-urban migration is that the cities of Soviet Central Asia have been dominated until recently both socioeconomically and culturally by Russians. For the rural members of the indigenous nations, migration to the cities within their homelands may have been perceived as migration from the homeland itself, since they must adapt not only to a relatively unfamiliar socioeconomic setting, but also to an alien sociocultural environment. According to Arutyunyan (1985: 30): "Restrained migration to cities of the Central Asian republics by members of the indigenous nationalities is explained, in particular, by the difficulty of linguistic adaptation to the urbanized regions, where the cultural environment to a significant degree is formed by the Russian population." In other words, lack of urbanization among the nations of Soviet Central Asia may say less about their social preparedness for an urban-industrial life than about their willingness to adapt culturally to a Russian environment.

If lack of knowledge of the Russian language has served as an impediment toward greater urbanization and social mobility among Central Asians in the past, increasing Russian fluency has been indicative of greater social mobilization. Though the data on Russian bilingualism are suspect, particularly for Uzbekistan, it is apparent that the indigenes of Soviet Central Asia, and especially the younger, more highly educated members of these nations, have become more fluent in Russian (Table 10.4). Does this mean that the indigenous nations are becoming Russified, that they are losing their sense of national self-consciousness and adopting a Russian identity? Hardly. Increasing fluency in Russian has been limited to learning the statewide lingua franca as a second language, while Central Asians overwhelmingly retain their mother tongue as the dominant or first language. Learning Russian appears to be a strategic choice made by indigenes preparing to become more competitive with Russians and other outsiders for high-status positions in the modernized, urbanized environment that Russians have historically dominated. Seen in this way, increasing fluency in the Russian language may actually serve to heighten national solidarity as it brings indigenes into closer contact and competition with the dominant Russians who stand in the way of further indigenous upward mobility (Kaiser, this volume).

Table 10.4 Central Asian nationalities with a free command of Russian as a second language, by urban and rural for 1970 and 1979 (percent and percentage point change).

Nation	Urban			Rural		
	1970	1979	% point change	1970	1979	% point change
Uzbek	34.5	60.9	26.4	7.8	44.5	36.7
Kazakh	57.4	65.7	8.3	36.1	64.1	28.0
Kirgiz	53.1	59.8	6.7	13.3	22.6	9.3
Tadzhik	32.3	46.0	13.7	9.5	23.4	13.9
Turkmen	32.0	41.9	9.9	7.9	17.5	9.6
Karakalpak	24.6	44.8	20.2	4.1	45.3	41.2

Source: Guboglo 1984: 114.

A number of comments may be made regarding trends in Russian bilingualism among indigenes in Central Asia (Table 10.4). First, the rate of Russian fluency closely corresponds to the proportion of Russians present, and is an indicator of the dominance of the Russian language in these areas.[6] Both Kazakhstan and Kirgiziya had a larger number of Russians than indigenes, and these two had the highest rate of

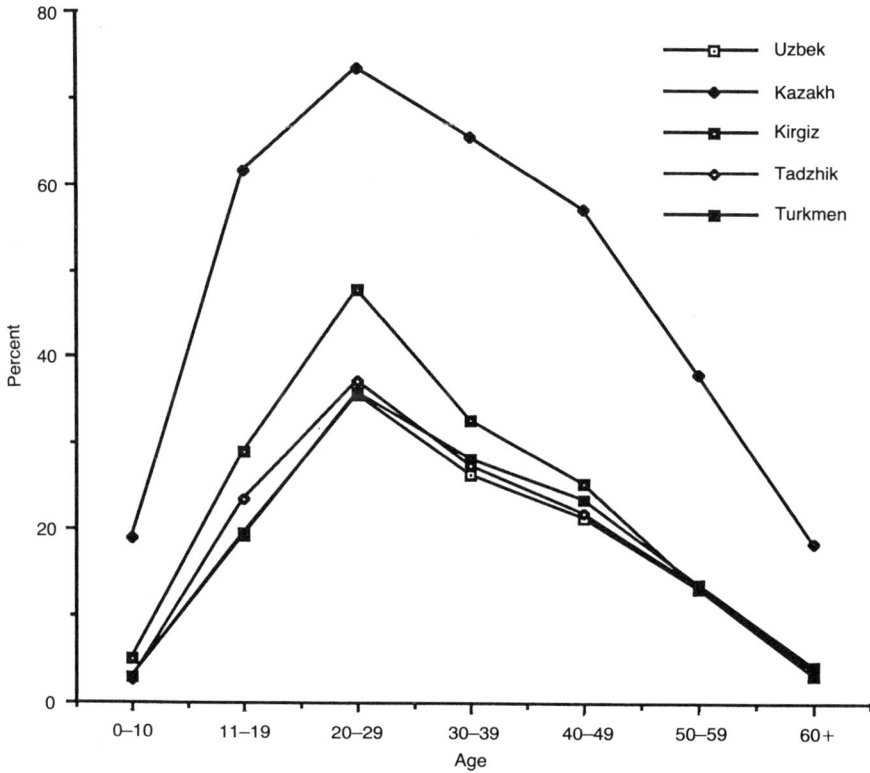

Figure 10.3 Free command of Russian as second language by nationality and age, 1970.
Source: TsSU SSSR 1973, vol. 4: 361–4, 378.

indigenous fluency in Russian. In a similar vein, the Russian population in Soviet Central Asia is concentrated in urban areas, and members of the indigenous nations residing in cities generally have a much higher rate of claimed Russian language fluency than their rural counterparts. The highest rates of Russian bilingualism in rural areas was found in Kazakhstan, and this is of course the republic that has the largest number of Russians residing in the countryside. A major influx occurred during the 1950s and 1960s as a consequence of the Virgin Lands Program. Second, while the urban rates are generally higher, the percentage point increase between 1970 and 1979 was higher among the rural indigenes for every nation except Tadzhiks. This corresponds to a more rapid rise in the level of educational attainment in the rural areas, and perhaps reflects the increasing desire of rural youth to become more socially mobile.

Rates of Russian bilingualism among the indigenous nations of Central Asia increase across the younger cohorts, reaching a peak among the 20–29-year-olds and declining thereafter (Figure 10.3). In part, this may be explained by the increasing rates of education, in that the study of Russian is a required part of the curriculum. However, the peak reached by the 20–29-year-olds also indicates that other factors influence the level of Russian fluency. First, the need for Russian increases among students enrolled in higher education, and as we have seen, the number of Central Asians at this level has increased markedly over time. In addition, according to surveys taken among non-Russians in the state, experience in the army, where Russian is the language of command and in which all 18–20-year-olds must serve, plays a role in increasing Russian fluency (Arutyunyan and Bromley 1986: 311).

Language data for the most recent census of 1989, while still not published fully, indicate the volatility of responses to this question. For every nation in the region, there was a marked decline in the rate of increase in Russian bilingualism, or even a registered decrease, during the 1980s (Table 10.5). A major question has been raised regarding the validity of these data, particularly the unbelievably high increase in Russian fluency among Uzbeks and Karakalpaks between 1970 and 1979. It was rumored that Rashidov, the party first secretary in Uzbekistan during this time period, was embarrassed by the low reported rates in the 1970 census and put pressure on census enumerators to ensure a more acceptable outcome in the 1979 census. It is interesting that the high rate of increase occurred among Karakalpaks as well as Uzbeks in 1979, and that the Russian fluency level for both these nations declined precipitously according to the 1989 census.[7]

For Uzbek bilingualism, alternative sources of information exist and do indicate that the 1970 census figures were probably too low. A survey

Table 10.5 Russian bilingualism by nation, 1970–89 (percent of each nation's population with a free command of Russian as a second language).

Nation	1970	1979	1989	Percentage point change 1970–9	1979–89	1970–89
Uzbek	14.5	49.3	23.8	34.8	−25.5	9.3
Kazakh	41.8	52.3	60.4	10.5	8.1	18.6
Kirgiz	19.9	29.4	35.2	9.5	5.8	15.3
Tadzhik	15.4	29.6	27.7	14.2	−1.9	12.3
Turkmen	15.4	25.4	27.8	10.0	2.4	12.4
Karakalpak	10.4	45.1	20.7	34.7	−24.4	10.3

Sources: Guboglo 1984: 105 (data for 1970, 1979); Goskomstat SSSR 1989a: 3 (data for 1989).

was taken at the end of the 1960s and beginning of the 1970s to measure actual competency in Russian as a second language. A comparison of the survey and census results indicates that the 1970 census figures on Russian bilingualism among Moldavians and Estonians in both the urban and the rural setting were fairly accurate; only rural Estonians showed a notable difference between "developed bilingualism" in the survey and the percentage claiming free command of the Russian language in the 1970 census (Table 10.6).

Table 10.6 Survey results of Russian bilingualism for urban and rural indigenes (select nations, percent).

Russian bilingualism[1]	Moldavians		Estonians		Uzbeks	
	Urban	Rural	Urban	Rural	Urban	Rural
Monolingual	4.1	13.9	8.6	15.8	12.9	32.9
Prebilingual	6.8	22.5	14.0	22.4	12.4	16.8
Incompletely bilingual	21.7	28.6	35.7	37.9	21.0	20.1
Completely bilingual	36.8	19.2	21.1	13.9	34.0	14.3
Postbilingual	29.2	10.7	16.4	8.2	9.0	5.5
Developing bilingual	28.5	51.1	49.7	60.3	33.4	36.9
Developed bilingual	66.0	29.9	37.6	22.1	43.0	19.8
Census results[2]	69.3	28.1	37.0	16.7	33.4	7.3

[1] Monolingual—"in general do not speak in a second language"; prebilingual—"speak a second language with great difficulty"; incompletely bilingual—"speak with some difficulty"; completely bilingual—"speak fairly freely"; postbilingual—"speak completely freely, think in the second language." Definitions from Guboglo 1984: 125.
[2] Census figures = number claiming Russian as a second language number claiming the respective indigenous language as the first language within each respective home republic. Since the survey was of national–Russian bilingualism, this is the most appropriate comparative measure.
Sources: survey results published in Guboglo 1984: 129; census figures from: Tsentral'noye Statisticheskiye Upravleniye SSSR 1973, vol. 4: 205–8, 277–8, 318–19.

The same close approximation of the two sets of data does not exist for Uzbeks residing in either urban or rural areas of Uzbekistan. The survey results, which attempt to gauge the actual level of Russian fluency, are clearly a better measure than the census figures, which make no effort to assess the validity of the responses to this subjective question. It is particularly interesting that the largest gap between survey and census figures was registered in the rural areas, and that this is also where the greatest increase between 1970 and 1979 occurred. Thus, it would appear that the 1970 census did underestimate the number of Uzbeks with a "free command" of the Russian language. Nonetheless, as the more recent results of the 1989 census make clear, the 1979 figures seriously overstated the actual degree of Russian bilingualism among Uzbeks and Karakalpaks.[8]

Female bilingualism

Equalization with indigenous males in Russian bilingualism may be seen as a further indicator of the breakdown of traditional value systems, and the trends among the Central Asians indicate that women as well as men are becoming better prepared to participate in the modernized sectors of society.

It is evident that while female rates of Russian fluency continue to lag behind those of male Central Asians, they were much closer to a ratio of 100 in 1979, which would signify equality (Table 10.7). The largest increases were again registered among Uzbeks and Karakalpaks; it may be the case that, to the degree that there was an undercount in 1970, it was particularly pronounced among indigenous women in Uzbekistan. At the same time, it is exceedingly unlikely that the female rates or ratios for 1979 accurately reflect the status of women with regard to their level of Russian fluency.[9] Though the data are flawed, they do correspond to the general picture of the status of women in Soviet Central Asia and indicate that, while equalization is occurring, female equality has not been achieved.

Table 10.7   Russian bilingualism among Central Asian nations, by sex, 1970–9 (percent claiming free command of Russian as a second language).

Nation	1970			1979		
	Male	Female	Female–male ratio	Male	Female	Female–male ratio
Uzbek	19.6	9.3	47.4	53.9	44.6	82.7
Kazakh	46.6	37.1	79.6	55.9	48.8	87.3
Kirgiz	24.2	14.2	58.7	34.8	24.0	69.0
Tadzhik	22.0	8.6	39.1	37.8	21.1	55.8
Turkmen	21.9	8.8	40.2	33.4	17.3	51.8
Karakalpak	14.5	6.2	42.8	43.8	40.3	92.0

Source: Guboglo 1984: 114.

Conclusions

Given the problems that exist with the data under consideration, it is difficult to argue convincingly whether or not the linguistic impediment to greater urbanization cited by Arutyunyan—a general lack of Russian language skills among the indigenous nations of Soviet Central Asia—has been overcome. It certainly seemed that this was the case, given the results of the 1979 census. However, the more recent census results of 1989 indicate that dramatic changes in the level of Russian bilingualism have not occurred during the past two decades. Comparing the figures of Russian fluency among the Central Asian nations with those for the country as a whole, it appears that only the Kazakhs exceed the all-union average of all non-Russians claiming "free mastery" of Russian as a second language, which stood at 49 percent in 1989 (Goskomstat SSSR 1989a: 3). The comparative status of these nations has apparently changed little in this measure since 1970, when the

all-union average stood at 37 percent of all non-Russians (Tsentral'noye
Statisticheskoye Upravleniye SSSR 1973, vol. 4: 20).

There is a second reason why the contemporary fluency in Russian
among the nations of Soviet Central Asia is difficult to utilize in
an assessment of indigenous social mobilization. The declining rate
of increase in Russian fluency, and even a decrease for some of
the indigenes, may also represent a new political reality in the region—
in the late 1980s Russians were losing the degree of sociocultural
and economic dominance they had enjoyed previously. The level of
nationalism and anti-Russian sentiments in the region has become much
more prominent during the past few years, and this should tend to
undermine the degree to which Russian is viewed as the language of
upward mobility. The present effort by national front organizations
throughout the region, and indeed throughout the country as a whole,
to enact laws which declare the indigenous language the lingua franca of
the republic is a clear statement about the local attitudes regarding the
status of their languages vis-à-vis Russian. However, at least in Soviet
Central Asia, the draft laws regarding the protection of the indigenous
languages state that bilingualism remains the desired objective, with
both Russians and indigenes in the region gaining mastery over one
another's language (JPRS 1989). In addition, the proportion of students
in general education who use Russian as their language of instruction
increased for all Central Asian republics except Tadzhikistan during the
1980s (Goskomstat SSSR 1989b: 5–7). This is a fairly strong indication
that Central Asian parents continue to view Russian as the language of
upward mobility.

While the data presented in this section and their relevance with
regard to social mobilization in Central Asia must be treated with
caution, it is possible to conclude that, in general, Russian bilingualism
is spreading, particularly among the younger generation, though the
rate of increase in this measure is not nearly so dramatic as the census of
1979 would have led us to believe. However, increasing mastery of the
Russian language by a growing number of young, upwardly mobile
indigenes is occurring simultaneously with rising national solidarity,
and the very need to learn the Russian language in order to become
more socially mobile may serve as a catalyst for rising anti-Russian
sentiments among this segment of the indigenous populations.

WORKPLACE MOBILITY

Trends in education and Russian bilingualism among the indigenous
nations of Soviet Central Asia indicate that for the younger members of

these nations the desire for upward mobility has increased dramatically in recent years. It is expected that the occupational structure is in the process of undergoing a rapid transition from a primarily agricultural work force to one with a much higher proportion in the secondary and tertiary sectors, and that in both these employment sectors Central Asians are increasingly represented in the higher-status or prestige occupations. The agricultural sector itself should be undergoing changes related to increased mechanization and a restructuring of the work force away from an emphasis on inputs of human energy.

In this section, we attempt to assess the degree to which the aspirations for upward mobility apparent among the younger indigenes are being accommodated in the workplace.[10] This analysis is made difficult by the lack of detailed occupation data by nation, age and sex. A number of general trends, however, may be established.

It is clear that during the period 1959–79, a shift in employment away from the "kolkhozy" and toward blue- and white-collar occupations has occurred for all of the indigenous nations in Soviet Central Asia (Table 10.8). Most of this shift was concentrated in the blue-collar sector, and the average annual growth in tertiary sector employment for the 1970–9 period actually declined relative to the 1959–70 growth rate for all but the Kazakhs.

Table 10.8   Class composition by nation, 1959–79 (percent).[1]

Nation	Kolkhozniki			Rabochiye			Sluzhashchiye		
	1959	1970	1979	1959	1970	1979	1959	1970	1979
Uzbek	65	45	32	27	39	50	8	16	18
Kazakh	40	13	8	43	65	64	16	22	28
Kirgiz	70	44	24	22	41	56	8	15	20
Tadzhik	74	48	30	18	37	55	8	15	15
Turkmen	69	51	45	22	32	39	9	17	16
Russian	24	12	6	54	63	63	22	25	31

[1] "Kolkhoznik" are agricultural workers on the kolkhozy; "rabochiye" is the rough equivalent of blue-collar employment, though agricultural workers on state farms or "sovkhozy" are also included in this "class"; "sluzhashchiye" is roughly equivalent to the service or tertiary sector.
Source: Arutyunyan and Bromley 1986: 55.

The degree to which this shift represents a movement away from agricultural employment is difficult to determine, since the blue-collar category includes agricultural workers on state farms or "sovkhozy". In 1959, the percentage of the "rabochiye" category engaged in industry ranged from a high of 23 percent for Turkmen to a low of 14 percent for Kazakhs (Arutyunyan and Bromley 1986: 55). And, since there has been a shift toward state farms during this time period, the actual shift away from agricultural employment is less than the figures would imply. According to Arutyunyan and Bromley (1986: 56), "as concerns the peoples of Central Asia, the agricultural population has remained relatively stable, even though the percentage of kolkhozniki markedly declined."

More recent data indicate that the agricultural segment of blue-collar workers has continued to increase only for the Turkmen during the 1980s, and that the percentage employed in industry increased for each nation (Table 10.9). Among white-collar employment, there has been a shift away from the traditional emphasis on the education sector, although the percentage in this category is still high among the nations of Central Asia relative to the other nations in the USSR. During the same time period, indigenes are comprising a growing proportion of the work force in each nonagricultural branch of the economy, and have attained majority status in most branches in Uzbekistan, Tadzhikistan, and Turkmeniya (Goskomstat SSSR 1988: 22–5).

Table 10.9 Distribution of blue- and white-collar employment by branch, 1977–87 (percent of employed and indigenes, excluding kolkhozniki).[1]

Branch	Uzbek 1977	Uzbek 1987	Kazakh 1977	Kazakh 1987	Kirgiz 1977	Kirgiz 1987	Tadzhik 1977	Tadzhik 1987	Turkmen 1977	Turkmen 1987
Industry	16	19	12	14	12	15	16	18	14	15
State farm	25	25	38	34	40	35	25	25	8	14
Transport	6	7	8	9	7	6	8	7	11	9
Construction	9	10	5	8	5	7	11	11	17	16
Trade	8	7	5	5	5	6	7	7	8	8
Housing/ communal services	2	2	1	2	1	2	3	3	3	3
Health	7	8	6	7	6	8	5	6	8	9
Education	15	14	12	11	13	12	14	13	16	15
Culture	2	1	2	2	2	2	2	2	3	2
Science	1	1	1	2	2	2	1	1	2	2
Government	3	2	4	3	3	3	3	2	4	3
Other	6	4	6	3	4	2	5	4	6	4

[1] Percentages are equal to the distribution of the blue- and white-collar employment among members of the indigenous nation living in their home republic, and total to 100 percent.
Source: Goskomstat SSSR 1988: 24–5.

For the nations of Soviet Central Asia, the shift away from "kolkhoz" employment in recent years has increased the relative weight of state farm employment only among Turkmen, where "sovhozniki" comprised a relatively small percentage of the blue- and white-collar work force in 1977. Employment in industry and construction has become relatively more important in the region. However, the shift to tertiary employment has been unimpressive during the contemporary period. Even though the indigenes make up a growing proportion of these branches, the percentage of the indigenous work force engaged in the service sector appears to have stagnated or even declined. Within the tertiary sector itself, there continues to be a relatively high percentage

Table 10.10 National composition of blue- and white-collar employment by branch, 1977–87 (percent of each branch comprised by indigenes living in home republic).

Branch	Uzbek		Kazakh		Kirgiz		Tadzhik		Turkmen	
	1977	1987	1977	1987	1977	1987	1977	1987	1977	1987
Industry	38	53	13	21	15	25	35	48	34	53
State farms	67	76	38	52	63	69	62	63	58	81
Transport	44	55	20	28	28	35	45	57	40	48
Construction	36	50	11	21	16	26	36	48	41	54
Trade	63	66	21	29	23	34	52	61	47	65
Housing/ communal service	52	55	17	23	19	30	45	56	41	53
Health	52	64	25	38	32	46	33	50	48	62
Education	61	69	36	43	44	43	51	58	56	67
Culture	61	63	36	42	46	46	55	56	64	70
Science	31	39	17	25	20	27	23	31	35	48
Government	46	57	34	40	35	42	42	51	45	51

Source: Goskomstat SSSR 1988: 22–3.

in the education branch, though this has declined in recent years (Table 10.10).

An examination of the science branch indicates that the indigenous nations of Central Asia have yet to dominate these high-status occupations in their respective republics. Except for Moldavians and Latvians, all other nations with Union republics comprise a majority in this sector. Other measures suggest that the growth of employment in this sector continues to lag behind the rest of the country. The proportion of scientific workers located in Soviet Central Asia actually declined between 1960 and 1986, even though this region has come to have a larger share of the younger Soviet population (Goskomstat SSSR 1987: 64). There has been a shift toward Central Asia in the proportion of specialists with higher and specialized secondary education, increasing from 8.6 percent of the total in 1960 to 11.5 percent in 1986 (Goskomstat SSSR 1987: 419). However, the rate at which specialists were graduating from higher educational institutions actually declined relative to the all-union average between 1970 and 1986 for all but Kazakhstan, and most of the increase was registered at the level of specialized secondary education (Goskomstat SSSR 1987: 555).

There would thus appear to be a growing disparity between aspirations, on the one hand, and the capacity for upward mobility among indigenes, on the other. However, the increasing dissatisfaction with one's life chances expected as a result of this apparent increase in relative deprivation has not been found in surveys designed to measure it. In response to the question "Has life become better with the course of time?", Uzbeks expressed relatively greater satisfaction with improvements in their work situation than did either Estonians or Russians. This held true for unskilled physical workers as well as for highly skilled physical and mental workers in both urban and rural areas, with the exception that Russian unskilled physical workers in urban areas were marginally more satisfied with improvements than was the same group of Uzbeks (Arutyunyan and Bromley 1986: 257).

A number of reasons may account for the relatively positive assessment of working life. First, upward mobility across generations has been dramatic. For example, while only 33 percent of 50–59-year-old urban Uzbeks and 13 percent of the 50–59-year-old rural Uzbeks are engaged in qualified or skilled mental and physical labor, 82 percent of urban 20–24-year-old Uzbeks and 58 percent of rural 20–24-year-old Uzbeks have such employment (Arutyunyan and Bromley 1986: 72). In terms of intergenerational change, conditions for the younger generation have definitely improved. More children whose parents ("fathers") are engaged in physical labor are employed in mental work at present than was the case a generation ago, and the proportion engaged in unskilled physical labor, particularly in urban areas, has declined precipitously (Table 10.11). While change in rural areas has been less dramatic, nearly one-third of the younger generation is employed as skilled physical or mental labor, compared to less than one-fifth of the parents' generation (Arutyunyan and Bromley 1986: 94).

Table 10.11 Intergenerational changes in employment, Uzbeks (percent).

| | Children | | | |
| | Respondents 18–39 | | Respondents > 40 | |
Fathers	Physical	Mental	Physical	Mental
Physical	61	39	82	18
Mental	30	70	55	45

Source: Arutyunyan and Bromley 1986: 95.

The intergenerational shift from physical to mental labor has accelerated over time, while children whose parents are employed in mental labor are currently more able to retain this status than they were in the past. It appears as though occupational change has been at a level sufficient to satisfy the aspirations for upward mobility among the younger generation, at least through the 1970s.

Of course, one may question what "satisfactory employment" means in Central Asia, and indeed there do appear to be differing assessments of the prestige or status of specific occupations between members of European nations, on the one hand, and those of the indigenous nations of the region, on the other. The level of satisfaction with work expressed by Uzbeks may be a reflection of their somewhat different attitudes about what constitutes a "good life." In a ranking of values in their personal life, urban Russians ranked "family" first, followed by "interesting work", "material well-being", "esteem of one's associates," and "tranquil life." Urban Uzbeks also ranked "family" first, but this was followed by "esteem of one's associates," "tranquil life," "material well-being," and "interesting work" (Arutyunyan and Bromley 1986: 251). From this ranking, it would appear that Uzbeks depend somewhat less on rewarding work for their psychic satisfaction. This accords with the more general finding that "the prestige of the rural way of life has been retained, and influences even the urban population of the indigenous nationalities" (Arutyunyan 1987: 4). These surveys also tend to support the view that modernization is a process of adaptation which continues to be influenced by traditional value systems.

Nonetheless, as Uzbeks and other Central Asians become more upwardly mobile, there is convergence between them and other nations as regards their value orientation. While the above ranking holds for Uzbeks engaged in unskilled physical labor, for those employed as highly qualified mental labor "interesting work" and "material well-being" are ranked third and fourth, after "family" and "esteem of one's associates," while "tranquil life" is ranked last (Arutyunyan and Bromley 1986: 255). In this way, social mobilization affects the attitudes of the population, resulting in a more universal outlook regarding satisfaction with employment. With this convergence, it is expected that occupations will be ranked as to their prestige or status in similar ways by all socially mobile individuals, regardless of national affiliation. This will in the future result in an intensification of competition for a limited number of prestigious occupations, and increase the international competition for these positions between Russians and other socially mobile outsiders, on the one hand, and socially mobilized indigenes, on the other. For this reason, it may be predicted that social mobilization in Soviet Central Asia will serve to exacerbate national problems in the region (Kaiser, this volume).

Status of women

Over the recent past, women have become more equally represented in blue- and white-collar employment in all but Tadzhikistan (Table 10.12). There has been a convergence among these republics toward the all-union average, which stands at about half the employment in these general "classes". However, the problems associated with these categories noted in the introduction to this section raise serious questions regarding the actual degree to which indigenous women in Soviet Central Asia have experienced a shift in occupations away from agriculture. Since the blue- and white-collar occupations are not listed separately, it is not possible to determine whether these increasing percentages represent anything more than the transition from kolkhozy to sovkhozy in the region.

Table 10.12 Female representation in blue- and white-collar
employment by republic, 1960–86 (percent and percentage point
change).

Republic	1960	1980	1986	% point change
USSR	47.2	51.2	50.8	3.6
Uzbekistan	38.6	42.8	43.1	4.5
Kazakhstan	37.6	48.7	48.7	11.1
Kirgiziya	40.6	48.4	48.7	8.1
Tadzhikistan	36.6	38.9	38.2	1.6
Turkmeniya	36.3	41.2	41.3	5.0

Source: Zhenshchiny i Deti v SSSR 1988: 60.

In addition to this concern, the lack of nation-specific data leaves
open the question of which women are experiencing upward mobility.
According to Michael Sacks (1982: 171–4, 1989), the apparent equalization between the sexes that occurred during the 1960s and 1970s was the
result of a larger number of Russian women entering the region, and not
necessarily a reflection of the improved status of the indigenous female
population.

There is a striking disparity between urban and rural areas in the
status of women engaged in mental labor (Table 10.13). In urban areas,
where nonindigenes are concentrated, there is a higher percentage of
women than men in mental labor, and in every republic in both years
the female–male ratio exceeds the statewide average. The reverse is true
in rural Central Asia, indicating that the indigenous women, who are
concentrated in rural areas, are experiencing little equalization with men.
The only two exceptions to this statement are Kazakhstan and Kirgiziya,
and both these republics between 1959 and 1970 experienced a massive
influx of Russians and other nonindigenes into both urban and rural areas.

In the occupational structure of Soviet Central Asia, there appears to
be a continuation of both national and sexual stratification in employment. Russian and other nonindigenous males occupy the highest-
status positions in heavy industry, while Russian women are employed
in the service sector and in light industry. Each of these is concentrated

Table 10.13 Mental labor by republic and sex, 1959–70 (female–male
ratios).

Republic	Urban		Rural	
	1959	1970	1959	1970
USSR	139.5	147.8	86.8	125.5
Uzbekistan	152.7	156.0	41.1	49.7
Kazakhstan	162.6	170.5	98.7	129.0
Kirgiziya	149.0	153.3	66.2	98.8
Tadzhikistan	155.5	156.0	33.3	32.6
Turkmeniya	164.7	155.3	29.9	37.6

Source: Tsentral'noye Statisticheskoye Upravleniye SSSR 1973, vol. 6: 6–13.

primarily in the cities of the region. Occupationally, indigenous men compete with Russian women, but this competition to date is more apparent than real, since the indigenous men are primarily employed in rural areas. The least competitive segment of the population is the indigenous women, who have remained in rural areas in unskilled and primarily agricultural occupations (Sacks 1982: 174–8, this volume).

The lack of employment equalization evident between 1959 and 1970 apparently continued during the 1970s. This was particularly true in the more highly skilled occupations across "classes," though in the white-collar sector women were more equal to men than in the kolkhoz and blue-collar sectors (Table 10.14).

Table 10.14 Employed having a higher or specialized secondary education by republic, 1970–9 (female–male ratios).

| Republic | Kolkhoz | | Blue collar | | | | White collar | | | |
| | | | Urban | | Rural | | Urban | | Rural | |
	1970	1979	1970	1979	1970	1979	1970	1979	1970	1979
USSR	52.8	65.7	85.4	84.4	85.7	83.6	83.7	84.7	104.3	90.7
Uzbekistan	12.8	18.4	79.0	70.3	30.6	33.8	85.2	83.7	83.0	87.8
Kazakhistan	31.4	55.0	83.7	81.3	78.9	81.6	82.3	84.1	92.7	85.3
Kirgiziya	17.6	31.6	93.8	82.5	63.3	71.0	87.3	87.9	94.0	91.6
Tadzhikistan	8.3	6.8	72.9	65.2	35.7	16.2	85.6	85.1	84.3	88.0
Turkmeniya	18.6	12.3	81.1	56.2	53.1	23.7	84.4	81.6	96.4	86.4

Source: Tsentral'noye Statisticheskoye Upravleniye SSSR 1984: 157–72.

One explanation for the movement away from equality which a number of the figures in the table represent is that an increasing proportion of the labor force was comprised by indigenes during the 1970s. This would appear to be particularly true for the urban blue-collar work force. On the kolkhoz, it is clear that women continue to be concentrated in unskilled manual labor throughout Soviet Central Asia. Aside from Kazakhstan and Kirgiziya, this also holds true for rural blue-collar labor, which is predominantly agricultural employment on state farms. As noted before, the two republics that are exceptions to this rule are also those with a large number of Russian and other nonindigenous agricultural workers. Finally, while there was little evidence of equalization in either the urban or the rural white-collar sector, the female–male ratios throughout Central Asia are approximately equivalent to those obtaining for the total Soviet Union. This indicates that with social mobilization and the movement toward higher percentages engaged in tertiary activity comes the improved status of women.

An examination of specialists with higher and specialized secondary education does indicate that among this more elite socioprofessional group, the status of women has improved markedly in most Central Asian republics (Table 10.15). Women made impressive gains at the level of specialists in Uzbekistan, Kazakhstan and Kirgiziya. However, as was stated previously, from this general category it is difficult to ascertain the degree to which indigenous women participated in this improved status. Most of the increase occurred between 1960 and 1970, a time when these republics experienced an influx of Russians and other non-indigenes. However, the continued shift toward female specialists during the 1980s may be an indication that indigenous women have also become competitive at this level of employment.

Table 10.15   Female specialists with higher and specialized secondary education by republic, 1960–86 (percent).

Republic	1960	1970	1980	1986	% point change
USSR	59.1	58.8	59.3	60.0	0.9
Uzbekistan	44.9	47.3	48.2	49.6	4.7
Kazakhstan	54.0	57.8	60.1	60.8	6.8
Kirgiziya	52.1	55.3	57.8	58.3	6.2
Tadzhikistan	43.9	43.4	41.6	42.3	−1.6
Turkmeniya	44.5	44.0	45.4	44.4	−0.1

Source: Goskomstat SSSR 1987: 419–20.

Women appear to be losing ground to men in Tadzhikistan and Turkmeniya. More generally, in terms of urbanization and other measures of social mobilization, Tadzhikistan and Turkmeniya are not experiencing the process of modernization at anywhere near the rate of the rest of the country; these most southerly republics of Soviet Central Asia appear as a backwater of underdevelopment.[11]

Conclusions

Upward mobility in the occupational structure in Soviet Central Asia has occurred over time, and at least through the 1970s was sufficient to reduce the degree of dissatisfaction among the indigenous nations. Given the rapid population growth in the region and the dramatic increases in educational attainment, it is unlikely that the degree of satisfaction with living conditions will continue. The gap between indigenous aspirations and capabilities will grow wider unless higher-status employment is created in the region or out-migration occurs. At present, out-migration of Russians and other nonindigenes is also serving to provide openings in higher-status occupations. However, it does not appear that the rate of job creation will be sufficient to satisfy indigenous demands. For a number of reasons examined in the following chapter, indigenous out-migration in the near future is not expected. And, as pointed out by Liebowitz (this volume), investment in the region is not occurring at a level sufficient to bring about a satisfactory expansion in high-status employment *in situ*. On the contrary, unemployment in the region appears to have risen sharply over the past few years.[12] Given these conditions, it appears that social mobilization among indigenes in Soviet Central Asia will result in a more intractable national problem in the future.

IMPLICATIONS

The indigenous nations of Soviet Central Asia evidence a pattern of social mobilization indicative of their intermediate position in the modernization process. Over the recent past, trends in education indicate that, far from remaining a tradition-bound, isolated rural community, the younger generation of Central Asians clearly aspires toward greater levels of participation in the modernized sectors of society. The social mobilization occurring among the younger genera-tion is not limited to men, but includes an increasing number of women. Equalization between the sexes in educational attainment is taking place and will have a significant impact not only on the economy of the region, but also on traditional family value systems, and will ultimately result in a fertility decline experienced elsewhere in the developed world.

In the workplace, agricultural employment has remained at about the same level over time. The lack of urbanization and industrialization of the indigenous work force is explained at least partially by the ethnocultural character of the urban and industrial environment, which has up to now been dominated by Russians. The degree of indigenous upward mobility in the past was apparently sufficient to satisfy the limited demands of the indigenes.

However, trends in education and Russian bilingualism indicate that there is a growing number of young indigenes aspiring to compete for high-status occupations currently held by Russians and other non-indigenes. Without a more rapid restructuring of the Central Asian economy away from the "cotton belt" and toward job creation in the manufacturing and service sectors, the rapid increase in expectations among younger indigenes cannot be satisfied and will lead to increasing perceptions of relative deprivation. Since the path to greater upward mobility is currently blocked both by a lack of adequate and appropriate investment from Moscow and by Russians and other nonindigenes occupying elite positions in the region, this relative deprivation is likely to serve as a catalyst for growing national tensions in Soviet Central Asia. Indeed, the recent reports of international violence in the region may be seen as a response to this growing sense of relative deprivation. In this way, increasing social mobilization has not created the conditions for solving the national problem through "Sovietization," but rather has contributed to the growing national divisiveness evident in the country as a whole, and in Soviet Central Asia in particular. This aspect of the consequences of social mobilization is the subject of the following chapter.

NOTES

1. Hereafter this region is referred to as simply Central Asia or Soviet Central Asia.

2. The implications of this view of modernization are far-reaching, and according to Agnew (1987: 64–5) are behind the declining importance attached to place-based explanations of societal processes: "When the 'modernizing' forces of society overpower the 'traditional' forces of community, place is overpowered too and continues to exist only as the location of nationally defined social activities."

3. Sovietization, defined by Aspaturian (1968: 159) as "the process of modernization and industrialization within the Marxist–Leninist norms of social, economic, and political behavior," has been presented as an alternative to "Europeanization" in the USSR. This is of course paradoxical, since communism and Marxism themselves are "Western" ideologies.

4. Equality of educational attainment here refers to the quantity of education received on average, and not to quality of schooling. While the conventional wisdom has it that Central Asian education is inferior, Jones and Grupp (1984: 164–7) find no empirical evidence to support this contention.

5. In 1985 the differential between urban and rural education rates ranged from a low of 58/1,000 population 10 years old and older with higher or secondary education for the Tadzhiks to a high of 157/1,000 for the Kirgiz (Statisticheskiye materialy 1986: 67).

6. This correlation is also found at the oblast level. Interestingly, the reverse also appears to be true—in regions where there are relatively few Russians, a larger percentage of the Russians claim fluency in the locally dominant (indigenous) language (Smirensky 1985: 31–41).

7. The Karakalpak Autonomous Republic lies inside Uzbekistan, and census takers in the Karakalpak ASSR would be under the same orders as those in the rest of Uzbekistan.

8. It is interesting to note that the rate of change in degree of Russian bilingualism varied widely among the Central Asian nations between 1970 and 1979 and also between 1979 and 1989, but that all five groups experienced approximately the same rate of increase between 1970 and 1989. This lends greater credibility to the 1989 figures, and helps to confirm the belief that the 1979 results were seriously flawed. However, given other measures of social mobility and interaction with Russians, 1989 percentages for Uzbeks in particular appear too low, possibly reflecting an overreaction to the artificially high level in 1979 and the relatively higher degree of nationalist mobilization among Uzbeks at present.

9. Unfortunately, at present the language data for 1989 have not been provided by sex.

10. For a comparative treatment of occupations in Central Asia and the RSFSR, and an examination of the positive and negative aspects of economic ties between these two regions, see Michael Sacks (1989).

11. This is reflected in a number of measures, including an abysmal infant mortality rate. Turkmeniya's 54/1,000 rate in 1989 was over twice as high as the state-wide average of 22/1,000 (Goskomstat SSSR 1990b).

12. The actual rate of increase is difficult to determine, since in the past unemployment was not acknowledged to exist in the state.

REFERENCES

Agnew, John 1987. *Place and Politics: The Geographical Mediation of State and Society*. Boston, Mass.: Allen & Unwin.

Anderson, C. Arnold 1966. The modernization of education. In Myron Weiner (ed.), *Modernization: The Dynamics of Growth*, pp. 68–80. New York: Basic Books.

Arutyunyan, Yu. 1985. Natsional'nyye osobennosti sotsial'nogo razvitiya. *Sotsiologicheskiye Issledovaniya* 3: 28–35.

Arutyunyan, Yu. 1987. Natsional'no-regional'naya spetsifika protsessov sblizheniya goroda i derevni v SSSR. *Sovetskaya Etnografiya* 1: 3–10.

Arutyunyan, Yu. and Bromley, Yu. (eds) 1986. *Sotsial'no-Kul'turnyy Oblik Sovetskikh Natsiy*. Moscow: Nauka.

Aspaturian, Vernon 1968. The non-Russian nationalities. In Allen Kassof (ed.), *Prospects for Soviet Society*, pp. 143–98. New York: Praeger.

Black, Cyril 1966. *The Dynamics of Modernization: A Study in Comparative History*. New York: Harper & Row.

Carrere d'Encausse, Helene 1981. *Decline of an Empire: The Soviet Socialist Republics in Revolt*. New York: Harper & Row.

Eisenstadt, S. N. 1966. *Modernization: Protest and Change*. Englewood Cliffs, NJ: Prentice-Hall.

Goldscheider, Calvin 1971. *Population, Modernization, and Social Structure*. Boston, Mass.: Little, Brown.

Goskomstat SSSR 1987. *Narodnoye Khozyaystvo SSSR za 70 let*. Moscow: Finansy i Statistika.

Goskomstat SSSR 1988. *Trud v SSSR: Statisticheskiy Sbornik*. Moscow: Finansy i Statistika.

Goskomstat SSSR 1989a. *Natsional'nyy Sostav Naseleniya*. Moscow: Finansy i Statistika.

Goskomstat SSSR 1989b. Obucheniye v dnevnykh obshcheobrazovatel'nykh shkolakh na Russkom yazyke; yazykakh drugikh natsional'nostey po soyuznym respublikam. *Press Vypusk*, 15 November (524): 5–7.

Goskomstat SSSR 1990a. Sostav studentov vysshykh uchebnykh zavedeniy po korennym natsional'nostyam na nachalo 1989/90 uchebnogo Goda. *Press Vypusk*, 16 March (105): 1–2.

Goskomstat SSSR 1990b. Koeffitsienty mladencheskoy smertnosti po soyuznym respublikam. *Press Vypusk*, 20 February (75): 1.

Guboglo, M. N. 1984. *Sovremennyye Etnoyazykovyye Protsessy v SSSR.* Moscow: Nauka.

Gurr, Ted Robert 1970. *Why Men Rebel.* Princeton, NJ: Princeton University Press.

Inkeles, Alex 1966. The modernization of man. In Myron Weiner (ed.), *Modernization: The Dynamics of Growth*, pp. 138–50. New York: Basic Books.

Jones, Ellen and Grupp, Fred 1984. Modernization and ethnic equalization in the USSR. *Soviet Studies* 36, 2: 159–84.

Jones, Ellen and Grupp, Fred 1987. *Modernization, Value Change and Fertility in the Soviet Union.* London: Cambridge University Press.

JPRS 1989. Republic language legislation. *JPRS-UPA* (Soviet Union: Political Affairs), 5 December (63).

Makarova, G. P. 1987 *Narodnyy Komissariat po Delam Natsional'nostey RSFSR, 1917–1923 gg.* Moscow: Nauka.

Sacks, Michael Paul 1982. *Work and Equality in Soviet Society: The Division of Labor by Age, Gender, and Nationality.* New York: Praeger.

Smirensky, Nicholas 1985. Moscow or Mecca: the modernization of the titular nationalities of Central Asia and their integration into Soviet Society. Master's thesis, Columbia University.

Statisticheskiye materialy 1986. *Vestnik Statistiki* 7: 67–79.

Tsentral'noye Statisticheskoye Upravleniye SSSR 1973. *Itogi Vsesoyuznoy Perepisi Naseleniya 1970 goda.* Vols 3–7. Moscow: Statistika.

Tsentral'noye Statisticheskoye Upravleniye SSSR 1984. *Chislennost' i sostav Naseleniya SSSR: po dannym Vsesoyuznoy Perepisi Naseleniya 1979 Goda.* Moscow: Finansy i Statistika.

Zhenshchiny i Deti v SSSR 1988. *Vestnik Statistiki* 1: 57–76.

Zinchenko, I. P. 1984. Natsional'nyy sostav naseleniya SSSR. In A. A. Isupov and N. Z. Shvartsera, (eds), *Vsesoyuznaya Perepis' Naseleniya 1979 Goda*, pp. 150–61. Moscow: Finansy i Statistika.

Zyuzin, D. I. 1983. Prichiny nizkoy mobil'nosti korennogo naseleniya respublik Sredney Azii. *Sotsiologicheskiye Issledovaniya* 1: 109–18.

Chapter 11 ———————————————————

Nations and Homelands in Soviet Central Asia

Robert J. Kaiser

INTRODUCTION

Over the past few years, Gorbachev's assessment of the national[1] question has undergone a dramatic metamorphosis, changing almost overnight from an issue largely ignored to "the most fundamental, vital question of (Soviet) society" (*Pravda* 1988: 2). Indeed, after a relative lull in overt demonstrations of nationalism, the past few months have seemed almost revolutionary. With officially sanctioned "glasnost" and "demokratizatsiya," nationalists have become increasingly willing and able to make demands for greater independence from Moscow. This nationalist dimension of the radical reform program instituted since 1985 was almost certainly unintended, and indicates that nationalism has continued to survive and even thrive in a political environment which has officially promoted "proletarian internationalism."

The national problem which has assumed center stage recently in the USSR, as in most other multinational, multi-homeland states, has a distinct territorial dimension. That is, the indigenous nations resident in the state lay claim to particular geographic regions as their respective national homelands. The boundaries of these national homelands are perceptual; their delimitation is often made by nationalists who are both looking back to a mythical golden age when the "nation" dominated vast land areas, and also looking forward to a time when the nation will once again regain sovereignty over the place it considers home.[2] The striving for such sovereignty is at the heart of national self-determination movements in the contemporary world political system, and the main strategy used by nationalists to gain control for the nation is national territoriality. In the USSR, national territoriality has become a dominant strategy among nationalists seeking greater independence from Moscow. Therefore, the meaning of homeland is central to an understanding of nationalism in the Soviet Union.

OF HOMELANDS AND NATIONAL TERRITORIALITY

A close, emotional attachment to a perceived homeland is a seemingly universal phenomenon among nations (Connor 1986). This nationalist affinity for the homeland is motivated by more than simply a desire to control resources. The more subjective "sense of place" (Agnew 1987: 28) which ties nation to homeland is also an important element in the formation and maintenance of national self-consciousness, which in turn revolves around a shared sense of common past and a belief in a common future.

A nation's sense of a shared past refers not only to a mythical common ancestry, but also to a common geographical birthplace (Connor 1986: 16). The homeland thereby becomes a part of the nation in the perceptions of its members; blood and soil have mixed for generations to create a united whole. This perceptual relationship is clearly evident in the contemporary literature of Soviet Central Asia. For example, the Uzbek poet Cholgan Ergash wrote of the homeland (Allworth 1973: 15):

> So that my generation would comprehend the Homeland's worth,
> Men were always transformed to dust, it seems.
> The Homeland is the remains of our forefathers
> Who turned into dust for this precious soil.

The sense of a primordial connection between nationalist and homeland is said to be as strong as that between nationalist and ancestors. This is clear in an admonition made by poet Turar Kodzhomberdiev to his fellow Kirgiz to "Remember, even before your mother's milk / You drank the milk of the homeland" (Allworth 1973: 16). The nationalistic perception that the homeland is the sacred geographic cradle of the nation is perhaps the most powerful example of what Soja (1971: 34) refers to as a "sense of spatial identity", which he cites as the first ingredient of human group territoriality.

In addition to this intimate "sense of spatial identity" binding nation to homeland, and deriving from it, nationalists also normally claim the right to exclusive use or control over their perceived homelands. Thus, according to Shibutani and Kwan (1972: 445), "the national land is often regarded as a group possession on which foreigners are interlopers." This exclusionary or proprietary outlook conforms to the more general "sense of exclusiveness" which Soja (1971: 34) cites as the second ingredient of human group territoriality. This "sense of exclusiveness" has certainly been enhanced by the federal structure of the USSR, and particularly by the naming of republics after the indigenous national communities. Kazakhstan, Uzbekistan, and so forth, literally mean "land of the Kazakhs, Uzbeks . . ."—in this way reinforcing the perception that the homeland belongs to one and only one nation. This conforms to a more general statement made by Connor (1986: 18): "Who but the Scots could have plenary claim to residence within Scotland, who but Germans to Deutchland, Kurds to Kurdistan (literally 'Land of the Kurds'), or Nagas to Nagaland?"

In multi-homeland states, this nationalistic exclusivity at minimum serves to inhibit inter-homeland mobility and international mixing, thereby seriously reducing the potential for international integration or assimilation. Beyond this, relations between nations (that is, international relations) are more likely to be confrontational than accommodative as indigenes compete with "outsiders" for the resources of the homeland. All nationally mixed areas are potential sites of conflict, since all territory in the state is claimed as one or another nation's homeland, and often two nations' claims overlap. Several of the international conflicts that have emerged in the USSR in recent years, including the most recent outbreak of violence between Kirgiz and Uzbeks in June 1990, involve at least in part international disputes as to the proper delimitation of the national homeland.[3]

The homeland also plays a critical role in the nation's belief in a common destiny. Clearly, this future-oriented aspect of national self-consciousness implies an active nationalism whose practitioners seek self-determination. According to Smith (1981: 187), "nationalism always involves a struggle for land." Clearly, the land over which nationalists are most likely to struggle is the ancestral homeland. National territoriality, defined as a strategy used by nationalists seeking to gain greater control over the destiny of the nation by asserting their exclusive right to sovereignty in the homeland,[4] represents a powerful force for political decentralization in any multinational, multi-homeland state.

In the USSR in recent months, national territoriality has become much more overt, and indicates a growing (or more vocal) intolerance of control from Moscow. The events in Transcaucasia over the status of Nagorno–Karabakh Autonomous Oblast are perhaps the most graphic illustration of this national problem. This incident clearly is not an isolated phenomenon. Indeed, one of the principal reasons for refusing to transfer the mostly Armenian Nagorno–Karabakh to Armenia is the fear that similar demands would then be raised throughout this multinational, multi-homeland state (Gorbachev 1989: 95–6). However, it is not merely the redrawing of internal borders which must concern Moscow, but also the increasing effort by nationalist elites to gain political control of the territory which they perceive as their respective homelands. In other words, national territoriality in the USSR, as in other multinational, multi-homeland states, is a strong centrifugal force which, in the absence of an equally powerful countervailing force from the center, will lead to the devolution of political power, moving the state toward the confederal end of the unitary–confederal continuum (Kaiser 1988). With "glasnost" and "demokratizatsiya," the center's ability to act as a counterweight has been greatly diminished, and consequently this aspect of national territoriality is becoming a predominant feature on the political landscape of the Soviet Union.

The "national question" in the USSR may thus be viewed as the competition between indigenous nationalists striving for greater sovereignty over their homelands and the central authorities pressing for greater territorial and international integration. This formulation is somewhat different from the typical portrayal of the national question—how to balance the interests of the dominant Russians with those of the subordinate non-Russians. However, it is a generally valid statement that Russians dominate the central organs of power, so that the "center-republic" interaction is at the same time a reflection of the Russian–non-Russian relationship.

Just as Moscow had until recently retained most of the political power within the USSR, the Russian nation had also continued to enjoy dominance throughout the state. However, through the constant national territorial pressure exerted by non-Russians, it was also true that the political system had evolved over time to one which was less unitary, though still not federal in essence. As a consequence, the national stratification system had become a two-tiered structure, with the Russians in a dominant position throughout the state, but also with the members of each nation attaining a privileged position within their own homeland (Kaiser 1988: 374). Since 1985, the balance has shifted much more radically in favor of indigenes seeking to control the destiny of their nations by gaining sovereignty over their respective homelands (that is, national territoriality).

In the remainder of this chapter, we consider these general aspects of the national question in the specific context of Central Asia. From a national perspective, this region extends to include the Kazakhs and their homeland (Map 11.1).[5] Indeed, the nations and homelands of this region do not end at the interstate border, but continue into Iran, Afghanistan, and China. However, given limited space and serious data availability and comparability problems at the interstate scale in this region, the present chapter examines the national situation within the Soviet borders of Central Asia only.

THE NATIONAL QUESTION IN SOVIET CENTRAL ASIA

In the West, Central Asia has often been viewed as the Kremlin's "Achilles' heel," as the region that will bring about the disintegration of

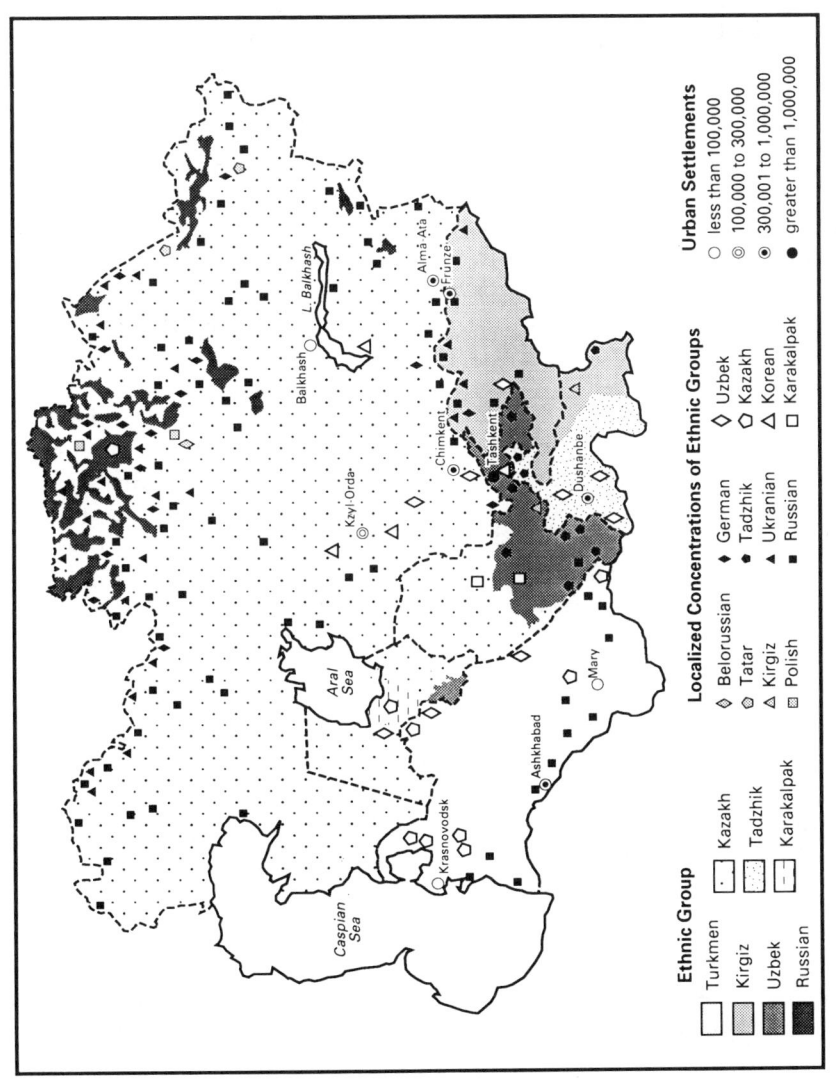

Map 11.1 Ethnic composition of Soviet Central Asia.

the "last empire." According to Bennigsen, the most ardent proponent of this view, "Muslim nationalism is probably more dangerous for the final stability of the USSR than the better known Ukrainian, Georgian, and Baltic nationalist movements" (1986a: 132). Kinship ties between the Turkic-Muslims of Soviet Central Asia and those across the border have led to numerous predictions of increased nationalism in the region, particularly during the past decade with the Soviet invasion of Afghanistan and the Islamic revolution in Iran. The recent international conflicts in Soviet Central Asia between members of the indigenous nation, on the one hand, and "outsiders," on the other, have provided growing evidence of indigenous resentment toward privileges (whether real or merely rumored) enjoyed by nonindigenes, in what could be seen as a much more activated "sense of exclusiveness." However, while the growing international violence in the region is certainly serious, it is not nearly so threatening to the future viability of the state as are the demands for greater national sovereignty in the homeland, which are heard primarily in Lithuania, Latvia, and Estonia, not in Uzbekistan or Kirgiziya.

Of course, the relative lack of demands for outright independence in Soviet Central Asia does not mean either that this more radical version of national territoriality cannot occur in the region. In order to help explain the reasons why Central Asia has not been at the forefront of the separatist movements in the USSR, as well as to assess the direction and pace of future political trends, a closer examination of national self-consciousness in the region is clearly warranted.

NATIONAL IDENTITY IN SOVIET CENTRAL ASIA

Among those analysts who predicted that Central Asia would be the downfall of the USSR, a question as to the importance of national self-consciousness has been raised. For example, Bennigsen (1979: 51) asserted that

> In Soviet Central Asia today, there are three levels of ethnic consciousness among Muslims: sub-national, supra-national and national. Both the sub-national and supra-national have long histories and are deeply rooted in the culture of the area. The national, in contrast, is primarily a Soviet creation imposed in 1924—designed to divide, and to secure Soviet control over Central Asia.

In this section, these three levels of consciousness are assessed in an effort to determine the relative importance of each. It should be noted at the outset that, since identifying the group to which one owes primary loyalty is subjective, there is no completely satisfactory objective measure available which would unerringly indicate which level of identity is most important. In addition, because of the lack of extensive Soviet studies on this issue, there is little evidence available to help sort out this controversy. The question of an overall primary loyalty itself is too simplistic, since humans have the capacity to feel loyalty of varying intensity to several groups at the same time, ranging from the family to the entire human race. In the Soviet context, a loyal Soviet citizen may also be a loyal Uzbek and a Muslim. It is only when two or more of these various levels of identity compete that the question of primacy becomes relevant.

Subnational ethnic identity

The nations of Soviet Central Asia and throughout the USSR as listed in the Soviet censuses are at least in part contrived. In addition to those individuals who identify themselves directly as members of a specific nation, census tabulators include in the total for each nation individuals identifying themselves as members of subnational ethnic groups supposedly belonging to that nation (Appendix 11.1).[6] During the postwar period, the number of nationalities enumerated dropped from 109 in the 1959 census to 101 in 1979, in an apparent effort to reinforce the perception that the national communities of the state were merging into one Soviet people. The increased number of national categories from 1979 (101) to 1989 (128) is more a reflection of the new political reality, which has undermined official support for international integration, than a real increase in the number of national communities resident in the USSR.

Since a nation is a self-conscious human collectivity, the practice cited above calls into question the extent to which the nations of Soviet Central Asia (and throughout the USSR) are real. The significance of this problem is unknown, since the population size of these subnational ethnic groups is not given. However, the continued presence of such subnational affiliations provides clear evidence that, for at least some individuals, primary self-consciousness resides at this subnational or ethnic level.

A number of these subnational groups are listed in the 1926 census. In general, their numbers were relatively small in 1926, and figures for some of these groups in 1959 indicate that they have not increased over time (Appendix 11.1). However, not all subnational categories were included in the 1926 census. Therefore, their contribution to the total number of individuals affiliated with any given nation must remain an unanswered question.

Studies of ethnic processes in Soviet Central Asia have been conducted during the past 20 years and indicate that consolidation toward the main indigenous nations of the region is accelerating (Gurvich 1982: 25). As mass education and communication have spread, local dialects and other "ethnographic peculiarities of former local and ethnic groups" have disappeared (Zhdanko 1972: 28). From the evidence at hand, it would appear that those individuals identifying most strongly with a subnational ethnic group make up a small percentage of the population of each nation in Central Asia. The consolidation process of incorporating these subnational elements more fully into the nations of the USSR is said to be essentially complete (Bromley 1983a: 9), though this is apparently less true for the nationalities residing in the Gorno–Badakhshan Autonomous Oblast of Tadzhikistan (Monogarova 1980). Thus, even though the subnational ethnic units have "deep roots" in the region (which is true of nearly all regions of the USSR, or the world for that matter), national self-consciousness appears to have superseded the subnational level of identity for the majority of Central Asia's inhabitants.

Islamic identity

The primacy of a supranational Islamic identity in Central Asia is often asserted by Western analysts, who see a pan-Islamic movement at the root of most problems facing the USSR in the non-Slavic south. One analyst went so far as to equate the nationalism displayed by Kazakhs in the 1986 riots in Alma-Ata with Islamic fundamentalism (Bennigsen 1986b: A18). Islam is seen as a unifying force superseding the national and subnational distinctions that exist in Soviet Central Asia. The development of a "Homo islamicus," immune to the forces of integration that other national communities in the USSR experience, was said to pose a direct and ominous challenge to the future of both the Soviet state and the dominant Russian nation (Carrere d'Encausse 1978: 246–50). The specter of an Islamic "nation" some 55 million strong, growing rapidly, and united in its opposition to the Orthodox infidels of the north, presents a far more threatening image than does the existence of dozens of smaller national communities whose members to some extent also consider themselves Muslims.

In the USSR, a state that is officially supportive of atheism, it is difficult to determine the strength of any religion. According to Lubin (1984: 7), "Soviet and Western estimates on the number of Muslim believers in Soviet Central Asia range from about 5 per cent of the area to about 65 per cent." Until recently, Islam has been tolerated to a greater extent than a number of other religions, mainly for the foreign relations benefits that may accrue to the Soviets as a result.

Nonetheless, freedom of religious worship is seriously circumscribed in Central Asia.

There have been attempts by Soviet sociologists to measure the strength of people's religious belief in Central Asia. In the Karakalpak ASSR in the early 1970s, for example, 25.8 percent of the indigenous population surveyed were categorized as "believers" and 21.5 percent as "atheists." The remainder of the population were categorized as intermediate in their belief, and included those "hesitants" who believe in Allah, as well as "nonbelievers" who nevertheless observe some religious rites and rituals. In addition, this study indicated that none of the population aged 18–30 was categorized as "believers," whereas 47.8 percent were classified as "atheists." For the population aged 30–42, the figures were 1.2 percent and 34.8 percent respectively. By comparison, 67.6 percent of the population above age 54 were "believers" and only 6.8 percent considered themselves to be "atheists" (Bazarbayev 1973: 50–3). This seems to indicate that a strong attachment to Islam is primarily a phenomenon of the past.[7]

This is not to suggest that Islam has no meaning for the nations of Central Asia, but rather that Islam does not provide the basis for the creation of an Islamic nation that would incorporate all of the nationally diverse individuals who also consider themselves Muslim to some extent (Akiner 1983: 4): "The Muslim population in the Soviet Union is certainly large, but . . . it cannot be said to have a corporate identity. It is formed of a great number of separate elements that have no direct links with one another."

In examining the phenomenon of pan-Islamism within a global framework, it is difficult not to be skeptical of the position taken by its advocates in the Soviet context. Both interstate struggles such as the Iraq–Iran War, and also civil war as in the case of Lebanon, seem to indicate that Islam is not a unifying force acting to create a human collectivity superseding all others. On the contrary, the global picture seems to verify the statement made by Rupert Emerson some 30 years ago (1960: 95–6): "The nation is today the largest community which, when the chips are down, effectively commands men's loyalty, overriding the claims both of the lesser communities within it and those which cut across it or potentially enfold it within a still greater society, reaching ultimately to mankind as a whole." Smaller and smaller groups throughout the world are proclaiming themselves nations deserving of their own states, while overarching identities or "pan-isms" appear to be fading in importance.

Is there something unique about the nations of Soviet Central Asia that would tend to place them outside this global norm? Again, according to Bennigsen (1979: 60–1), the nations of Central Asia were created by the Soviet Union in 1924 in an effort to "divide and conquer" the emerging Muslim "nation." Since the indigenous peoples of Central

Asia had attained no self-consciousness as members of these created nations, they were and continue to be artificial collectivities.

There is some question as to the degree of national self-consciousness which existed at the turn of the century. While the small but vocal intellectual elite, including the literate Muslim clergy, was supportive of pan-Turkism and pan-Islamism, the vast majority of the population were illiterate, isolated, and therefore probably less aware of their membership in these more broad-based human collectivities. Their level of self-consciousness was most likely limited to the subnational ethnic groups discussed above (Matley 1973: 141), making it difficult to imagine that there was a need to "divide" the population in the region further in order to "conquer" it.

On the other hand, it would be incorrect to say that no Uzbek, Tadzhik, Turkmen, Kirgiz, or Kazakh ethnic identity existed at the turn of the century. These self-names are found in the literature from earlier centuries (Krader 1971: 54–68). Rather, it would be more accurate to state that members of these ethnic groups had not yet attained national self-consciousness, and that this came only later. Since the turn of the twentieth century there has been a coalescence around these national communities, which was undoubtedly influenced by the creation of officially recognized home republics during the 1920s and 1930s, particularly in the case of the Tadzhiks (Rakowska–Harmstone 1970: 76–9).[8]

Interest in religion in Soviet Central Asia and elsewhere in the USSR has increased along with rising nationalism in recent years. There is nothing surprising in this, particularly since religion was suppressed by the state in the past. However, it is apparent that Islam has not served to unify the international Muslim brethren in the region. Indeed, the recent violence in Central Asia has not only been between Muslim and non-Muslim (for example, Tadzhiks against Armenians in Dushanbe), but also among the Turkic Muslims themselves (for example, Kirgiz against Uzbeks in Osh). More relevant than religious affiliation in understanding these international conflicts is the indigenous nations' "sense of exclusiveness" and the threats posed to this by "outsiders," regardless of whether these "outsiders" are Muslims or not.

THE NATIONS OF SOVIET CENTRAL ASIA

A dramatic transition in the national composition of the population of Central Asia has occurred in the postwar period (Table 11.1). The indigenous nations made substantial gains over the past 30 years, increasing from 55.3 percent of the population in 1959 to 68.6 percent in 1989.

Table 11.1   National composition of Soviet Central Asia and Kazakhstan, 1959–89 (absolute, percent, percentage point change, and average annual percent change).

Nation	1959	%	1989	%	% Point change 1959–89	Ave. annual % change 1959–89
Uzbek	5,973,147	26.0	16,520,080	33.6	7.6	3.4
Kazakh	3,232,403	14.1	7,476,295	15.2	1.1	2.8
Tadzhik	1,385,835	6.0	4,162,524	8.5	2.5	3.7
Turkmen	985,643	4.3	2,672,174	5.4	1.1	3.3
Kirgiz	962,001	4.2	2,482,210	5.1	0.9	3.2
Karakalpak	168,274	0.7	416,152	0.8	0.1	3.0
Russian	6,213,830	27.0	9,516,229	19.4	–7.6	1.4
Ukrainian	1,034,965	4.5	1,234,556	2.5	–2.0	0.6
Tatar[1]	779,840	3.4	1,178,698	2.4	–1.0	1.4
German[2]	985,623	3.0	1,134,097	2.3	–0.7	0.5
Korean	212,472	0.9	320,189	0.7	–0.2	1.4
Belorussian	121,596	0.5	239,765	0.5	0.0	2.3
Uygur	92,974	0.4	258,952	0.5	0.1	3.4
Azerbaydzhan	102,169	0.4	187,273	0.4	0.0	2.0
Jewish[3]	147,495	0.6	137,445	0.3	–0.3	–0.2
Armenian	47,066	0.2	111,064	0.2	0.0	2.9

[1] Tatar figure for 1989 includes Crimean Tatars, a new census category as of 1989.
[2] German figures are for 1970 and 1989, not 1959 and 1989.
[3] Jewish figure for 1989 includes Jews of Central Asia, Georgia and the mountains—three new census categories added to the general category "Evrey." Between 1979 and 1989, the number of Central Asian Jews increased dramatically, while those in the general category declined precipitously, indicating that both reidentification and out-migration were probably taking place.
Sources: Tsentral'noye Statisticheskoye Upravleniye SSSR 1962: vols 4, 5, 11, 12, 14: table 53 (for 1959); Goskomstat SSSR 1989a: 64–70, 87–92, 94–5 (for 1989).

Much of the shift in favor of the indigenous nations can be attributed to their much higher rate of natural increase (Bondarskaya and Darskiy 1988). For example, in 1979 the average number of children per family in Uzbekistan was 6.1 for Uzbeks and only 2.3 for Russians. This differential was found throughout Central Asia (Arutyunyan and Bromley 1986: 120). The national differentials in natural increase have recently been enhanced by a growing net out-migration of Russians and other nonindigenes, a significant reversal of a long-standing trend (Arutyunyan and Bromley 1986: 21–22; Anderson and Silver 1989; Goskomstat SSSR 1989b: 61–70; Rowland, this volume). In addition to sharply declining economic conditions in Central Asia during the 1980s, the recent rise in overt national territoriality among indigenes is undoubtedly helping to accelerate the pace of this nonindigenous out-migration.

Obviously, trends for the region as a whole mask significant intraregional variations in national composition. The majority of each nation's members resides within the home republic (for example, Uzbeks in Uzbekistan). The percentage of the national community living within its home republic actually increased for all six Central Asian nations in both intercensal periods between 1959 and 1979, except for the Kirgiz between 1970 and 1979 (Table 11.2).

Table 11.2   Population distribution of the Central Asian nations, 1959–89 (percent).

Nation	% in home republic				% in Central Asia			
	1959	1970	1979	1989	1959	1970	1979	1989
Uzbek	83.8	84.0	84.9	84.6	99.4	99.2	99.3	99.0
Kazakh	77.2	79.9	80.7	80.3	89.3	90.8	91.9	91.9
Kirgiz	86.4	88.5	88.5	88.0	99.3	99.2	99.0	98.1
Tadzhik	75.2	76.3	77.2	75.1	99.2	99.1	99.2	98.7
Turkmen	92.2	92.9	93.3	92.9	98.4	98.5	98.7	98.3
Karakalpak	90.4	92.2	92.9	91.8	99.0	98.6	98.2	98.3

Sources: Tsentral'noye Statisticheskoye Upravleniye SSSR 1962, vol. 16: 206–8 (for 1959), 1973, vol. 4: 321–4 (for 1970), 1984: 138–41 (for 1979) Goskomstat SSSR 1989 (for 1989).

However, between 1979 and 1989, a reversal of this trend toward greater indigenous concentration occurred, and the decrease was particularly large for the Tadzhiks. Since it is unlikely that members of the Central Asian nations are experiencing greater levels of natural increase outside the homeland than in, the deconcentration occurring in the 1980s is most likely the result of a net out-migration among indigenes. Nevertheless, the rate of dispersal was very slight, and the level of concentration of these nations in their respective home republics in 1989 remained extremely high, indicating that even under increasingly adverse economic conditions, the nation–homeland tie remains a strong force holding the indigenous population in place.

A large percentage of the individuals not residing in the home republics are found in the remaining union republics of Central Asia (Table 11.2). At a more refined geographic scale, it is also apparent that a majority of these individuals are located in oblasts adjacent to their own home republic. For example, 88.7 percent of the Uzbeks living outside Uzbekistan in 1979 resided in oblasts bordering the Uzbek Union Republic.[9] This provides at least indirect confirmation that there is a difference between the borders of the union republics and the dimensions of the homelands as perceived by the nations themselves. This also sets the stage for international conflicts over competing claims to land at the spatial fringes of each indigenous nation. There have already been international disputes between Uzbeks and Tadzhiks over this issue, and this was undoubtedly a factor in the recent international violence along the border between Kirgiziya and Uzbekistan.

Between 1979 and 1989, the deconcentration of Central Asians occurred not only at the republic scale, but also within the region as a whole. That is, the limited net out-migration that probably occurred during the 1980s was not to other republics in Central Asia, but to other regions of the USSR. However, as with the level of concentration in the national homeland, the vast majority of each nation's membership remains in Central Asia.

The degree to which each nation numerically dominates its respective home republic varies substantially within the region (Table 11.3). In the most nationally homogeneous republics–Uzbekistan and Turkmeniya— the indigenous nations comprise over two-thirds of the population. The only other nation close to this figure is the Tadzhiks. This is potentially significant, since the new law on secession calls for a popular referendum in which a two-thirds majority votes in favor of independence (*Pravda* 1990: 2).[10]

Table 11.3 National composition of the population by republic, 1959–89 (percent).

Republic	Indigenous nation				Russians			
	1959	1970	1979	1989	1959	1970	1979	1989
Uzbekistan[1]	62.1	65.5	68.7	71.3	13.5	12.5	10.8	8.3
Kazakhstan	30.0	32.5	36.0	39.7	42.7	42.4	40.8	37.8
Kirgiziya	40.5	43.8	47.9	52.3	30.2	29.2	25.9	21.5
Tadzhikistan	53.1	56.2	58.8	62.2	13.3	11.9	10.4	7.6
Turkmeniya	60.9	65.6	68.4	71.9	17.3	14.5	12.6	9.5
Karakalpak ASSR	30.6	31.0	31.1	32.1	4.5	3.6	2.4	1.6

[1] The figures for Uzbekistan include the Karakalpak ASSR.
Sources: Tsentral'noye Statisticheskoye Upravleniye SSSR 1962, vol. 16: 206–8 (for 1959), 1973, vol. 4: 321–4 (for 1970), 1984: 138–41 (for 1979); Goskomstat SSSR 1989a (for 1989).

In-migration of individuals from other parts of the USSR has occurred over time; the most significant nation in this regard is the Russians. The Russians in Central Asia are spatially concentrated in Kazakhstan, where they comprised a plurality of the republic's population until the 1980s. They represent a smaller proportion of the total population in the other republics, and have experienced a relative decline in numerical strength throughout Soviet Central Asia. The Russian population declined absolutely for the first time between 1979 and 1989 in Uzbekistan, Tadzhikistan, and Turkmeniya, while Ukrainians declined absolutely in Kazakhstan, Kirgiziya, and Turkmeniya during the same period. This represents a major reversal of a long-standing trend of net in-migration for Russians, and given the economic and political conditions in Central Asia (and throughout the non-Russian periphery), this net out-migration is likely to continue accelerating in the near future.

In the past, the level of international tensions between indigenes and "outsiders" was limited by the spatial concentration of non-indigenes in the cities of the region. Even today, the majority of Russians in each of the union republics live in urban areas, whereas all of the Central Asian nations have remained predominantly rural (Table 11.4). This geographically distinct settlement pattern is less true in Kazakhstan, where Russians and others migrated to rural areas in large numbers during the Virgin Lands Program.

Table 11.4 Urbanization in Soviet Central Asia and Kazakhstan, 1959–79, by republic and nation (percent).

Republic	Total 1959	1970	1979	1989	Indigenes 1959	1970	1979	Russians 1959	1970
Uzbekistan[1]	34.1	36.7	41.2	40.2	20.1	22.8	29.0	83.7	89.1
Kazakhstan	43.7	50.3	53.9	57.2	24.3	26.3	32.0	59.0	69.1
Kirgiziya	33.7	37.4	38.7	38.2	11.0	14.5	18.0	57.8	65.9
Tadzhikistan	32.6	37.1	34.9	32.6	19.6	25.5	25.0	86.9	93.8
Turkmeniya	46.2	47.9	48.0	45.4	26.3	31.7	33.0	94.5	95.7
Karakalpak ASSR	27.2	35.5	42.3	48.1	19.6	30.0	n.a.	80.9	88.0

[1] The Karakalpak ASSR figures are not included with Uzbekistan.
Sources: Tsentral'noye Statisticheskoye Upravleniye SSSR 1962, vol. 16: 206–8 (for 1959), 1973, vol. 4: 321–4 (for 1970), 1984: 138–41 (for 1979) Goskomstat SSSR 1989 (for 1989); 1979 figures for the titular nations provided in Arutyunyan and Bromley 1986: 38.

Most of the indigenous nations registered gains in their level of urbanization between 1970 and 1979, though none became predominantly urban. There was a "ruralization" of the population in Central Asia excluding Kazakhstan during the 1980s. However, this does not mean that indigenes were moving back to the countryside in large numbers, but rather the population in the countryside grew more rapidly than in the cities. Urban growth has nonetheless been quite rapid in Central Asia, and is mainly to be accounted for by growing numbers of indigenes residing in cities. Indeed, with the exception of Kazakhstan and Kirgiziya, Central Asia's urban areas appear to be undergoing a rather rapid transition from Russian enclaves to indigenous pluralities (Table 11.5). While in 1959 only Uzbekistan's and Karakalpakiya's urban areas contained more indigenes than Russians, in 1970 the indigenous members of Uzbekistan, Tadzhikistan, Turkmeniya, and the Karakalpak ASSR surpassed the Russian percentage in the urban regions. This transition has apparently accelerated during the 1970s and 1980s, with a rapid increase in the number of indigenes residing in cities coupled with Russian out-migration from urban areas (Arutyunyan and Bromley 1986: 27–28).

Table 11.5 National composition of the urban population by republic, 1959–79 (percent and percentage point change).

Republic	Indigenes 1959	1970	1979	% point change 1959–70	1970–9	Russians 1959	1970	% pt chg 1959–70
Uzbekistan	37.2	41.1	48.0	3.9	6.9	33.4	30.4	−3.0
Kazakhstan	16.7	17.1	21.0	0.4	3.9	57.6	58.4	0.8
Kirgiziya	13.2	16.9	23.0	3.7	6.1	51.8	51.4	−0.4
Tadzhikistan	31.8	38.6	42.0	6.8	3.4	35.3	30.0	−5.3
Turkmeniya	34.7	43.4	47.0	8.7	3.6	35.4	29.0	−6.4
Karaklapak ASSR[1]	22.0	26.2	n.a	4.2	n.a.	13.4	8.9	−4.5

[1] In the Karakalpak ASSR, Uzbeks and Kazakhs collectively comprised 53.9 percent of the urban population in 1959 and 56.5 percent in 1970.
Sources: Tsentral'noye Statisticheskoye Upravleniye SSSR 1962, vol. 16: 206–8 (for 1959), 1973, vol. 4: 321–4 (for 1970), 1984: 138–41 (for 1979); 1979 figures for the titular nations provided in Arutyunyan and Bromley 1986: 38.

The Russians have clearly lost ground to the indigenous nations in numerical terms; as the nations of Central Asia continue to experience rapid urban growth (but not urbanization), they are displacing the Russians as the majority group—though not necessarily as the dominant nation—in the cities of Soviet Central Asia. There is no reason to foresee a reversal of this trend in the near future, and it will lead over time to ever greater contact and competition between the indigenous nations and the Russians.

The implications of this trend are far-reaching. As the nations of Central Asia become more socially and geographically mobile, national territoriality is likely to become intensified, since it is much more likely that an indigene's "sense of exclusiveness" will become activated under the conditions of increasing international interaction that are found in cities. In line with this, recent Soviet surveys have found that the intelligentsia, the most socially mobilized sector of a nation's population, is also the most nationally self-conscious (Arutyunyan and Bromley 1986: 414). Surveys of the language in which parents desired their children to be instructed reflect this relationship (Table 11.6).

Table 11.6 **Language of instruction desired for one's children, Uzbeks in Uzbekistan, by socioprofessional group (percent).**

| Group | Language of instruction desired: | | | |
| | Urban | | Rural | |
	Uzbek	Russian	Uzbek	Russian
Unskilled physical labor	66.7	18.2	66.4	16.6
Middle skilled physical	57.9	26.3	71.0	18.8
Highly skilled physical	46.0	38.0	66.0	24.0
Office workers	44.1	40.7	63.4	23.1
Middle skilled specialists	44.9	32.6	50.2	35.1
Highly skilled specialists	37.2	38.7	54.2	28.3
Mid-level managers	52.9	41.1	54.6	35.4
High-level managers	62.5	25.0	61.2	27.8

Source: Arutyunyan and Bromley 1986: 309.

The "U"-shaped curve in evidence indicates that it is the least and most socially mobile members of the nation that are most concerned with the maintenance of the tangible attributes of the nation. Socioprofessional elite Uzbeks living in both urban and rural areas expressed a strong desire for their children to be instructed in Uzbek, even though Russian has been recognized throughout the USSR as the language of upward mobility.[11]

The high level of national self-consciousness among the elite which may be inferred from these data is not a phenomenon unique to Uzbeks. Indeed, a national resurgence is occurring throughout the "modernized" world. With social mobilization and even international equalization, individuals from different national communities become more competitive for the same limited social, economic, and political resources (Nielsen 1985). In a multi-homeland state, the indigene's "sense of exclusiveness" makes this competition particularly intense between elite indigenes, on the one hand, and "outsiders" holding high-status positions, on the other. The old argument that indigenes are not capable of managing their own lives and their homelands' development no longer accords with reality, undermining the legitimacy of continued dominance by "outsiders."

With rising social mobilization, then, the prospects for international integration or the merger into one Soviet people throughout the USSR have become all the more remote, while the more traditional forms of identity are weakened. This indicates that members of indigenous nations actively resist acculturation and assimilation while participating more intensely in the "modernized" sectors of society, and not that Islam or some other traditional cultural attribute makes members immune to both social mobilization and international integration. A closer examination of the acculturation and assimilation processes indicates that members of nations living in their own home republics are not participating to any significant degree. Rather, it is more frequently those nonindigenous "outsiders"—who must accommodate themselves to the national territoriality of the indigenous nations—who undergo a process of international integration. This is true not only in Soviet Central Asia, but throughout the USSR, and indeed in multi-homeland states generally. In the following section we explore the influence of the homeland and national territoriality on the pace and direction of international integration in Central Asia.

THE HOMELAND FACTOR IN INTERNATIONAL INTEGRATION

International integration refers to the processes of drawing together ("sblizheniye") and merger ("sliyaniye") of nations into a unified (that is, Soviet) people. In Central Asia, this process could also refer to the development of an Islamic "nation" from the Uzbeks, Kazakhs, Kirgiz, and so on. It is often viewed as a multistage process beginning with acculturation, or the adoption of one or several of another nation's tangible attributes (such as language), and ending with the psychological assimilation of an individual from one nation to another. Exogamy, or international marriage, marks an intermediate stage in the process, since it reflects the acceptance of former "outsiders" into the most intimate societal unit, and in the process serves to internationalize the family.

While international integration does normally proceed in stages, acculturation is not a sufficient condition for the occurrence of assimilation. According to Connor (1972: 341–2): "An individual (or an entire national group) can shed all of the overt cultural manifestations customarily attributed to his ethnic group and yet maintain his fundamental identity as a member of that nation. Cultural assimilation need not mean psychological assimilation."

The benefits of national belonging in the USSR are relatively high and have increased over time in one's national homeland, but decrease when one leaves. In general, individuals living outside their national homelands experience a much higher degree of international integration than those who remain in the homeland. This spatial differential indicates that the cost of retaining one's national identity, particularly on an intergenerational basis, increases dramatically outside the national homeland.

The rate of assimilation also appears affected by the national self-consciousness of individuals who leave the homeland. While those who voluntarily leave acculturate quite readily, those who have been forced to emigrate resist the assimilative pressures of their new surroundings. Those who choose to leave the homeland appear to be more nationally marginal (that is, value their national identity less highly) than those who do not, and this characteristic itself accentuates the spatial differential in the rate of international integration between members of the nation living inside and outside the homeland (Kaiser 1988: 366–7).

The complex relationship between location and the international integration process is briefly sketched below. Each "stage" of the process is examined separately, beginning with acculturation as measured by the adoption of another nation's language as one's own, followed by rates of international marriage, and concluding with the national identity of children from international families.

Acculturation

Acculturation in the Soviet context has often been equated with linguistic "Russification", which in turn has often been viewed as a major step in the complete "Russification" of the non-Russian population.[12] Linguistic Russification is used as one measure of acculturation, since the adoption of the tangible attributes of another nation as one's own is the hallmark of this preliminary stage in the process of international integration. However, as was noted above, acculturation does not signify the loss of national self-consciousness, nor is it alone a sufficient condition for psychological assimilation at some later date. Linguistic Russification, even where it is found to be extensive, may be the terminal point in the process of international integration, and even this tentative step may be reversed at some later date.

Linguistic assimilation to the indigene's language (for example, to Uzbek in Uzbekistan) may also occur. This aspect of acculturation has rarely been explored, but is included in this analysis in an effort to assess the relative status the indigenous language enjoys in the national homeland. If linguistic "Russification" may be equated with accultura- tion toward the dominant nation in the state, linguistic assimilation to the indigenous language is indicative of the preferential position attained by the indigenous nation within its homeland.

Linguistic Russification

Learning Russian as a second language has not led to linguistic "Russification" in Central Asia. Indeed, the introduction of a question on second language fluency in the 1970 census appears to have resulted in a decline or an extremely minimal increase in linguistic "Russifica- tion" between 1959 and 1970 for members of each nation who resided in the national homeland. This decline or low growth was of course more apparent than real, and indicates that the choice of first language is not necessarily made solely on the basis of fluency level, but may be influenced by the international climate at the time of the census (Silver 1987b: 79–80).

Spatial variations in the rate of linguistic Russification among Central Asians do exist (Table 11.7). Rates of linguistic Russification are much lower in rural areas, and are highest among members living in urban areas outside their home republic. Since members of Central Asia's nations are concentrated in rural homeland areas, the higher rates outside affect only a small proportion of the population. In addition, even this outside segment of the national community does not indicate a strong tendency toward linguistic Russification. It should be recalled that the majority of indigenes living outside the officially delimited

Table 11.7 Linguistic Russification, 1959–79, by nation and location (percent).[1]

Nation	In home republic			Outside home republic		
	1959	1970	1979	1959	1970	1979
Uzbek	0.3	0.3	0.4	1.3	1.5	1.6
Kazakh	0.8	1.1	1.4	2.8	3.7	4.7
Kirgiz	0.2	0.2	0.4	0.8	1.0	1.5
Tadzhik	0.4	0.4	0.6	1.0	1.3	1.6
Turkmen	0.5	0.6	0.7	2.3	2.9	4.0
Karakalpak	0.1	0.1	0.2	1.7	3.2	3.7

[1] 1989 census figures for linguistic Russification are not yet available.
Source: Tsentral'noye Statisticheskoye Upravleniye SSSR 1962: vols. 4, 5, 11, 12, 14, 16 (for 1959), 1973, vol. 4 (for 1970), 1984: 71–136 (for 1979).

home republics reside in oblasts adjacent to it. Thus, residence outside the home republic, which normally results in much higher rates of acculturation in the USSR, has not had the same effect on the nations indigenous to Soviet Central Asia (Kaiser 1988: 255–75).

Linguistic Russification in the region is occurring primarily among nonindigenes (for example, Ukrainians, Belorussians, Germans, and Jews). The languages of these nonindigenous non-Russians are for all intents and purposes useless in this setting, and there are few support mechanisms in place that would encourage their retention.[13] For these nonindigenes, Russian has become the most attractive alternative. Among these non-Central Asians, linguistic Russification is most advanced in the republics and oblasts where Russians are concentrated (Table 11.8).

Table 11.8 **Native language retention rates (NL), 1979–89, and linguistic Russification (LR), 1979 (percent).**

Republic	Ukrainian NL 1979	Ukrainian NL 1989	Ukrainian LR 1979	Belorussian NL 1979	Belorussian NL 1989	Belorussian LR 1979	German NL 1979	German NL 1989	German LR 1979	Jew[1] NL 1989
Uzbekistan	45	50	55	n.a.[2]	56	n.a.	n.a.	48	n.a.	26
Kazakhstan	41	37	59	38	35	62	65	54	35	19
Kirgiziya	38	34	62	n.a.	35	n.a.	72	63	28	17
Tadzhikistan	44	50	56	n.a.	50	n.a.	71	60	29	19
Turkmeniya	55	53	45	n.a.	56	n.a.	n.a.	52	n.a.	29

[1] Jews in 1989 were separated into four categories: "Evrey" = Western Jews; Mountain Jews; Georgian Jews, and Central Asian Jews. The figures given above are for Western Jews only in 1989. Since no comparable figure exists for 1979, only 1989 data are presented.
[2] N.a. = data for these nations in these republics were not available for 1979.
Sources: Tsentral'noye Statisticheskoye Upravleniye SSSR 1984: 71–136. (for 1979); Goskomstat SSSR 1989a (for 1989).

Ukrainians chose their native language more frequently in 1989 than in 1979 in Uzbekistan and Tadzhikistan, and these were the two Central Asian republics in which the Ukrainian population grew relatively rapidly. This indicates that the population growth was caused primarily by the in-migration of Ukrainians from Ukraine, where native language retention rates are much higher. In contrast to this situation, the German population experienced a high rate of decline in native language retention between 1979 and 1989, and this was matched by a slow rate of growth or actual decline in the German population. It may be surmised that the more nationally self-conscious Germans are leaving Soviet Central Asia, many for emigration to Germany.

Acculturation toward the Russian nation has been minimal in Soviet Central Asia among the indigenous nations throughout the postwar period, and there is no indication that this is likely to change in favor of "Russification" in the near future. Indeed, recent trends in the region indicate that even the minimal inroads made by linguistic Russification over the past 30 years are likely to be reversed in the near future, as indigenes become more assertive with regard to the status of their own languages in their own home republics. For nonindigenous non-Russians, linguistic Russification has proceeded much further; this may continue as the more nationally self-conscious members of these groups leave for "home" (for example, Germans). On the other hand, the Russian language is experiencing a serious erosion in its level of prestige in Central Asia and elsewhere in the non-Russian periphery, and as this occurs the number of nonindigenes willing to acculturate to the Russian nation should decline. Linguistic assimilation to the indigenous languages would in this case become a relatively more attractive alternative.

Linguistic assimilation to the indigenous languages

While the indigenous Central Asian nations are not linguistically Russified, neither are Russians living in Central Asia linguistically assimilating to the indigenous languages of the region. They are, however, increasingly claiming mastery of the indigenous languages as their second language, which provides evidence that even the dominant Russians feel the need to converse in the locally dominant language when they are outside the national homeland. This is especially true when they are the minority population in a given region (Smirensky 1985: 31–41).

The non-Russians residing in a republic that is not their national homeland are linguistically assimilating to the indigenous language to a greater extent, though this process of acculturation is also not extensive (Table 11.9). Indeed, the percentage of non-Russian nonindigenes claiming the indigenous language as their mother tongue actually declined between 1959 and 1970, indicating that when nonindigenes were given the option of claiming the locally dominant language as a

Table 11.9 Nonindigenes claiming fluency in the indigenous language as either a first or second language, 1970–79, by republic (percent).[1]

Republic	Including Russians		Non-Russians	
	1970	1979	1970	1979
Uzbekistan	12.4	13.5	17.8	20.2
Kazakhstan	1.3	1.2	1.8	2.1
Kirgiziya	2.5	2.7	3.5	4.2
Tadzhikistan	8.4	10.4	10.5	12.9
Turkmeniya	6.9	8.6	10.3	12.8
Karakalpakiya	4.5	11.7	4.4	11.5

[1] 1989 census figures are not yet available.

Sources: Tsentral'noye Statisticheskoye Upravleniye SSSR 1973, vol. 4 (for 1970), 1984: 71–136 (for 1979).

second language in 1970, they did so. This occurred mainly in rural areas among members of Central Asian nations living outside their home republics but inside their national homelands (that is, in oblasts adjacent to the home republic).

Aside from Kazakhstan and Kirgiziya, a fairly large proportion of the nonindigenous population is claiming fluency in the indigenous languages of Central Asia, and this percentage increased in every republic between 1970 and 1979. While not yet truly competitive with Russian, the indigenous languages do appear to be gaining in usage among nonindigenes. For a certain segment of the nonindigenous population, it is not enough to know one's native language and Russian; to succeed in another nation's homeland it is also necessary to know the indigenous language.

The choice of first language among members of Central Asian nations is clearly the language of that nation, whether the individuals live in urban areas or rural, or whether they are living within their homelands or outside. Even in cities where the Russian population is in the majority, such as Alma-Ata, the indigenous population over-whelmingly chooses its own language as the first language. If linguistic assimilation is viewed as a preliminary stage in the process of inter-national integration, it is apparent that none of the indigenous nations is on its way to disappearing into some larger national (for example, Russian) or supranational (for example, Soviet or Islamic) people. Indeed, just the opposite appears to be the case today, as each indigenous nation presses for the enactment of legislation guaranteeing its own native language the status of lingua franca in its own home republic (JPRS 1989). With the attempt to establish the predominance of their own languages in their own home republics, the nations of Central Asia are clearly asserting that they have the right as indigenes to preferential treatment in their respective homelands. To the degree that they are successful in the use of this national territoriality strategy, nonindigenes will feel increasing pressure to leave the region or accom-modate themselves to the new reality (that is, acculturate to the indigenous nations).

INTERNATIONAL MARRIAGE PATTERNS

Soviet authors view exogamy or international marriage as playing a major role in the merger of all national communities into one Soviet people (Susokolov 1987: 24–5). Extensive intermarriage and the resultant internationalization of the family are said to pave the way to complete international integration through the "natural" assimilation of the children and grandchildren of international families (for example, Bromley 1983b: 204). However, while this statement of the relationship between exogamy and assimilation appears valid, it is too one-dimensional to capture the complex nature of the process. First, the rate of exogamy does not, as has often been assumed, increase in a linear fashion over time, nor is there a simple causal relationship with social mobilization. Beyond this, the direction of the international integration process itself is not necessarily toward the dominant nation (that is, "Russification"), but rather is strongly influenced by national territoriality.

There exists a clear differential between the rate of international families in urban and rural areas, both within Central Asia and in the USSR as a whole (Table 11.10). In rural areas the population is generally less concentrated and more nationally homogeneous, affording fewer opportunities for international marriage. The rural population of Kazakhstan and Kirgiziya is more nationally mixed, and this is reflected in a higher incidence of exogamy. The urban areas in Central Asia are much more nationally diverse, and higher rates of international marriage are found in these settings.

Table 11.10 International family rates, 1959–79, by republic (percent of all families, by urban and rural location).

Republic[1]	Urban				Rural		
	1959	1970	1979		1959	1970	1979
Total USSR	15.1	17.5	19.2		5.8	7.9	9.2
Uzbekistan	14.7	18.4	16.2		4.7	5.7	4.4
Kazakhstan	17.5	23.8	23.9		11.9	17.1	15.9
Kirgiziya	18.1	24.1	21.6		9.2	10.6	9.1
Tadzhikistan	16.7	22.3	21.5		5.5	6.5	5.5
Turkmeniya	14.9	20.0	17.7		2.5	3.4	3.0

[1] Data for Karakalpakiya were not available for 1959 and 1979; therefore this autonomous republic is not included. Uzbekistan figures include Karakalpakiya.

Sources: Isupov 1964: 38 (for 1959); Tsentral'noye Statisticheskoye Upravleniye SSSR 1973: vol. 7: 272–303 (for 1970), 1984: 298–301, 312–15, 318–19 (for 1979).

It is clear that a decline in the rate of international families occurred between 1970 and 1979. Over time, the population in the Central Asian republics is becoming more nationally homogeneous; members of the indigenous nations are comprising a greater percentage of the population entering into marriage; and indigenes clearly intermarry at a lower rate than nonindigenes. These factors have combined to cause a decline in the proportion of families which are international, even though an absolute increase in the number of international families was registered throughout Central Asia between 1970 and 1979, except in rural Uzbekistan (Table 11.10).

According to data recently released for the first time, patterns of exogamy vary substantially across national communities in Soviet Central Asia, with the lowest rates not surprisingly found among members of the indigenous nations (that is, Uzbeks in Uzbekistan, and so on) (Table 11.11). However, members of Central Asian nations living outside their respective home republics engage in exogamy at rates comparable to Russians living in the region. The highest rates of intermarriage occur among nonindigenes whose homelands lie outside the region (for example, Ukrainians and Belorussians) (Table 11.12).

Table 11.11   **Rate of exogamy for indigenes and Russians by republic, 1978–88 (percent).**

| | Indigenes | | | | Russians | | | |
| | Urban | | Rural | | Urban | | Rural | |
Republic	1978	1988	1978	1988	1978	1988	1978	1988
Uzbekistan	8.2	6.7	2.3	2.4	24.0	26.3	30.1	21.0
Kazakhstan	7.0	3.1	2.0	2.6	21.1	23.4	29.9	34.9
Kirgiziya	5.7	5.8	2.0	2.6	16.3	19.3	22.8	26.0
Tadzhikistan	12.7	10.8	4.2	4.3	26.0	28.5	35.4	40.8
Turkmeniya	10.0	7.8	1.8	2.0	26.2	30.7	35.8	54.1

Source: Goskomstat SSSR 1989c: 204–321.

Table 11.12   **Rate of exogamy for nonindigenes,[1] by republic, 1978–88 (percent).**

| | Other Turkic-Muslims | | | | Other nonindigenes | | | |
| | Urban | | Rural | | Urban | | Rural | |
Republic	1978	1988	1978	1988	1978	1988	1978	1988
Uzbekistan	31.9	34.3	17.4	20.0	89.3	91.4	90.5	89.8
Kazakhstan	46.5	49.9	25.2	23.8	80.9	85.5	71.0	78.1
Kirgiziya	8.4	11.6	2.9	9.1	76.0	86.9	65.1	81.0
Tadzhikistan	27.6	28.7	8.1	9.1	n.a.	n.a.	n.a.	n.a.
Turkmeniya	39.2	64.1	38.0	80.4	n.a.	n.a.	n.a.	n.a.

[1] Uzbekistan: "other Turkic-Muslims" = Tatars, Kazakhs, Tadzhiks, and Kirgiz; "other nonindigenes" = Ukrainians. Kazakhstan: "other Turkic-Muslims" = Tatars and Uzbeks; "other nonindigenes" = Ukrainians and Belorussians. Kirgiziya: "other Turkic-Muslims" = Uzbeks; "other nonindigenes" = Ukrainians. Tadzhikistan: "other Turkic-Muslims" = Uzbeks; no "other nonindigenes" given. Turkmeniya: "other Turkic-Muslims" = Tatars; no "other nonindigenes" given.

Source: Goskomstat SSSR 1989c: 204–321.

For every indigenous nation except the Kirgiz, the urban rate of exogamy actually declined between 1978 and 1988, and this is most likely an indication of the growing size of the indigenous population in the urban areas. However, this may also be seen as evidence that the direct relationship normally assumed to exist between social mobilization and international integration needs to be questioned. Indeed, an examination of these data indicates that there is less difference between urban and rural rates of exogamy within each national community than there is between indigenes, on the one hand, and nonindigenes, on the other, in either an urban or a rural setting.

    Survey data must be used to determine which nations are intermarrying with one another. In general, three types of international marriage are of interest: (a) that between a Russian and a member of the indigenous nation; (b) that between a Russian and a member of a nonindigenous nation; and (c) that between members of two Central Asian nations. In assessing trends in these three categories, where possible we have compared the actual rate of intermarriage with the theoretical probability of its occurrence.[15]

A number of international marriage surveys have been conducted in various localities within Soviet Central Asia during the past three decades (for example, Kozenko and Monogarova, 1971; Evstigneev 1974; Kalyshev 1984; Tolstova 1985. A review of these and other surveys of international marriage patterns leads one to the conclusion that the propensity for exogamy is greatest for members of nations who are dispersed and living outside their homelands. Primarily this refers to the nonindigenous nations in Central Asia such as the Ukrainians, Belorussians, and Germans. The members of these nations most often marry Russians, and represent the population most likely to become "Russified." Members of the indigenous nations intermarry at less than the theoretical probability, and this holds true for intermarriage between Central Asians and "outsiders" or between members of different Central Asian nations. For members of the Central Asian nations living outside their respective home republics, intermarriage occurs more frequently, most often with members of the indigenous nation (for example, Uzbeks marrying Tadzhiks in Tadzhikistan). However, the actual rate of this type of intermarriage rarely exceeds the theoretical probability for its occurrence. Overall, the evidence provided in these surveys indicates that the indigenous nations of Central Asia are maintaining an endogamous barrier to international integration, and that neither "Russification" nor homogenization into one Muslim nation is occurring in the region.

NATIONAL TERRITORIALITY AND "NATURAL" ASSIMILATION

As noted above, the importance of exogamy is that over the course of time, measured in generations, international marriage leads to the "natural" assimilation of the population. Particularly important in this regard is the national identity chosen by children of international families.[16]

When members of Central Asian nations do intermarry, the children of such marriages most often identify with the indigenous nation, provided that one of the parents is a member of that nation (Vinnikov 1980: 37): "In Uzbekistan in the preponderant majority of Uzbek–Russian, Uzbek–Tatar, and Uzbek–Turkmen families teenagers choose the Uzbek nationality; in Turkmenia—in Turkmen–Russian and Turkmen–

Uzbek [families]—[they choose] Turkmen. This also takes place in Tadzhikistan and Kirgiziya."

This dominant trend in "natural" assimilation means that international integration results in the "indigenization" of the population in each national homeland over time. This occurs even in the case of Russian–indigene families, which actually result in a loss to the Russian nation through the national identification of these children with the indigenous nation. Other surveys have shown that the preference for identification with the homeland nation is a nearly ubiquitous feature of international integration in the USSR.

The children of marriages between Russians and non-Central Asians most frequently identify themselves as Russians. The tendency for Russian national identity to be preferred by children from Russian–nonindigenous couples is found throughout the USSR. The losses registered to the potential growth of the Russian population due to Russian–indigene exogamy is more than compensated for by the gains resulting from exogamy with Ukrainians, Belorussians, Germans, and so forth.

The indigenous nations of Soviet Central Asia are clearly not on the path of international integration, toward either an Islamic or a Soviet nation. On the contrary, indigenes in this region and elsewhere in the USSR have been relatively successful in promoting their own nations to a privileged position in their own homelands, and "outsiders" increasingly must accommodate themselves to this new political reality. The nations who do appear to be losing adherents over the course of generations are the Ukrainians, Belorussians, Germans, Poles, and other nonindigenes living in Russian-dominated areas of Central Asia. As was noted previously, the Belorussians, Ukrainians, and Jews have declined absolutely in the region in recent years. While part of this decline was undoubtedly due to out-migration of these nations' members during this time period, part of it is apparently also due to the "natural" assimilation toward the Russian nation.

CONCLUSIONS

It is clear that the social and geographic mobilization occurring among the indigenes in Soviet Central Asia has not resulted in their merger into one Soviet people. Traditional belief systems and lifestyles have given way to a more "modern" outlook with an emphasis on higher education and upward mobility. However, this has not led to a weakening of national self-consciousness. On the contrary, the social, economic, and political developments that have taken place during the post-World War II era in Soviet Central Asia and elsewhere in the USSR have led to greater national self-awareness and assertiveness. Since educational attainment among the urban working population of the Turkic-Muslim nations is on a par with that of the Russians, and since nations in their own homelands perceive that they should be the ones to benefit from any development that takes place on this territory, the privileged position of the Russians and other "outsiders" will increasingly be challenged by mobilized members of the indigenous nations in the region. If the nationally stratified system does not prove flexible enough in accommodating the pressure for upward mobility among a growing segment of the indigenous population, international tensions and conflict will increase.

Before 1985, it appeared that the system was able to accommodate the pressure for upward mobility in Soviet Central Asia. However, population growth, along with rising educational attainment and rising expectations among the indigenous nations, presents the state with the prospects of an ever-growing indigenous elite seeking to gain greater control of its homeland (that is, national territoriality). The problems of the region will almost certainly grow if economic development does not keep pace with population growth and social mobility. We have already seen this in the outbreaks of violence by indigenes against "outsiders" who are perceived as threatening the indigenous claim to preferential status in their respective homelands.

Out-migration is often viewed as one solution to the developing imbalance between economic development, on the one hand, and population growth, on the other (Rowland, this volume). If a state policy to promote this exists at all, it is clearly a failure, for the indigenous nations of Soviet Central Asia have remained highly concentrated in their respective home republics. This is true even while underemployment and actual unemployment in the region are escalating rapidly.

To date, out-migration from one's homeland for all but the dominant Russians has been impeded by the multi-homeland nature of the state and the national stratification system which provides preferential treatment for the indigene in his or her homeland, but not outside.[17] To leave the homeland means to forgo the privileges gained by being a member of the indigenous nation. Indeed, given the national territoriality exhibited in Central Asia, we would predict that, as an alternative to out-migration, indigenes may make more vocal demands for a larger share of investment funds in order to bring about greater regional development.

Out-migration will occur when spatial differentials in economic opportunities become great enough to overcome the impediment created both by the emotional attachment of indigenes to the homeland, and also by the concrete benefits derived from residence in the homeland. For most Central Asians, this threshold clearly has not yet been reached. It is the Russians and other nonindigenes who are currently leaving the region, and they, of course, have no such threshold to overcome.

Out-migration, above and beyond its economic desirability, may be seen by Soviet policy makers as a solution to the national question. Once outside the homeland, individuals tend to become more rapidly acculturated. Indeed, members of Central Asian nations living in the cities of the RSFSR have acculturated to a much greater extent than have their co-nationals residing in the national homeland or more generally in Central Asia. However, the number of Central Asians living in cities in the RSFSR is quite small, and this group may not be representative of the nation as a whole. That is, those who have left the homeland voluntarily appear to be more nationally marginal than individuals who remain in the homeland. Larger migration streams from Central Asia to the RSFSR, especially if the migration is perceived as being forced—due either to a direct policy or to a lack of economic opportunities in the homeland resulting from central investment decision making—are unlikely to result in the same acculturation rates. Individuals out-migrating in response to what may be perceived as coercion from above may in fact become more nationalistic as a result. Also, due to the potential for a nativistic backlash among indigenes in the areas of destination, massive movement of any national community to another's homeland will in all likelihood increase rather than decrease the level of international tensions in the state. Thus, while out-migration may occur as a response to differential economic opportunities brought about by regional differences in demographic and socioeconomic development patterns, it should not be assumed that this will help to resolve the national question.

The growing international tensions and conflicts in Soviet Central Asia are serious, but not life threatening to the USSR. National territoriality in the region appears to be motivated by a more localist "nativism," with indigenes reacting to perceived nonindigenous affronts to their "sense of exclusiveness." Reflective of this, there are relatively few national front organizations in the region, and those that do exist such as "Birlik" ("Unity") appear more interested at present in obtaining a higher status for the nations' cultural attributes (for example, language) than in mobilizing the national membership behind the goal of independence (Brown 1990). One of the main reasons why Western analysts missed so badly is that their predictions of Central Asian revolt were based on the assumed existence of an internally cohesive Muslim nation hostile to the Slavic, socialist, and Orthodox north. This assumption was apparently based more on wishful thinking than on solid empirical evidence. Rather than one "Homo islamicus," Soviet Central Asia is home to several relatively underdeveloped nations whose socioeconomic and geopolitical position argues in favor of continued participation in a more decentralized USSR.

However, this is not to argue that the nations of Central Asia are incapable of mobilizing mass-based support for self-determination, including secession from the USSR. Should Moscow prove unwilling or unable to meet demands for greater levels of investment in the region, and should other nations gain their independence from the USSR, nationalists in Soviet Central Asia would certainly be capable of rallying the national membership and pressing for greater political sovereignty. Their ability to do so has increased, not decreased, over time as each nation's membership has become more socially mobilized and at the same time more aware of themselves as members of nations deserving of a privileged place in their own perceptual homelands.

APPENDIX: THE NATIONS OF CENTRAL ASIA AND AFFILIATED SUBNATIONAL GROUPS

Nation	Subnational group[1]	1926 census	1959 figures[2]
Uzbek (Ozbak)		3,904,622	6,015,416
	Kipchak[3]	33,502	
	Kurama	50,079	
	Tyurk[3]	537	
	Barlas, Durmen, Kangly, Kaluk, Katagan, Kirk, Kongrat,[3] Lakay, Mangyt, Mugul, Musobozor, Nayman,[3] Saray, Turkman,[3] Chagatay,[3] Yuz		
Kazakh		3,968,289	3,621,610
	Nogay[3]	36,274	38,583
	Kipchak[3]		
	Aday, Alban, Argyn, Bersh, Beskalmak, Dulat, Zhagalbayly, Zhalair, Zhapps, Kazak,[3] Kerey, Kongrat,[3] Nayman,[3] Oshakty, Sary-Uysun, Sreyl, Suan, Tabyn, Tama, Tortkara, Uak, Sherkesh, Shekty, Ysty		
Kirgiz (Kyrgyz)		762,736	968,659
	Kalmak Issyk-Kulya[4]	2,405	2,500 (3,587)
Tadzhik (Todzhik)		978,680	1,344,939*
	Yagnob	1,829	2,000
(Pamirskiy Tadzhik)			
	Bartangi		3–4,000
	Ishkashimtsy		500
	Vakhanchi		6–7,000
	Oroshortsy		1.5–2,000

Rushansty		7–8,000
Sarykol'tsy		5,000
Khuftsy		1–1,500
Shugnantsy		20,000
Yazgulyamtsy		1.5–2,000

Badzhuvtsy, Barvoztsy, Gorontsy, Gundtsy, Darvozi, Mundzhantsy, Kharduri, Chagatay,[3] Shakhdarintsy

Turkmen	763,940	1,001,585

Alili, Anevli, Ata, Gandyry, Gokleny, Emreli, Yomudy, Karadashly, Mukry, Nokhurly, Salory, Saryki, Sakhary, Syrkhy, Teke, Trukhmeny, Tyurk,[3] Chovdury, Ersari, Shikh

Karakalpak (Kalpak, Karolpak)	146,317	172,556

[1] Changes between the 1969 and 1988 *Slovari:*
 Uzbek—Saray and Kirk are new names found under heading; Chagatay, Nayman, and Turkman whose native language is Uzbek are also included in this for the first time.
 Kazakh—all Naymans were included in this category; now they are divided between Kazakhs and Uzbeks on the basis of language.
 Kirgiz—no changes.
 Tadzhik—all Chagatay were included in this category; now they are divided between Tadzhiks and Uzbeks on the basis of language. Darvozi is an added ethnonym in 1988.
 Turkmen—Anevli, Gandyry, Sakhary, and Syrkhy are new ethnonyms found here; Seyit, Makhtum, Myudzhevuri, and Khodzha were deleted from the list.
[2] Figures for subnational ethnic groups in 1959 are based on language affiliation rather than national self-identity, and thus may overstate or underestimate the size of these groups.
[3] Subnational groups divided according to native language:
 Kipchak and Kongrat are divided betwen Uzbeks and Kazakhs (the number listed for the Kipchaks in 1926 was for the entire group's membership). This division is a recent decision; the 1959 *Slovari* listed Kypchaks and Kongrats as subnational groups within the Uzbek nation.
 Tyurk is split between the Uzbeks and Turkmens on the basis of language. The figure cited for 1926 is for "Tyurks of Fergana and Samarkand."
 Turkmans who speak Uzbek are an addition to the Uzbek listing. This ethnonym was not found under the Turkmen or any other heading, and was a new category as of 1988.
 Kazak is split between Kazakhs, Russians, and Ukrainians on the basis of language, but refers to Cossacks.
 Nogay is split between Kazakhs and Nogays on the basis of language. This appears to be a recent decision as well, in that the 1959 *Slovari* does not seem to allocate this group on the basis of language.
 Nayman is divided between Uzbeks and Kazakhs on the basis of language. This is a recent decision; the 1969 *Slovari* listed all Naymans under the Kazakh nation.
 Chagatay is divided between Uzbeks and Tadzhiks on the basis of language. This is a recent decision; the 1969 *Slovari* listed all Chagatays with the Tadzhik nation.
[4] Kalmaks of the Issyk Kul' are not listed in the *Slovari* for 1959, 1969, or 1988. Information on this subnational group is found in Zhukovskaya (1980: 157–66). The figures given correspond to the following years: 2,405 in 1917; 2,500 in 1959; and 3,587 in 1970.
Sources: Subnational listing—Goskomstat SSSR 1988; comparisons were made with Tsentral'noye Statisticheskoye Upravelniye (TsSU) SSSR 1959, 1969.
 1926 data—TsSU SSSR 1929, vol. 17.
 1959 data—TsSU SSSR 1962, vol. 16: table 53; this citation is relevant for number of Uzbeks, Kazakhs, Nogays, Kirgiz, Tadzhiks, Turkmens, and Karakalpaks in 1959; note that the figure given for the Tadzhiks has been reduced by 52,000, the maximum estimate for the subnational groups listed.
 1959 data—Vinogradov 1968, vol. 1; information for subnational groups listed under Tadzhiks; information on the Pamir Tadzhik groups also found in Monogarova 1980: 125.

NOTES

1 The term "nation" and its derivatives (national, nationalism, international) refer to self-conscious human collectivities such as Uzbeks, Russians, and so forth, not to states.

2. This perceptual delimitation of the national homeland is almost certain to be somewhat different from the official delimitation of the union republics and autonomous regions that comprise the federal structure of the USSR. For a discussion of the latter, see Schwartz (this volume).

3. A recent map published by the US State Department (Dillon 1990) provides a cartographic catalog of these international border disputes in the USSR.

4. This definition of territoriality follows the work of Sack (1983, 1986), who defined this concept "as an attempt by an individual or group to affect, influence, or control people, phenomena, and relationships, by delimiting and asserting control over a geographic area" (1986: 19).

5. The reader is cautioned that other authors in this volume define the region differently, and this may result in data comparability problems.

6. The latest listing of these subnational groups, in preparation for the 1989 census, enumerated over 700 potential ethnographic groups (Goskomstat SSSR 1988), from which 128 "census nations and nationalities" appeared in 1989 (Goskomstat SSSR 1989a). Twenty-seven of these 128 are "nationalities of the north" ("narodnosti severa") which are not enumerated separately in the figures for the entire USSR, and appear only in specific administrative units of the RSFSR.

7. Since the pronouncement of strong religious beliefs was likely to have a negative impact on an individual's career, even young people who considered themselves "believers" may have been unlikely to declare this in surveys. Thus, while the survey findings cited above are generally considered to be valid, they probably understate the degree of belief among the young to some extent.

8. The timing of the consolidation process in Central Asia is not unique. According to Vakar (1956), a mass-based Belorussian national consciousness did not exist prior to the interwar period. Even in France, there existed little French self-consciousness in rural areas prior to the beginning of the twentieth-century (Weber 1976). The depth of the national "roots" in Soviet Central Asia is apparently not atypical of European experience.

9. The national composition of the Soviet population at the oblast level is not yet available for 1989.

10. However, at present the significance of this is potential only, since the nations of Central Asia have not become politically mobilized behind the goal of independent statehood to date.

11. The interest in maintaining or strengthening the native language among the socioprofessional elite may be read as more than mere nostalgia. Clearly, language may also be used as an instrument in the struggle with "outsiders" for the resources of the homeland, including high-status occupations. Having one's native language recognized as the dominant language in the homeland provides the upwardly mobile indigene with a competitive edge over "outsiders". This may be one of the main reasons why every nation with a union republic and a

number of those with autonomous republics have either passed or drafted legislation declaring the native language (e.g. Uzbek in Uzbekistan) the lingua franca of the republic (JPRS 1989). In this way, it appears that indigenes who have attained high-status positions in their homelands become more nationally assertive in the process.

12. The growing percentage of non-Russians who claim fluency in Russian as a second language has also been viewed by some Western analysts as a significant step in the process of assimilation to the Russian nation. However, this view appears incorrect; learning Russian as a second language is related more to the process of social mobilization than to international integration (Clem 1980: 44, Kaiser, this volume). Nevertheless, assigning Russian the role of lingua franca in the USSR is certainly advantageous for the dominant Russians, and cannot be viewed as a neutral "a-national" aspect of social mobilization.

13. RUKH, the Ukrainian national front organization, advocates the right to "national–cultural autonomy" for Ukrainians living outside Ukraine as a way of overcoming the population at risk of acculturating (1989: 30).

14. International family rates (the measure provided in the censuses) are cumulative, and therefore not identical to international marriage rates, which are annual.

15. The theoretical probability measures the probability that a member of one nation will marry a member of another nation, given the national composition of men and women entering into marriage in a given location in a given year. This measure is useful for comparative purposes, since it factors out the regional differences in national representation of the population entering into marriage. For a discussion of the construction and use of the theoretical probability, see Gantskaya and Debets (1966).

16. In the Soviet Union, each individual must declare his or her national identity at the age of 16, at which time he or she is issued an internal passport. While children from endogamous families do not have a choice, children from international families may choose the national identity of either parent. Published surveys of these data are used as the basis for the analysis in this section.

17. The exception of the Russian nation may be more apparent than real, since as the dominant nation Russians often lay claim to the entire state territory as their rightful "home."

REFERENCES

Agnew, John 1987. *Place and Politics: The Geographical Mediation of State and Society.* Boston, Mass.: Allen & Unwin.

Akiner, Shirin 1983. *Islamic Peoples of the Soviet Union.* London: Kegan Paul International.

Allworth, Edward 1973. *Regeneration in Central Asia. In Edward Allworth (ed.), The Nationality Question in Soviet Central Asia*, pp. 3–18. New York: Praeger.

Anderson, Barbara and Silver, Brian 1989. The changing ethnic composition of the Soviet Union. *Population and Development Review* 15, 4: 609–56.

Arutyunyan, Yu. and Bromley, Yu. (eds) 1986. *Sotsial'no-Kul'turnyy Oblik Sovetskikh Natsiy.* Moscow: Nauka.

Bazarbayev, Zhumanazar 1973. *Sekulyarizatsiya Naseleniya Sotsialisticheskoy Karakalpakii.* Nukus: Karakalpakstan.

Bennigsen, Alexandre 1979. Several nations or one people? Ethnic consciousness among Soviet Central Asian Muslims. *Survey* 24, 3: 51–64.

Bennigsen, Alexandre 1986a. Soviet minority nationalism in historical perspective. In Robert Conquest (ed.), *The Last Empire,* pp. 131–50. Stanford, Calif.: Hoover Institute Press.

Bennigsen, Alexandre 1986b. The long shadow of anti-Soviet rioting. *New York Times,* 30 December: A18.

Bondarskaya, G. and Darskiy, L. 1988. Etnicheskaya differentsiatsiya rozhdaemosti v SSSR. *Vestnik Statistiki* 12: 16–21.

Bromley, Yu. 1983a. Etnograficheskoye izucheniye sovremennykh natsional'nykh protsessov v SSSR. *Sovetskaya Etnografiya* 2: 4–14.

Bromley, Yu. 1983b. *Ocherki Teorii Etnosa.* Moscow: Nauka.

Brown, Bess 1990. The role of public groups in perestroika in Central Asia. *Radio Liberty: Report on the USSR* 26 January (RL 46/90): 20–5.

Carrere d'Encausse, Hélene 1978. *Decline of an Empire: The Soviet Socialist Republics in Revolt.* New York: Harper & Row.

Clem, Ralph 1980. The ethnic dimensions of the Soviet Union, parts I and II. In Jerry Pankhurst and Michael Sacks *Contemporary Soviet Society,* pp. 11–62. New York: Praeger.

Connor, Walker 1972. Nation-building or nation-destroying? *World Politics* 24, 3: 319–55.

Connor, Walker 1986. The impact of homelands upon diasporas. In Gabriel Sheffer (ed.), *Modern Diasporas in International Politics,* pp. 16–46. London: Croom Helm.

Dillon, Leo 1990. Ethnicity and political boundaries in the Soviet Union. Map No. 0858 published by the Office of the Geographer, US Department of State (State INR/GE).

Emerson, Rupert 1960. *From Empire to Nation.* Boston, Mass.: Beacon Press.

Evstigneev, Yu. 1974. Interethnic marriages in some cities of northern Kazakhstan. *Soviet Sociology* 13 (winter): 3–16.

Gantskaya, O. and Debets, G. 1966. O graficheskom izobrazhenii rezul'tatov statisticheskogo obsledovaniya mezhnatsional'nykh brakov. *Sovestskaya Etnografiya* 3: 109–18.

Gorbachev, M. 1989. *Izbrannyye Rechi i Stat'i.* Vol. 6. Moscow: Politizdat.

Goskomstat SSSR 1988. *Slovari Natsional'nostey i Yazykov.* Moscow.

Goskomstat SSSR 1989a. *Natsional'nyy sostav naseleniya.* Moscow: Finansy i Statistika.

Goskomstat 1989b. Naseleniye. *Statisticheskiy Press-Bulleten'* 4: 49–79.

Goskomstat SSSR 1989c. *Naseleniye SSSR 1988: Statisticheskiy Yezhegodnik.* Moscow: Finansy i Statistika.

Gurvich, I. 1982. Osobennosti sovremennogo etapa etnokul'turnogo razvitiya narodov Sovetskogo Soyuza. *Sovetskaya Etnografiya* 6: 15–27.

Isupov, A. 1964. *Natsional'nyy Sostav Naseleniya SSSR.* Moscow: Statistika.

JPRS 1989. Republic language legislation. JPRS–UPA (Soviet Union: Political Affairs) 63 (5 December).

Kaiser, Robert 1988. National territoriality in multinational, multi-homeland states: a comparative study of the Soviet Union, Yugoslavia and Czechoslovakia. PhD disseration, Columbia University.

Kalyshev, A. 1984. Mezhnatsional'nyye braki v sel'skikh rayonakh Kazakhstana. *Sovetskaya Etnografiya* 2: 71–7.

Kozenko, A. and Monogarova, L. 1971. Statisticheskoye izucheniye pokazateley odnonatsional'noy i smeshannoy brachnosti v Dushanbe. *Sovetskaya Etnografiya* 6: 112–18.

Krader, Lawrence 1971. *Peoples of Central Asia*, 3rd edn. Bloomington, Ind.: Indiana University Press.

Lubin, Nancy 1984. *Labour and Nationality in Soviet Central Asia*. Princeton, NJ: Princeton University Press.

Matley, Ian 1973. Ethnic groups of the Bukharan state ca. 1920 and the question of nationality. In Edward Allworth (ed.), *The Nationality Question in Central Asia*, pp. 134–42. New York: Praeger.

Monogarova, L. 1980. Evolutsiya natsional'nogo samosoznaniya pripamirskikh narodnostey. In R. Dzharylgasinova and L. Tolstova (eds), *Etnicheskiye Protsessy u Natsional'nykh Grupp Sredney Azii i Kazakhstana*, pp. 125–35. Moscow: Nauka.

Nielsen, Francois 1985. Toward a theory of ethnic solidarity in modern societies. *American Sociological Review* 50 (April): 133–49.

Pravda 1988. Rech' general'nogo sekretarya TsK KPSS M. S. Gorbacheva na plenume TsK KPSS 18 Fevralya 1988 goda. 18 February: 2.

Pravda 1990 Zakon SSSR: o poryadke resheniya voprosov, svyazannykh c vykhodom soyuznoy respubliki iz SSSR. 7 April: 2.

Rakowska-Harmstone, Teresa 1970. *Russia and Nationalism in Central Asia: The Case of Tadzhikistan*. Baltimore, Md: Johns Hopkins, University Press.

RUKH 1989. *RUKH: Program and Charter*. Ellicott City, Md: Smoloskyp Publishers.

Sack, Robert 1983. Human territoriality: a theory. *Annals of the Association of American Geographers* 73, 1: 55–74.

Sack, Robert 1986. *Human Territoriality: Its Theory and History*. Cambridge: Cambridge University Press.

Shibutani, Tamotsu and Kwan, Kian 1972. *Ethnic Stratification: A Comparative Approach*. New York: Macmillan.

Silver, Brian 1986. The ethnic and language dimension in Russian and Soviet censuses. In Ralph Clem (ed.), *Research Guide to the Russian and Soviet Censuses*, pp. 70–97. Ithaca, NY: Cornell University Press.

Smirensky, Nicholas 1985. Moscow or Mecca: the modernization of the titular nationalities of Central Asia and their integration into Soviet society. Master's thesis, Columbia University.

Smith, Anthony 1981. States and homelands: the social and geopolitical implications of national territory. *Millenium: Journal of International Studies* 10, 3: 187–202.

Soja, E. 1971. *The Political Organization of Space*. Washington, DC: Association of American Geographers.

Susokolov, A. A. 1987. *Mezhnatsional'nyye Braki v SSSR*. Moscow: Mysl'.

Tolstova, L. 1985. Natsional'no-smeshannyye braki u sel'skogo naseleniya Karakalpakskoy ASSR. *Sovetskaya Etnografiya* 3: 64–72.

Tsentral'noye Statisticheskoye Upravleniye (TsSU) SSSR 1929. *Vsesoyuznaya Perepis' Naseleniya 1926 Goda*. Vol. 17. Moscow: Izdaniye Soyuza SSR.

TsSU SSSR 1959. *Slovari Natsional'nostey i Yazykov*. Moscow: Gosstatizdat.

TsSU SSSR 1962. *Itogi Vsesoyuznoy Perepisi Naseleniya 1959 Goda*, 16 vols. Moscow: Gosstatizdat.

TsSU SSSR 1969. *Slovari Natsional'nostey i Yazykov*. Moscow: Statistika.

TsSU SSSR 1973. *Itogi Vsesoyuznoy Perepisi Naseleniya 1970 Goda*, 7 vols. Moscow: Statistika.

TsSU SSSR 1984. *Chislennost' i Sostav Naseleniya SSSR: po Dannym Vsesoyuznoy Perepisi Naseleniya 1979 Goda*. Moscow: Finansy i Statistika.

Vakar, N. 1956. *Belorussia: The Making of a Nation*. Cambridge, Mass: Harvard University Press.

Vinnikov, Ya. 1980. Natsional'nyye i etnograficheskiye gruppy Sredney Azii po dannym etnicheskoy statistiki. In R. Dzharylgasinova and L. Tolstova (eds), *Etnicheskiye Protsessy u Natsional'nykh Grupp Sredney Azii i Kazakhstana*, pp. 11–42. Moscow: Nauka.

Vinogradov, V. V. (ed.) 1968. *Yazyki Narodov v Pyati Tomakh*. Leningrad: Nauka.

Weber, Eugen 1976. *Peasants into Frenchmen: The Modernization of Rural France 1870–1914*. Stanford, Calif.: Stanford University Press.

Zhdanko, T. 1972. Natsional'no-gosudarstvennoye razmezhevaniye i protsessy etnicheskogo razvitiya u narodov Sredney Azii. *Sovetskaya Etnografiya* 5: 13–29.

Zhukovskaya, N. 1980. Issyk-Kul'skiye Kalmaki (Sart-Kalmaki). In R. Dzharylgasinova and L. Tolstova (eds), *Etnicheskiye Protsessy u Natsional'nykh Grupp Sredney Azii i Kazakhstana*, pp. 157–66. Moscow: Nauka.

Index

Note: Page numbers in italic refer to figures; page numbers in bold refer to tables.